"This book on *Advanced Wind Turbines*, organized in 11 chapters, is a good reference for those wanting to better understand the aerodynamics driving the design and performance of vertical axis wind turbines. The authors have written it from the perspective of industrial applications focused on the conditions specifically for small city-states that are not naturally endowed with sufficiently strong wind. It is well written and is very pleasant to read, covering the most important issues about wind turbines and a very detailed treatment of dynamic stall phenomena. It is a veritable mine of numerical and experimental data, useful for research and professional work. It proved to be a useful resource throughout the design process for a client's unique wind energy project.

The optimization of the Vertical Axis Wind Turbine (VAWT) of Darrieus type for low wind speed condition has been well presented in Chapters 2 to 6 using the concept of Hybrid design — Savonius-Darrieus rotor. Savonius rotor generates higher starting torque as the blades are drag driven, hence these machines can be wisely integrated with Darrieus turbine to enhance the performance at low wind speeds. A good technical evaluation of the Floating Offshore Wind Turbines (FOWT) is presented in Chapters 7 to 10. I would like to point out the impressive number of updated references in the wind turbine field, which provide a very rich source of documentation to the reader.

The authors Palanisamy MohanKumar, Krishnamoorthi Sivalingam and Teik-Cheng Lim were involved in the development of an upgraded hybrid wind turbine that was operating in unique wind conditions and using non-conventional rotor designs. The software and libraries were customized to give an excellent overview of the rotor performance and important design criteria. The simulations and experimental results provided were pivotal in understanding how to achieve optimum performance. In my opinion, this level of process simulation validated with experiments is an essential tool in designing advanced wind turbines.

I would strongly recommend this book as a good reference for scientists, engineers and any companies worldwide working on wind turbines."

Ion Paraschivoiu, P.Eng., Ph.D.
Full Professor, Polytechnique Montréal, Canada
President, IOPARA Inc.

I.Paraschivoiu

"As the world grapples with the urgent need to transition to cleaner and more sustainable forms of energy, wind power has emerged as one of the leading contenders in the race to mitigate the impacts of climate change. Wind turbines are becoming more efficient, reliable, and cost-effective, making them a viable option for powering homes, businesses, and entire communities.

This book explores the cutting-edge technologies and innovations that are driving the advancement of wind turbines. From the latest designs in turbine blades and addressing complex phenomena on various wave and wind motions causing the aerodynamics loads on the rotor and towers to new control systems, the book provides a comprehensive overview of the state of the art in wind energy.

Written by leading experts in the field, the book covers topics such as:

- The fundamentals of Vertical Axis Wind Turbines (VAWT), such as Darrieus wind turbines, design optimization for low wind speed energy and the design principles of wind turbines
- The latest advances in turbine blade and tower design, including innovations in materials, aerodynamics, and manufacturing
- The role of Floating Offshore Wind Turbines (FOWT) and its advancements

- The key aerodynamic aspects of FOWT with various motions caused by waves and wind
- The techno-economics of large-scale offshore wind turbines/farms and FOWT including Remaining Useful Life (RUL) prediction using physics-based methods, and the costs and benefits of large-scale wind farms

This book is intended for students, researchers, and professionals in the fields of wind energy, renewable energy, and sustainable development. It provides a comprehensive and up-to-date overview of the latest advances in wind turbine technology, and offers insights into the key challenges and opportunities facing the industry.

Overall, the book aims to inspire and inform readers about the exciting and rapidly evolving world of wind power, and to encourage further research and innovation in this critical area of energy development."

Sundararajan Natarajan, Ph.D.
Assistant Professor, Indian Institute of Technology-Madras, India

Advanced

WIND TURBINES

Other Titles by Teik-Cheng Lim

An Introduction to Electrospinning and Nanofibers
Nanosensors: Theory and Applications in Industry, Healthcare and Defense
Advances in Therapeutic Engineering
Auxetic Materials and Structures
Mechanics of Metamaterials with Negative Parameters
A Partially Auxetic Metamaterial Inspired by the Maltese Cross

Advanced WIND TURBINES

Palanisamy MohanKumar
Nanyang Technological University, Singapore

Krishnamoorthi Sivalingam
Surbana Jurong Pte Ltd, Singapore

Teik-Cheng Lim
Singapore University of Social Sciences, Singapore

World Scientific

NEW JERSEY • LONDON • SINGAPORE • BEIJING • SHANGHAI • HONG KONG • TAIPEI • CHENNAI • TOKYO

Published by

World Scientific Publishing Co. Pte. Ltd.

5 Toh Tuck Link, Singapore 596224

USA office: 27 Warren Street, Suite 401-402, Hackensack, NJ 07601

UK office: 57 Shelton Street, Covent Garden, London WC2H 9HE

Library of Congress Cataloging-in-Publication Data

Names: MohanKumar, Palanisamy, author. | Sivalingam, Krishnamoorthi, author. | Lim, Teik-Cheng, author.

Title: Advanced wind turbines / Palanisamy MohanKumar, Nanyang Technological University, Singapore, Krishnamoorthi Sivalingam, Surbana Jurong Pte Ltd, Singapore, Teik-Cheng Lim, Singapore University of Social Sciences, Singapore.

Description: New Jersey : World Scientific, [2024] | Includes bibliographical references and index.

Identifiers: LCCN 2022061353 | ISBN 9789811272486 (hardcover) | ISBN 9789811272493 (ebook) | ISBN 9789811272509 (ebook other)

Subjects: LCSH: Wind turbines. | Wind power.

Classification: LCC TJ828 .M64 2024 | DDC 621.4/5--dc23/eng/20230111

LC record available at https://lccn.loc.gov/2022061353

British Library Cataloguing-in-Publication Data

A catalogue record for this book is available from the British Library.

For any available supplementary material, please visit
https://www.worldscientific.com/worldscibooks/10.1142/13311#t=suppl

Desk Editor: Shaun Tan Yi Jie

Typeset by Stallion Press
Email: enquiries@stallionpress.com

Preface

As the world continues to shift towards cleaner and more sustainable sources of energy, wind power has emerged as one of the most promising options. Wind turbines have become increasingly popular in recent years, as they offer a reliable and affordable way to generate electricity without relying on fossil fuels. In fact, wind energy has the potential to become the largest source of renewable energy in the world.

Although there is no lack of monographs on sustainable energy, the authors feel that the outlook from industry is somewhat less prominent when compared against the more academic viewpoint. With this in mind, the authors have attempted to produce a wind turbine book from the perspective of the industry in contrast to the more academic bent. A secondary motivation for this book arises from the conditions faced by small city-states that are not naturally endowed with sufficiently strong wind. After the customary Introduction, Chapters 2 to 6 mainly address the optimization of Darrieus wind turbines for low wind speed condition. Another limitation faced by small island states in their wind energy endeavor is the limited coastline to cater for wind turbines, due to competition from trade (the use of limited coastline for ports). To address such drawbacks, wind turbines in some jurisdictions have pushed the location of their wind turbines further away from the shore, hence introducing floating offshore wind turbines (FOWT). A holistic technical evaluation of the FOWT is addressed in Chapters 7 to 10. Finally, a set of concluding remarks is furnished in Chapter 11. Figures where a reference number is indicated at the end of the caption are reproduced from published works, with kind permission from the respective publishers. Due to fast-changing technological developments, any recommendations put forth is confined to only the near future.

Whether you are an engineer, investor, policymaker, or simply interested in renewable energy, this book offers a valuable resource on the state of wind power and its future prospects. We hope it will serve as a useful guide and reference for those working in the field, and inspire further innovation and progress in the wind power industry.

Palanisamy MohanKumar
Krishnamoorthi Sivalingam
Teik-Cheng Lim

About the Authors

Dr. Palanisamy MohanKumar is an expert in renewable energy technologies, thermal management of electrical systems, and precision machine design. Currently, he is the technical manager for the Mechanical and Thermal division of Rolls Royce Electrical (RRE) at Rolls Royce Singapore Pte Ltd. He is specialized in wind energy, tidal energy, resource assessments, and various energy storage technologies with over 18 years of experience in design and deployment of these systems. He has developed multiple wind and tidal turbines, especially for low wind and tidal resources that have been deployed in and around Singapore. He specialized in wind turbine aerodynamics and related computational fluid dynamics which earned him an industrial PhD for the development of vertical axis wind turbines. Previous roles include Research and Development engineer specializing in the design of CNC machines, leading to a Masters in product design at Nanyang Technological University, Singapore. As a research engineer at the Energy Research Institute, Singapore, he developed battery energy storage systems, compressed air energy storage, and thermal energy storage for microgrid applications. As a research fellow at the National University of Singapore, he devised thermal management systems for medical devices. In RRE, currently, he is heading an international team to develop battery systems for aircraft and utility-scale energy storage and thermal management of power converter systems.

Dr. Krishnamoorthi Sivalingam is an expert in energy transition (including wind energy). Currently he is Senior Manager at Surbana Jurong Pte Ltd, Singapore, managing the Energy Transition portfolio of the company's Energy and Industrial division to enable industries and various infrastructure-related projects to create a sustainable pathway to achieve net-zero carbon emission targets via renewable energy, energy efficiency with combined cycle gas turbines, and electrification. He was previously a Technical Manager at Rolls Royce Singapore Pte Ltd, Engineering Manager at NTUtive of Nanyang Technological University, Singapore, and also worked in Lloyds Register, Siemens, Vestas, UTAC, Delphi Automotive Systems, and the Indian Government R&D Center. He was deeply involved in several prominent multi-million-dollar projects and possesses vast experience in project, people, proposal and technical management in his 23 years of industrial experience. He is an expert in renewable energy systems including wind energy, energy storage systems, hydrogen fuel cells, electrification, and electro-thermo-mechanical aspects of other energy engineering systems. He has authored/co-authored nearly 30 papers in conferences and journals. He has also contributed to 21 patents and made more than 150 invention disclosures. He has conducted numerous market analyses, concept and feasibility studies, and managed various stakeholders, standards committees and regulatory authorities.

He had directly contributed to and managed several significant projects, such as aerospace-related products development, wind (onshore and offshore including floating offshore wind), thermo-mechanical systems design, development, battery energy storage systems, CoolerTop thermal systems for megawatt-scale wind turbines and water treatment systems. Moreover, he has vast experience in dealing with fluid flow machinery and thermal systems at various levels including computational methodologies, analytical solutions, and final design of the product or project. He is proficient in the application of artificial intelligence, machine learning, Internet of Things, data science, digital twin, and predictive analytics to engineering systems. He is also a digital transformation leader in engineering applications.

Associate Professor Teik-Cheng Lim is Head of PhD (Engineering) and Master of Enginering Programmes at the Singapore University of Social Sciences, Singapore. He won a Faculty of Engineering Annual Book Prize for his undergraduate studies at the National University of Singapore (NUS) and was subsequently awarded a research scholarship to pursue his PhD at NUS. He has written and edited 6 books, namely *A Partially Auxetic Metamaterial Inspired by the Maltese Cross* (Cambridge University Press, 2022), *Mechanics of Metamaterials with Negative Parameters* (Springer, 2020), *Auxetic Materials and Structures* (Springer, 2015), *Advances in Therapeutic Engineering* (Taylor and Francis, 2012), *Nanosensors* (Taylor and Francis, 2011), and *An Introduction to Electrospinning and Nanofibers* (World Scientific, 2005). He has served as competition judge for the Singapore Science & Engineering Fair (SSEF) organized by the Singapore Science Centre, and Energy Innovation Challenge organized by the Institution of Engineers, Singapore. He also served as external examiner for PhD theses at the University of Malta, Hong Kong Polytechnic University and Indian Institute of Technology, and has been teaching part-time for Harbin Institute of Technology, China.

Contents

Chapter 1

Introduction

1.1 A Case for H-Darrieus Wind Turbine

The rising energy demand along with fossil fuel depletion has shifted the focus towards renewable energy resources. The quest for clean energy has increased significantly in recent years in the wake of rising global temperatures and their impact on humanity. Decentralized or distributed energy generation is increasingly popular for their flexibility in power generation and to cater to local demands with negligible transmission losses.[1]

Small-scale power generation is economical compared to large-scale centralized power plants with large network and distribution infrastructures of the olden days. The transmission and distribution losses outweigh the efficiency gains of large-scale power generation. The decentralized power generation encompasses many small energy sources along with inexpensive energy storage systems. Power supply and demand can be matched effectively by shifting the automated loads to off-peak periods with independent energy sources assuring the power security.

Wind energy is popular among various renewable energy sources such as solar, hydroelectric, geothermal and biomass. Wind energy plays a major role in electricity generation with a total installed capacity of 468.8 GW as of 2016.[2] Small wind turbines (SWTs) are indispensable devices of such systems which can coexist and are compatible with other renewable and non-renewable energy sources. The nominal power capacity of these turbines are less than 50 kW[3] and are widely installed in locations in the vicinity of demand compared to utility-scale wind turbines that are installed on sites with high wind energy potential. SWTs are often installed in low wind speed sites characterized by unsteady and highly turbulent flows for significant portions of the time.

Hence the immediate challenge is to develop SWTs that are efficient in energy capture at a reasonable level of economics. Though a large variety of wind energy conversion machines were proposed in the early days, the most commonly employed are either Horizontal Axis Wind Turbines (HAWTs) or Vertical Axis Wind Turbines (VAWTs). Each type has its own merits and demerits, but HAWTs are preferred for their high efficiency especially for utility-scale power generation. The distinct advantages of VAWTs such as insensitivity to the wind direction, high reliability, low noise and ease of maintenance with power train close to the ground level hold a niche market in the wind turbine arena. Apart from electricity generation, VAWTs are preferred over HAWTs for unique applications such as desalination, water pumping and water body aeration to name a few. The resurgence of VAWTs as a standalone device demands energy extraction in low wind speeds. H-Darrieus, being the most preferred and widely commercialized turbine, is plagued with poor self-starting capability and generally recognized to require external assistance to incite the rotation.[4] Acknowledged disadvantages include poor performance at low Reynolds number flows, pulsating torque for every rotation and lack of aerodynamics over speed-regulating features. Past endeavors on H-Darrieus achieved marginal success over the above said issues by adding complexity and cost. The past attempts successfully lifted the H-Darrieus out of self-starting issues, yet reduced performance at low wind speed is still grappling its development.[5]

10 kW HAWT rated at 11 m/s has recorded a cut in wind speed of 2.5 m/s,[6] whereas H-Darrieus with rated wind speed of 12 m/s displays a cut of 4.5 m/s.[7] The above comparison exposes the extent of reduced annual yield by H-Darrieus for similar specification with HAWT in the market. Wind speed of less than 5 m/s prevails over tropical regions around the world and hence reducing the cut in wind speed has significant influence on energy yield. To enable the Darrieus turbine to generate significant power compared to HAWT, the H-Darrieus configuration has to be revisited with thorough understanding on existing strategies. Though the configuration of H-Darrieus is simple in construction it is complicated in aerodynamics due to inconsistent angle of attack[8] and varying wind velocity, which demands extensive computational optimization and field test for any new concept implementation.

1.2 Overview of Floating Offshore Wind Turbine

The renewable energy sector has been paying particular attention to offshore wind technology given that a better quality of wind away from land leads to a stronger, more consistent, and more easily predictable method of power generation. So far, these benefits have been leveraged through the installation of fixed-bottom wind turbines in shallow waters of up to 30 m in depth. Attempts to install wind turbines with monopile or gravity-based foundations in deeper waters where the depth is over 60 m, however, is not economically feasible as the cost factor is not linear. Furthermore, many countries lack access to shallow water. Nevertheless, floating platform technology from the offshore oil and gas sector can be modified in order to place wind turbines in those deep waters. This means that wind turbines can be used more extensively and that they have a minimal impact on the horizon. As new technologies are developed in order to assist with simplified maintenance and installation the cost of energy will fall, thus making floating offshore wind turbines (FOWTs) more attractive from an investment perspective.[9]

There are a number of hurdles that need to be overcome, before FOWTs can be applied to commercial scale usage. One of the critical hurdles revolves around the rotor which, for onshore wind turbines, is stationary. On a FOWT, however, the floating platform can cause significant rotor plane shift and oscillation, which creates a layer of complexity. This results in a change to the rotor plane which, in turn, causes a transient flow field both on and behind the rotor. FOWTs face a serious problem in the form of the motions which are induced by waves and currents.

The control system also needs to be designed specifically to have the ability to respond to both changes in wind speed and direction as well as any motions which are induced by waves and currents. Harsh environmental conditions such as winds, currents, typhoons, storms, cyclones, and more can conspire to make it difficult for an FOWT to float in the upright position. The change in rotor plane will create a gyro effect on any rigid body with a wheel. This effect impacts FOWTs too, yet the influence upon its structure and control system has not been adequately studied. The aerodynamic, mooring dynamic, and hydrodynamic effects combine to shape the operating and failure conditions of an FOWT. Butterfield *et al.*[10] stated that the combined effects of these factors have not yet been studied in sufficient detail for floating systems. Due to their limited capacities, the previous dynamic models developed to predict various behaviors failed.[10]

Given that myriad factors such as aerodynamics, hydrodynamics, mooring line systems, and structural dynamics as shown in Figure 1.1 are brought together, using numerical simulations to study the multi-physics phenomena of an FOWT is highly challenging. Computational Fluid Dynamics (CFD)

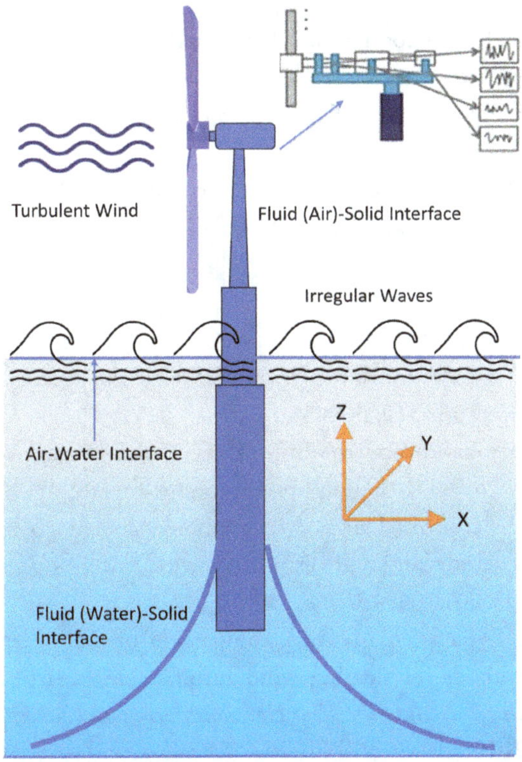

Figure 1.1 Floating offshore wind turbine — various loads at different major wind turbine components.

methodologies and computation power have improved in recent decades, yet the simulation of a fully-coupled FOWT still represents a significant challenge. More attempts have been made in recent years which place a strong emphasis upon the development of viable CFD methodologies for design and operation analysis. When it comes to certification, however, it is not possible to rely exclusively upon readings from high-fidelity tools like CFD and Finite Element Analysis due to the large number of simulations which are required in order to verify the design. The scope for this research is limited to application of aerodynamics of FOWT in engineering numerical tools and validation of the same.

References

1. Lasseter, R. H. Microgrids and distributed generation. *Journal of Energy Engineering* **2007**, *133*, 144–149.
2. http://www.gwec.net/wp-content/uploads/vip/GWEC_PRstats2016_EN_WEB.pdf.
3. Gipe, P. *Wind Energy Basics*. Chelsea Green Pub. Co. **2009**.
4. Worasinchai, S.; Ingram, G. L.; Dominy, R. G. The physics of H-Darrieus turbine starting behavior. *Journal of Engineering for Gas Turbines and Power* **2015**, *138*, 062605.
5. Baker, J. Features to aid or enable self starting of fixed pitch low solidity vertical axis wind turbines. *Journal of Wind Engineering and Industrial Aerodynamics* **1983**, *15*, 369–380.
6. http://bergey.com/products/wind-turbines/10kw-bergey-excel.
7. http://www.ropatec.it/index.php/en/prodotti-ropatec-4/196-sa40-en.
8. Ågren, O.; Berg, M.; Leijon, M. A time-dependent potential flow theory for the aerodynamics of vertical axis wind turbines. *Journal of Applied Physics* **2005**, *97*, 104913.
9. Sebastian, T; Lackner, M. A. A comparison of first-order aerodynamic analysis methods for floating wind turbines. *48th AIAA Aerospace Sciences Meeting Including the New Horizons Forum and Aerospace Exposition.* Orlando **2010**.
10. Butterfield, C. P.; Musial, W; Jonkman, J. Overview of offshore wind technology. *Chinese Renewable Energy Industry Association WindPower Shanghai Conference.* Shanghai **2007**.

Chapter 2

State-of-the-art Technologies for Low Wind Speed Operation

2.1 Types of Wind Turbine

Myriad configurations of wind energy conversion devices have been proposed throughout history as shown in Figure 2.1,[1] the most widely used being HAWTs and VAWTs. Numerous variations exist in both HAWTs and VAWTs based on the application and required output. The primary driving forces, i.e., lift and drag are created in all possible ways for efficient energy conversion. The lift force is created by airflow over an airfoil or through Magnus effect, where a spinning cylinder creates circulation and in turn generates lift. A recent addition to the novel rotors is the wind kite from Makani power. Apart from efficient rotors, the research on wind turbines focuses on the development of flow acceleration devices such as shroud, diffuser and guide vanes to create negative pressure behind the rotor and improve the aesthetics. The three-bladed design with its axis parallel to the incoming wind dominates the field of utility-scale power generation for its high efficiency. Demerits include the requirement of yaw mechanism to orient to wind direction and poor performance in turbulent winds.

VAWT being the main focus of the current study has its rotation axis perpendicular to the incoming wind. The niche advantages of VAWT over HAWT are insensitivity to the wind direction and ease of maintenance as the drive train components are placed close to the ground. VAWTs are further classified as lift-based and drag-based machines.

The lift-based machines employ airfoils arranged vertically and were invented by Georges Jean Marie Darrieus in 1931. A straight two-bladed Darrieus turbine is commonly referred to as H-Darrieus as it resembles the letter "H". Straight multi-bladed Darrieus is referred to as Giromill rotor. Research in the early days was focused on curved-bladed or troposkein blades. The efficiency of the Darrieus turbine ranges from 35% to 40%. Advantages of Darrieus turbines are lower footprint, quiet operation, bird friendly and easier blade fabrication. Notable drawbacks

Figure 2.1 Types of wind turbine.[1]

are poor starting capability and high cut in wind speed resulting in low annual energy yield. Darrieus wind turbines are mostly preferred for power generation.

Savonius turbine was invented by Sigurd Johannes Savonius in 1922 and basically consists of two vertical half cylinders. Though the efficiency of Savonius rotor (~15%) is lower than those of HAWTs (~45%) and VAWTs (~35%), the advantages such as simple in construction and a good starting torque at lower wind speeds attracted researchers. Being the most studied wind turbine, Savonius turbines are preferred for urban installations, street lighting, water pumping, and rooftop applications for their performance in turbulent wind conditions and good low wind speed performance. The major types of wind turbines are shown in Figure 2.1.

2.2 Aerodynamics of Darrieus Rotor*

The aerodynamics encompassing the Darrieus rotor is indeed more complex than its horizontal axis counterparts. These cross-flow devices are able to extract energy from any wind direction paving the way for complex flow field. This section sheds light on rotor dynamics that has major influence on the performance of Darrieus turbines.

2.2.1 *Tip-speed Ratio and Angle of Attack*

The blades of H-Darrieus turbine encounter inconsistent Angle of Attack (AoA) for every rotation. The AoA has strong correlation with tip-speed ratio (TSR) and the azimuthal position. When the rotor is at rest, the freestream wind will approach the rotor from any direction and the AoA can vary from 0^0 to 360^0. As the TSR increases the AoA reduces due to increase in the tangential velocity, resulting in higher relative velocity

* Parts of this section are adapted with permission from Kumar, P. M.; Rashmitha, S. R.; Srikanth, N.; Lim, T.-C. Wind tunnel validation of double multiple streamtube model for vertical axis wind turbine. *Smart Grid and Renewable Energy* **2017**, *8*, 412–424.

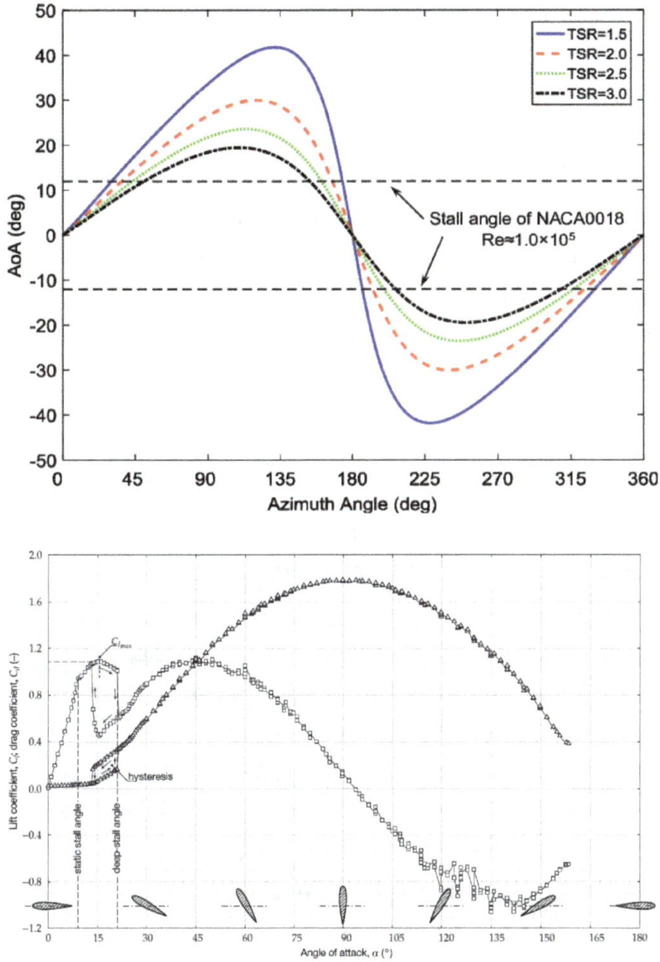

Figure 2.2 Cl and Cd of NACA 0018 (left) and AoA vs azimuthal angle (right).[1,2]

than freestream wind velocity. The variation of AoA with azimuthal position and TSR is shown in Figure 2.2 (bottom).[2] The AoA for the given TSR and the azimuthal position can be derived by Eq. (2.1).

$$\alpha = \tan^{-1}\left(\frac{\sin\theta}{\mathrm{TSR} + \cos\theta}\right) \tag{2.1}$$

The aerodynamic coefficients of NACA 0018 at Reynolds number (Re) 5×10^5 obtained from the wind tunnel experiment are shown in Figure 2.2 (top).[1] It can be noticed that the drag increases drastically for AoA at more than 15^0. Because of this large AoA and with almost negligible lift, the Darrieus suffers from poor self-starting capability. The starting torque is further reduced due to laminar bubble separation, a typical behavior of low Re flow. At higher TSR, both the Re and AoA are favorable for torque generation. Hence TSR is a critical factor in determining the performance of Darrieus rotors.

2.2.2 *Dynamic Stall*

The rapid change of AoA induces dynamic stall, a complex phenomenon associated with unsteady aerodynamics. The aerodynamic loads for static and dynamic pitching airfoils are different for the same AoA and Re, and hence it is imperative to accurately predict the effect of dynamic stall. The impact of dynamic stall on the performance of Darrieus rotors is based on the frequency at which the AoA changes and in turn on the TSR. As the TSR reaches the maximum, the blades operate under dynamic stall in which the flow is entirely different from the static airfoil flow. As the airfoil AoA moves past the static stall, the flow separation starts from the leading edge and tends to roll up. With further increase in the AoA, the vortex grows in size and travels towards the leading edge with substantial increase in lift. A trailing edge vortex tends to grow in the opposite direction. Finally the leading and trailing edge vortices separate from the airfoil forming a pair of counter-rotating vortices in the downstream wake as visualized by Fujisawa and Shibuya.[3] The sequence of dynamic stall occurrence in an airfoil is depicted in Figure 2.3.[1]

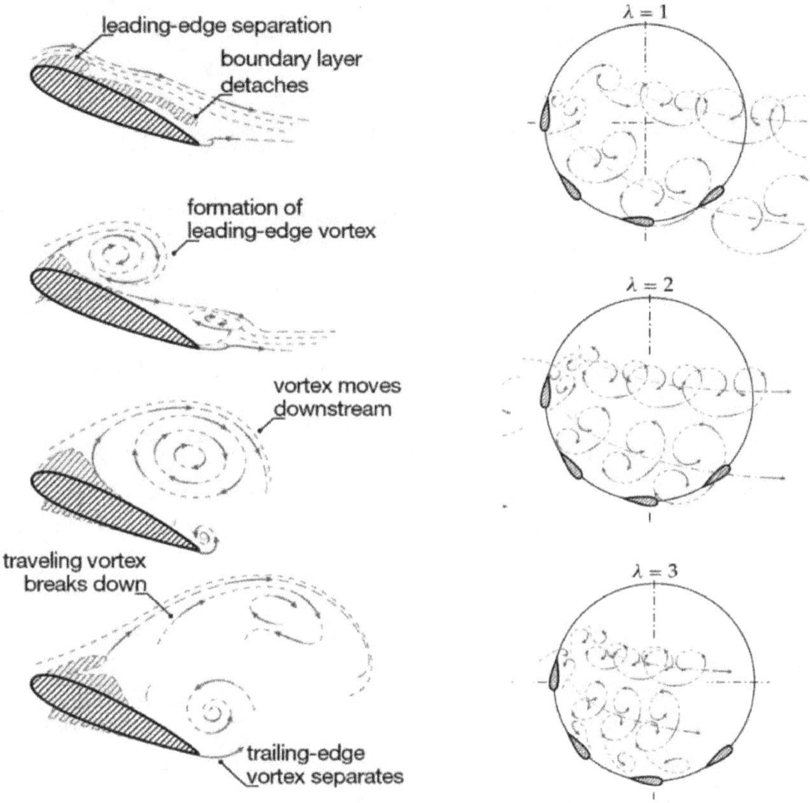

Figure 2.3 Sequence of dynamic stall occurrence on an airfoil.[1]

2.2.3 *Aerodynamic Models*

Several models have been proposed in the past to predict the power performance and aerodynamic loads for Darrieus turbines by Wilson and Lissaman,[4] Templin,[5] Strickland,[6] Holme,[7] and Paraschivoiu.[8] Every model has its assumptions, limitations, and inaccuracies. Hence a brief comparative analysis of the existing models with their mathematical expressions is performed in search of a suitable model to optimize the proposed adaptive hybrid Darrieus turbine (AHDT).

For any analytical model, the procedure to compute the performance is described as:

1. Calculating the local AoA and the relative velocity corresponding to the blade orbital position;
2. Calculating the induced velocity accounting for the blade-wake interaction;
3. Choosing the suitable mathematical model along with the mathematical expressions;
4. Calculation of blade tangential forces and the normal forces;
5. Pre-stall airfoil characteristics at different Re;
6. Aspect ratio consideration to account for the blade tip loss effects;
7. Dynamic stall model to account for the unsteady effects;
8. Flow curvature model to account for the circular blade motion.

2.2.3.1 *Cascade Model*

The cascade model was prescribed by Hirsch and Mandal[9] for VAWT analysis, which is widely used for turbomachinery with several blades. In this model, the blades are arranged in cascade on a planar surface, with the distance equal to the circumference of the rotor divided by number of blades as illustrated in Figure 2.4.

The local relative velocity and the AoA are calculated for the upwind and downwind configurations before unwinding into cascade configuration. Two induction factors are introduced to calculate the induced velocity at upstream designated as v_u and downstream designated as v_d:

$$\frac{v_u}{v_0} = \left(\frac{v_1}{v_0}\right)^{k_i} \tag{2.2}$$

$$\frac{v_d}{v_0} = \left(\frac{v_2}{v_1}\right)^{k_i} \frac{v_1}{v_0} \tag{2.3}$$

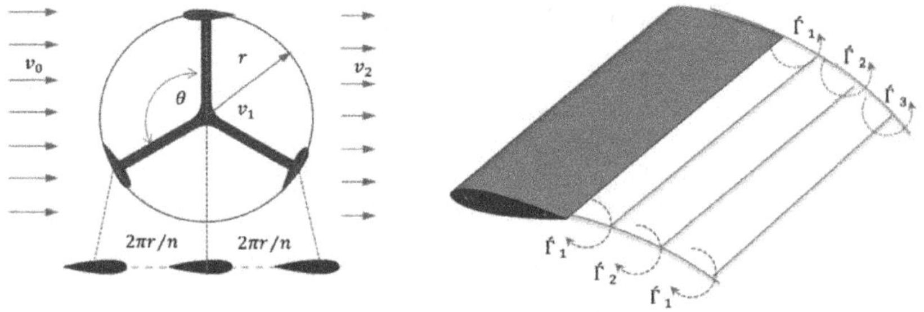

Figure 2.4 Cascade model (left) and vortex model (right).

$$k_i = 0.425 + 0.332\left(\frac{Nc}{R}\right) \tag{2.4}$$

$$Q = \rho h R^2 \int_0^{2\pi} (W_o^2 \sin\alpha_0 \cos\alpha_0 - W_i^2 \sin\alpha_i \cos\alpha_i)\, d\theta \tag{2.5}$$

with h as the height of the turbine, R as the radius of the turbine and c as the airfoil chord. The overall torque Q generated by the rotor can be deduced from Eq. (2.5). The cascade model is able to predict the performance of the rotors with high solidities and also at high TSR quite accurately compared to other mathematical models. Other geometric characteristics such as blade pitch, flow curvature effect and aspect ratio effects can be readily integrated into this model. To improve the predicting capability the dynamic stall model was included along with the flow curvature effects.[10]

2.2.3.2 *Vortex Model*

The vortex model is based on the influence of vorticity on the wake of the blade in which a bound vortex filament replaces each blade element as shown in Figure 2.4. From the Helmholtz theorem,[11] the strength of the bound vortex filament is equal to the trailing edge tip vortex. The velocity-induced dv at any point in the flow field by a single vortex filament of length l and at a distance r can be deduced by the Biot-Savart law as given by Eq. (2.6).

$$dv = \frac{\Gamma}{4\neq} \frac{dl \times r}{|r^3|} \tag{2.6}$$

The undisturbed freestream velocity added to the induced velocity by the entire vortex filaments comprise the fluid velocity at any point in the flow field. The relationship between the lift L per unit span on a blade segment and the bound vortex strength Γ_B is given by the Kutta-Joukowski law as in Eq. (2.7).

$$L' = \rho v \Gamma_B \tag{2.7}$$

The relation between the bound vortex strength, the lift coefficient C_l and the local relative fluid velocity U_r is given as

$$\Gamma_B = \frac{1}{2} C_l C U_r \tag{2.8}$$

The vortex model was improved in recent years by incorporating dynamic stall effects by Strickland *et al.*,[12] while flow curvature effects were included by Migliore *et al.*[13] Though vortex model results have better correlation with experimental values, significant efforts are needed to reduce the computational time. Potential flow assumption in the wake and the viscosity effects are simplified through empirical force coefficients, which are likely to contribute to inaccuracies in the performance prediction.

2.2.3.3 *Single Streamtube Model*

The kinetic energy in the wind invariably creates thrust in the incoming direction. The change in kinetic energy between the upstream wind v_0 and downstream wind v_2 is the useful power P extracted from the wind. Wind rotors can be conveniently considered as a permeable disk as shown

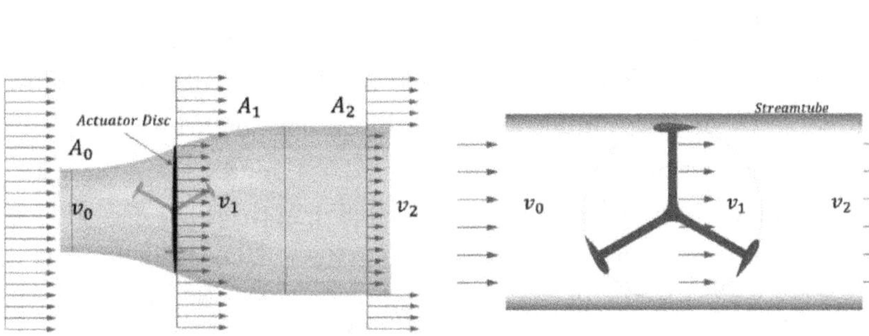

Figure 2.5 Actuator disc model (left) and single streamtube model (right).

in Figure 2.5, where the kinetic energy from the wind is converted into thrust F_D and the torque by the blades. In the single streamtube model as shown in Figure 2.5, the Darrieus rotor is represented as an actuator disk inside the streamtube as proposed by Templin[5] to calculate the performance characteristics. The assumptions of actual disk theory such as irrotational and incompressible upstream flow, and constant flow over the entire disk are also applicable for the streamtube model. The single streamtube model is the forerunner to incorporate the actuator disk in the Darrieus aerodynamic model and most of the successful models are based on this assumption. The upstream velocity v_0 is reduced at the disk by an induction factor a ($a < 1$). Rotor characteristics such as airfoil stall, solidity and aspect ratio are accounted for in this model and can effectively predict the performance for lightly loaded turbine. As the solidity and the TSR increase, this model tends to over-predict the performance due to uniform flow assumption for the entire disk.

$$F_D = mx(v_0 - v_2) \tag{2.9}$$

$$v_1 = (1 - a)v_0 \tag{2.10}$$

$$v_2 = (1 - 2a)v_0 \tag{2.11}$$

$$v_1 = \frac{v_0 + v_2}{2} \tag{2.12}$$

$$P = 2\rho v_0^3 a(1 - a)^2 \tag{2.13}$$

Since the upstream and downstream wind velocities vary considerably, double actuator disk model was introduced as shown in Figure 2.7. Additional induction factor a' has been introduced for the second actuator disk. Though this model is more sophisticated than the single disk model, it is a mere assumption that the downstream flow will be fully developed before it passes through the second disk.

$$v_1 = (1 - a)u_0 \quad (2.14)$$

$$v_2 = (1 - 2a)u_0 \quad (2.15)$$

$$v_3 = (1 - 2a)(1 - a')u_0 \quad (2.16)$$

$$v_4 = (1 - 2a')(1 - 2a)u_0 \quad (2.17)$$

2.2.3.4 *Multiple Streamtube Model*

An improved version of the single streamtube model was introduced by Wilson and Lissaman.[4] The swept volume of the turbine was divided into multiple parallel streamtubes as shown in Figure 2.6 which are aerodynamically independent. The blade element and the moment theory are applied for each streamtube. Rotor power and torque are obtained by averaging the values from each streamtube.

The model proposed by Wilson and Lissaman[4] accounts for only the lift force to calculate the induced velocity. Later Strickland, Webster and

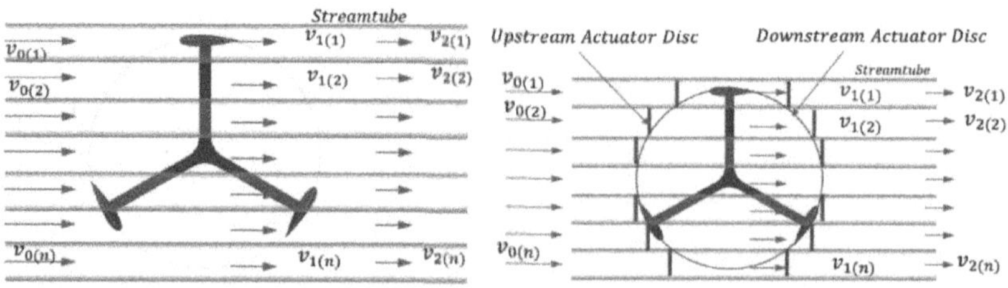

Figure 2.6 Multiple streamtube (left) and double multiple streamtube (right) models.

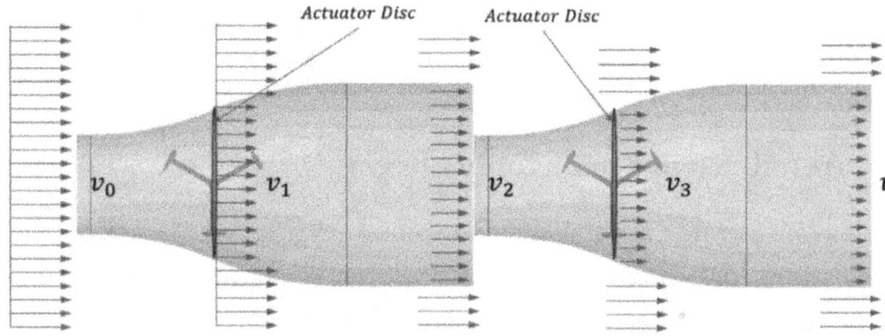

Figure 2.7 Double actuator disc model.

Nguyen[12] included viscous drag for similar calculations at the expense of slower convergence of the model. The multiple streamtube model (MST) was further improved by Muraca *et al.*[14] by the addition of support struts, blade aspect ratio, and solidity. The effect of Re was incorporated by Sharpe.[15] The MST is suitable for the power production of lightly loaded turbines of low solidity.

2.2.3.5 *Double Multiple Streamtube Model*

The MST was further extended by Paraschivoiu[8] by adding two actuator disks for upwind and downwind halves of the rotor and their respective induced velocities on each disk as shown in Figure 2.6. The induced velocities can be calculated either by constant or variable interference factors as a function of azimuthal angle. The proposed double multiple streamtube (DMST) uses two interference factors for upwind and downwind leading to double iteration. The torque C_Q for the upwind and the torque C'_Q for the downwind half are added to obtain the power coefficient C_p of the turbine. The current DMST model accounts for the secondary effects including streamtube expansion, blade geometry, rotating tower and the effect of struts and spoilers. Boeing–Vertoll stall model was the recent addition. A good agreement with the experimental data has been observed for different solidity and high TSR. By far the

DMST model predicts more accurately than any other available mathematical models.

2.3 Aerodynamics of Savonius Rotor

The Savonius rotor manifests in various configurations based on the applications such as water pumping or power generation. As the application demands different requirements, it may be either higher drag or higher speed; the Savonius geometrical parameters as shown in Figure 2.8 can be modified and optimized without notable decrease in the power performance. Due to its high starting torque, these machines are often integrated into Darrieus rotors as starters. The maximum efficiency of Savonius turbine is reported to be 21%–24%[16] at optimum TSR of 1.2–1.5. The following section explains the influence of geometrical parameters on power performance of the Savonius rotor.

2.3.1 *Effect of Aspect Ratio*

The aspect ratio can be defined as the ratio between the heights to the diameter of the rotor. It was concluded through experimental results that

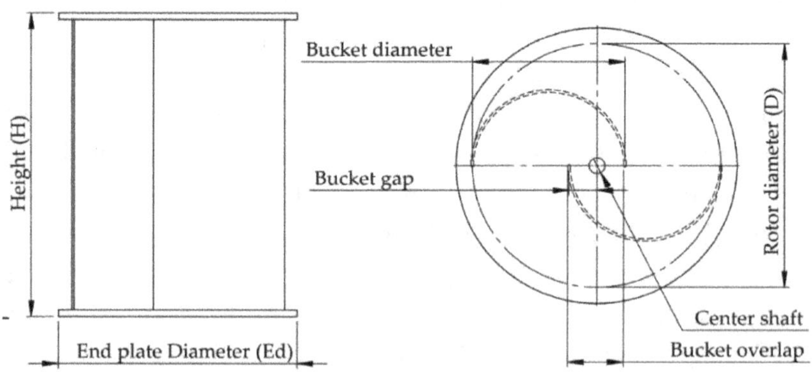

Figure 2.8 Geometric parameters of a conventional Savonius rotor.

the efficiency increases with the aspect ratio.[16] But a helical Savonius rotor with aspect ratio of 0.88 was found to yield better results than turbines with 0.93 and 1.17 aspect ratio.[17] Though higher aspect ratio of 6–8 is favorable for power performance, a rotor of lower aspect ratio is preferred for structural reasons. The performance of low aspect ratio can be significantly improved with end plates.

2.3.2 *Effect of End Plates*

Past research shows that the end plates significantly improve power performance by preventing the fluid flowing over the blades and maintaining the pressure difference between the concave and convex sides over the height of the rotor.[18] Sivasegaram[19] conducted experimental investigation on the effect of end plates and found that optimum end plate diameters is 1.1 times the rotor diameter. Jeon *et al.*[20] studied the effect on end plates on a helical Savonius rotor and found that with upper and lower end plates, the performance increased by 36% compared to without end plates. End plates with larger diameter tend to increase the rotor inertia and have to be carefully chosen for the given rotor diameter.

2.3.3 *Effect of Overlap Ratio*

There are conflicting judgments on the overlap ratio. According to Kamoji, Kedare and Prabhu,[17] the maximum Cp decreases with increase in overlap ratio, whereas the experimental study by Alexander and Holownia[16] reveals that the efficiency increases with the overlap ratio for the tested values of −0.07 to 0.22. He added that large bucket spacing let the flow diverge before concentrating on the returning bucket. The study by Mabrouki *et al.*[21] indicates that the maximum mechanical power was obtained at overlap ratio of 0.3. Mojola[22] conducted a field test on the Savonius rotor with seven different overlap ratios and concluded that 0.25 is the optimum value.

2.3.4 *Effect of Number of Blades*

The number of blades plays a critical role in power performance of Savonius turbines. With increase in number of blades, the static and dynamic torque fluctuations can be minimized. From an efficiency perspective, two blades offer maximum performance than three or four blades as verified experimentally by Mahmoud *et al.*[23] A maximum Cp of 0.24 was reported for two-bladed turbine at Re = 8.64×10^5 by Blackwell *et al.*[24] For rotors with three or four blades, the starting torque characteristics are better than two-bladed rotors. High torque application such as water pumping demands more blades whereas electricity generation requires fewer blades for higher speed. On the downside increased blade number will create wake for the following blades.

2.3.5 *Effect of Multi-Staging*

The number of stages does not have significant effect on the power performance as claimed by Kamoji, Kedare and Prabhu,[17] yet its influence is more pronounced on the static torque characteristics. 90^0 phase shift between the stages improves the directional starting with respect to incoming wind. Marginal improvement in power output was noted by Mahmoud, El-Haroun, Wahba and Nasef[23] between two-stage and single-stage rotor. The double-stage rotor has higher static torque leading to higher static torque coefficient.

2.3.6 *Effect of Accessories*

Ogawa and Yoshida[25] have investigated the effects of the deflecting plate and its parameters to improve the performance of the Savonius wind rotor. The results show that the rotor power is higher by approximately 30% than the conventional Savonius rotor without the deflecting plates.

2.3.7 *Effect of Blade Shape*

Saha *et al.*[26] have investigated the twisted blade in a three-bladed rotor system in a low-speed wind tunnel and compared its performance with semicircular blades. They found out that the twisted blade showed a maximum of Cp = 13.99%, whereas the semicircular blade showed a Cp = 11.04%.

To increase the efficiency, helical Savonius was proposed by Damak *et al.*[27] A comparative study was conducted to compare the performance with the conventional rotor. Maximum Cp of helical Savonius achieved was 0.2, whereas the conventional Savonius rotor can achieve only 0.16 at Re of 116,000.

2.3.8 *Effect of Reynolds Number*

The Re defines the flow behavior on the Savonius buckets and it is a vital parameter for turbine performance. Cp and Re have linear correlation. Kamoji, Kedare and Prabhu[17] reported that when the Re increases from 0.8×10^5 to 1.5×10^5, the performance increases by 19%. Akwa, Vielmo and Petry[18] also reported that the static moment coefficient also increases significantly with Re.

2.3.9 *Effect of Tip-speed Ratio*

TSR is an imperative parameter from a structural perspective, as the key factors such as number of blades, number of stages and other geometric parameters are based on it. Blackwell *et al.*[28] reported that the optimum TSR for maximum Cp is between 0.7 and 0.9. Whereas for the turbine with deflecting plate, the TSR reaches as high as 1.08. Also it was found that the TSR decreases with increase in overlap ratio.[17]

2.4 State-of-the-art Technologies for Starting and Low Wind Speed Operation*

The search for a feasible solution to enable Darrieus turbines to generate optimal power at low wind speed has persisted for decades. Over the years, countless efforts have been made through experimental, computational and field tests to achieve significant progress in understanding self-starting physics of a Darrieus turbine. The attempts substantially improved the starting behavior, yet Darrieus turbines are not preferred for low wind speed sites for their low annual energy output. Another limiting factor for their not being widely preferred is the lack of aerodynamic power regulation such as furling or blade pitching as in the case of HAWT.

The objective of the current literature search is to identify or propose a feasible solution that will enable the Darrieus rotor to generate power rather than self-start. The existing strategies are critically examined with the constraints on cost, ease of maintenance, lower complexity and high reliability. The concept behind each strategy is discussed in detail with a sketch of the Darrieus turbine incorporating the same. Cp of each strategy is compared with the conventional turbine to quantify the performance loss or gain. The chapter ends with a comparative table listing the strategies along with their merits and demerits.

2.4.1 Airfoil Characteristics

The airfoil geometric features and their aerodynamic characteristics have a profound effect on the turbine performance and starting characteristics. The airfoil characteristics evaluated in this text pertains to small fixed pitch straight-bladed Darrieus turbine of power capacity less than 20 kW

* Parts of this section are adapted with permission from Mohan Kumar, P.; Sivalingam, K.; Lim, T.-C.; Ramakrishna, S.; Wei, H. Strategies for enhancing the low wind speed performance of H-Darrieus wind turbine—part 1. *Clean Technologies* **2019**, *1*, 185–204.

that may typically operate at the rated wind speed of 10~12 m/s. Though a high C_{ll}/C_d ratio is preferred for any airfoil design, a maximum C_{ll}/C_d over a wider range of AoA is the desired characteristic for VAWTs. The chosen airfoil should be able to generate sufficient lift both in the positive angle of incidence during the upstream half and the negative angle of incidence in the downstream half.

The airfoil exhibits lowest drag for a narrow AoA termed as "drag bucket" by Islam *et al.*[29] A wider drag bucket is a desirable property to maintain the performance over larger AoA. Laneville and Vittecoq[30] identified that stall angle at low Re has strong influence on the performance at low wind speed and self-starting. A stalled blade at low TSR fails to incite the rotation there by delaying the acceleration of the rotor. Hence an airfoil that has wider stall angle at low Re is preferred. Deep stall behavior has detrimental effect on the performance and should be postponed to higher AoA. Timmer and Van Rooij[31] found that a linear correlation exists between the thickness of the airfoil and deep stall angle. Jasinski *et al.*[32] studied the effect of surface roughness on the performance of airfoil in low Re and found that a high surface roughness at the leading edge will induce turbulent flow over the entire airfoil. Hence a significant weightage factor on surface roughness should be accounted for while selecting an airfoil.

Most of the fixed pitch Darrieus turbines are installed in urban environments, where noise is one of the deciding factors as suggested by Shepherd *et al.*[33] The noise in the straight-bladed turbine is generated when the laminar separation bubble leaves the trailing edge by inducing vibrations. Kato and Seki[34] claimed that the pitching moment coefficient of the airfoil should be higher for increased power performance. His study shows that high Cp is achieved when $C_m = 0.05$ rather than $C_m = 0$. The C_m value is higher for the camber blades whereas for the symmetric blades it oscillates around zero. Myriad airfoils were proposed for Darrieus turbine in the past decades with the objective of promoting self-starting and to reduce the manufacturing cost. Following are the prerequisites for selecting an airfoil:

i. Operating wind speed range — Re range.
ii. Fixed pitch or pitch-able blades.
iii. Straight blades or helical blades.
iv. Installation environment — urban or rural location.
v. Self-start or assisted start such as motor start.
vi. Drive train arrangement — resistive torque and efficiency.
vii. Manufacturing method of blades — aluminum extrusion or moulding.

The need for optimized airfoil dedicated for VAWT arose in the late 1970s when wind energy emerged as a key renewable source after the energy crisis and there was a desire to improve the performance of Darrieus turbine to be on par with the HAWT. Following is the list of prospective airfoils developed so far as illustrated in Figure 2.9.

2.4.1.1 *WSU 0015*

The airfoil was developed by Snyder and Furukawa[35] at Witchita State University (WSU) in an effort to postpone the stall angle and to attain peak performance at lower TSR. The experimental comparison of WSU 0021 with NACA 0012 reveals that the synthesized WSU 0021 achieved a peak Cp of 0.25 at TSR 4, whereas the NACA 0012 can achieve 0.2 at TSR 5.

2.4.1.2 *NACA 00XX*

Symmetrical NACA profiles were preferred during early days of VAWT development due to readily available test data that are meant for aviation purposes. Initially NACA 0012 was employed in smaller capacity turbines. As the turbine size grows bigger, NACA 0018[36] is preferred for

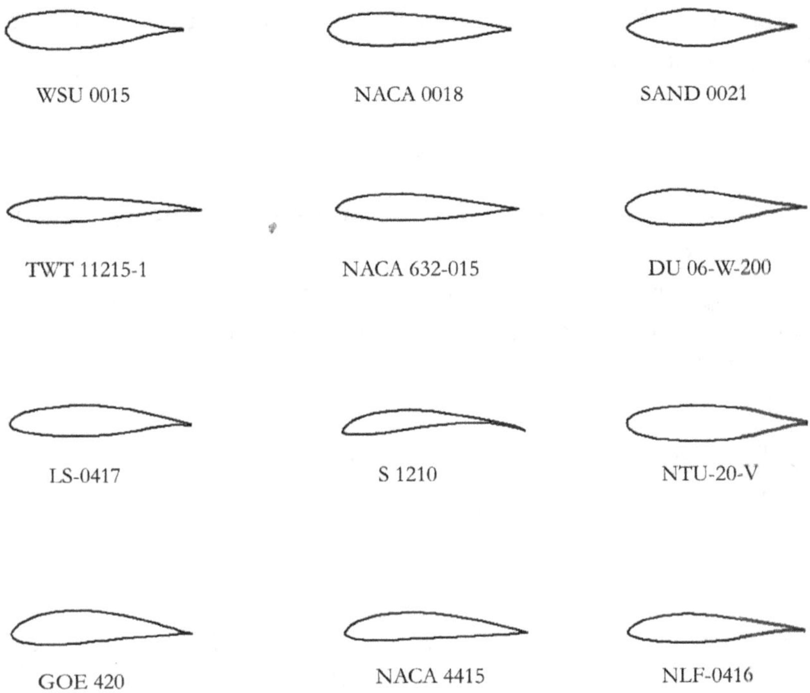

Figure 2.9 Prospective airfoils for VAWT.

the structural stiffness. Several NACA 00XX profiles are evaluated for higher AoA and Re range suitable for larger VAWTs.

2.4.1.3 *SAND 00XX/XX*

Sandia National Laboratories optimized early NACA profiles for sharp stall and wider drag buckets at high AoA. The resulted airfoils are designated as SAND 0015/47, SAND 0018/50 and SAND 0021/50.[37] The last two digits indicate the location of laminar flow on the chord. The new airfoils are less sensitive to surface roughness and can be seamlessly integrated with other airfoils for performance improvement. The 500 kW test turbine comprises both NACA 0021 at the root for structural rigidity and SAND 0018/50 at the equatorial region of Troposkein blade for aerodynamic performance.

2.4.1.4 *TWT 11215-1*

The TWT airfoil series is an outcome from Tokai University research on VAWT development. The airfoil TWT 11215-1 was developed by Kato *et al.* employing single streamtube model with the objective of gaining large pitching moment and minimal drag coefficient about zero lift line. The airfoil has a negative camber on the leading edge and a positive camber on the trailing edge, giving rise to an S-shaped camber line. The experimental results show that an increase in Cp was noted with increase in the pitching moment.

2.4.1.5 *NACA 6 Series*

NACA 64 and 63 series airfoils were optimized by Migliore[38] at West Virginia University and compared with traditional NACA 00XX airfoils. The results show a higher Cp at higher TSR than traditional airfoils. Subsequent testing revealed that the performance increases with the airfoil thickness, a desirable property from the structural perspective. NACA 63_2015, corrected for flow curvature effects, outputs 20% more annual energy than NACA 0015.

2.4.1.6 *ARC Series*

The ARC series of airfoils are conceived by National University of Athens. The series was devised as a potential solution for flow curvature effect on Darrieus turbine performance. It was found that the airfoil behaves differently when it is subjected to curvilinear flow and rectilinear flow. To counter the effect of flow curvature, a virtual camber and virtual angle of incidence are introduced on the symmetric airfoil in relation to the turbine radius. The airfoil camber value should be tailored for the given turbine parameters as c/R changes. Zervos[39] designated the flow curvature corrected airfoil of NACA 0015 as ARC 0015. An added advantage of ARC airfoils is the ability to diminish the unsteady loading, thereby extending the fatigue life of blades.

2.4.1.7 *DU 06-W-200*

The DU series of airfoils were developed by Claessens[40] at Delft University with the aim of improving both the aerodynamic and structural characteristics. The airfoil DU 06-W-200 was conceived in an effort to increase the thickness of the airfoil without compromising the maximum lift coefficient. The optimized airfoil is 20% thick and 0.8% camber. An extensive wind tunnel comparison with NACA 0018 shows that the DU 06-W-200 airfoil performs similar to NACA 0018 in the negative AoA, while an increased lift coefficient is reported at positive AoA. The deep stall occurs with reduced drop in lift coefficient at higher AoA. The laminar separation bubble of DU 06-W-200 does not extend up to the trailing edge as in the case of NACA 0018. The DMST validation shows 8% increment in Cp and 108% improvement with high surface roughness compared to NACA 0018.

2.4.1.8 *LS-0417*

The LS series were developed by McGhee and Beasley[41] at NASA for low-speed general purpose aviation. The symmetric airfoil with 17% thickness was compared with NACA 0012, 0015 and 0018. The results demonstrate that instantaneous Cp, normal and tangential force coefficients are smoothly distributed with smaller peaks. The LS-0417 was compared with GOE 420, NACA 4415 and NLF-0416 for Re 1×10^5 to 3×10^5.[29] The LS-0417 performed better than the all the investigated airfoils and later become the basis for the evolution of prospective airfoils.

2.4.1.9 *S1210*

S1210, high lift and low Re airfoil was considered by Islam, Ting and Fartaj[29] for VAWT from the self-starting perspective. Experimental verification indicates that a high static torque coefficient ensures the self-starting capability. The maximum Cp of 0.32 was obtained for the rotor solidity of 0.1, after which a significant drop is noted. Singh *et al.*[42] compared S1210 with NACA 0015 and found that the S1210 airfoil recorded a lower Cp due to reduced torque generation on the downstream half. The 7.2% camber is attributed for the poor downstream performance. Though S1210 may be appropriate for self-starting, 12% thickness will pose a challenge on achieving required structural stiffness.

2.4.1.10 *NTU-20-V*

The NTU-20-V airfoil developed by Kumar *et al.*[43] at Nanyang Technological University, Singapore is a recent addition to the VAWT airfoil list. The objective is to facilitate the manufacturing of inexpensive blades with extended fatigue life for smaller-capacity turbines. With the constraints on thickness and lift coefficient, genetic algorithm optimization was carried out to generate the new airfoil profile that can accommodate standard shapes of carbon fiber material from NACA 0018 surface pressure distribution. The conceived airfoil is 20% thick at 37.5%c. A minimum thickness of 17% is extended over 40%c. The computational comparative study with NACA 0018 on VAWT for the Re range 1.2×10^5 to 2.7×10^5 displayed almost the same Cp as the parent airfoil, but at a higher TSR.

2.4.2 *Camber and Symmetric Airfoils*

The most popular choice of airfoils for Darrieus turbines are symmetrical NACA profiles such as NACA 0015, 0018 and 0021 as they are experimentally validated for the range of Re that a VAWT typically experiences while in operation. Though it seems logical and efficient to use symmetrical airfoils as the suction and pressure side are reversed during the upwind and downwind rotations, the cambered airfoils are preferred from the self-starting perspective, especially for the fixed pitch VAWT as evaluated by Rainbird *et al.*[44] The airfoils should generate sufficient lift at low Re of 10^3 to 10^4 to overcome the resistive and inertial torque of the turbine to self-start. The symmetrical airfoils exhibit poor performance in this range and thus fail to start at low wind speeds. In contrast, a cambered airfoil is able to generate moderate lift in this range of Re, inducing rotation without significant abatement in power production at high winds. The cambered airfoils can be mounted in concave out or concave in (concave side facing the rotor center) configuration as shown in Figure 2.10. Since the power output is proportional to the cube of wind speed, the upstream half of a rotor encounters high velocity compared to downstream half and hence it is wise to select a configuration that performs better in upstream half. To enhance the self-starting with cambered airfoils, the concave out configuration extracts maximum power due to positive angle of incidence in upstream half, whereas for the concave in configuration, the airfoil encounters negative angle of incidence in upstream half leading to reduced power output. Though cambered airfoils generate moderate drag on the downstream half due to negative incidence angle, the superior performance due to positive angle

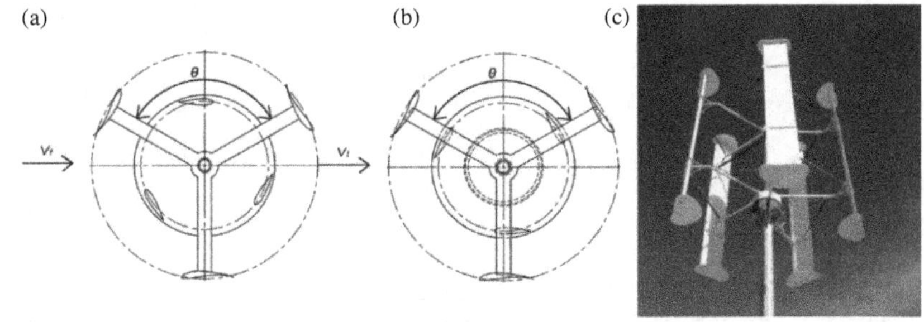

Figure 2.10 (a) Concave in. (b) Concave out. (c) Turbine with cambered blades. Image (c) is reproduced with permission from https://windsailenergy.com/our-vawt-models/p5000-ab-vertical-wind-turbine-vawt/.

of incidence in the upstream half will result in the higher overall performance at low Re. Wortmann FX63-137 airfoil of more than 5% camber was investigated by Islam *et al.*[45] comparing with NACA 0012. The outcome is that the airfoil with strong laminar separation bubble phenomena before stall and lower post-stall drag are desirable features for a fixed pitch VAWT. Zervos[39] compared NACA 63-015 with NACA 0012, 0015 and 0018 and found that cambered airfoil has smooth distribution of normal and tangential forces over a wide range of azimuthal angles compared to instantaneous peak forces by symmetrical airfoils, which extends the longevity of the blades and structures by reducing the fatigue stress. Kirke[46] suggested that the cost of manufacturing a symmetrical and cambered airfoil is almost the same with added minor tooling charges.

Several studies in the past were carried out experimentally to ascertain the performance of cambered over symmetrical airfoil leading to the development of special airfoils for fixed pitch VAWT. NACA 0018 was theoretically compared with the Gottingen series (GOE 460, GOE 676, GOE 738) of varying camber % and it was concluded that power output of the cambered airfoil was inversely proportional to camber %. A highly cambered S1210 airfoil was compared with NACA 0015 by Healy[47] who found that the maximum Cp occurs at lower TSR for cambered airfoils vs symmetrical airfoils as shown in Figure 2.11.[46] The cambered airfoils display larger stall angle due to the requirement of large adverse pressure

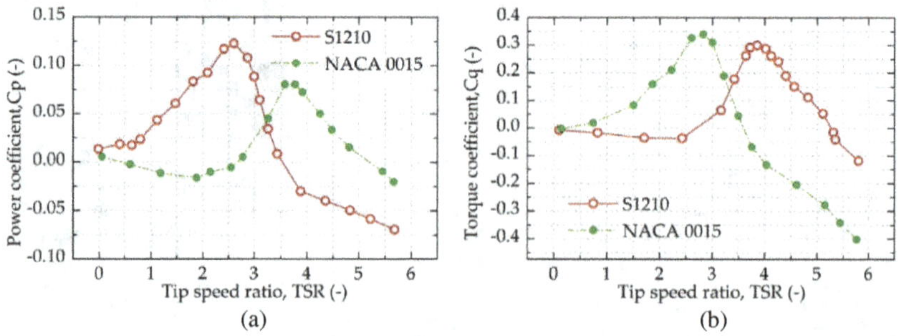

Figure 2.11 (a) Cp vs TSR and (b) Ct vs TSR for S1210 and NACA 0015 sections.

gradient either on suction or pressure side of the airfoil. Another notable advantage of cambered airfoil is higher pitching moment and less sensitivity to roughness compared to symmetrical airfoils. From the overall turbine performance, cambered airfoil performs superior to symmetric airfoils especially for the turbines operating at low TSR. For the self-starting and low wind speed operation, cambered airfoils are the better choice and have been commercially implemented as in Figure 2.10(c).

2.4.3 *Solidity*

Solidity is defined as the ratio between the total blade area to the swept area. The solidity can be achieved either by increasing the number of blades as shown in Figure 2.12(a) or by increasing the chord length as displayed in Figure 2.12(b) for a given radius of the rotor. Solidity is a critical factor in influencing the rotor characteristics and the immediate effect is on the TSR. Increasing the solidity of the Darrieus turbines has a negative impact on the aerodynamic performance. Ågren *et al.*[48] stated that the decrease in the peak power performance is attributed to the interaction of blade with the wake generated by the preceding blade, thereby increasing the AoA and leading to the earlier stall of the following blades. Despite a notable reduction in the Cp, high solidity rotors are favored for their desirable characteristics. Mohamed[49] confirmed

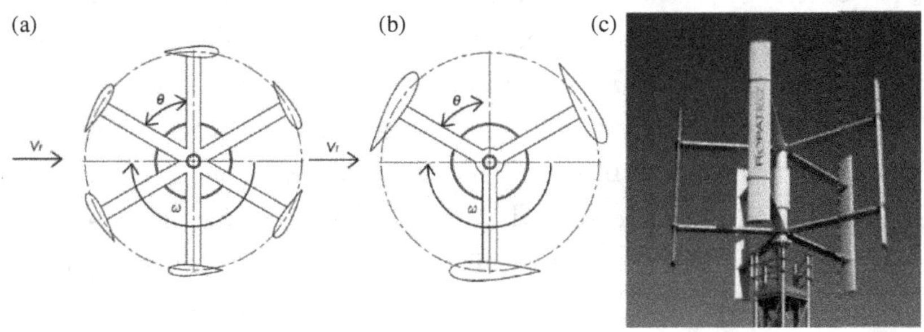

Figure 2.12 (a) Increased number of blades. (b) Increased chord. (c) High solidity turbine. Image (c) is reproduced with permission from https://www.ropat.it/language/it/energia-rinnovabile-impianti-eolici-ad-asse-verticale/.

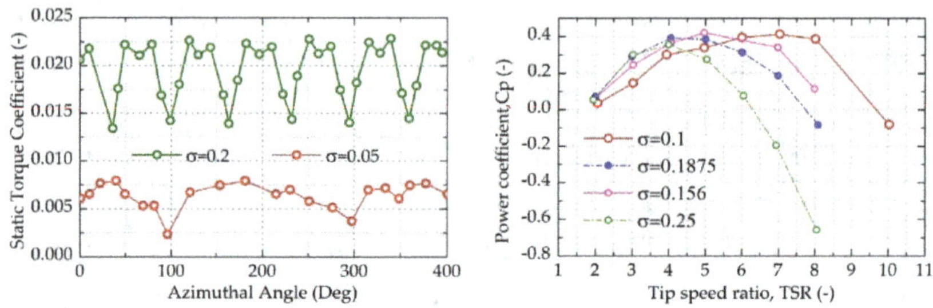

Figure 2.13 Effect of solidity on C_{Ts} (left) and Cp (right).

experimentally and numerically that increasing the number of blades generates marked increases in the static torque and high dynamic torque at low wind speeds. Increasing the solidity from $\sigma = 0.05$ to $\sigma = 0.2$, the static torque coefficient was increased by a factor of 4 as shown in Figure 2.13(a) and the variation of Cp in Figure 2.13(b). Hence to generate higher starting torque, more number of blades are preferred. Additionally, a high solidity rotor is able to operate at a wider TSR, extracting the energy from a larger wind speed spectrum and outputting higher annual energy. Nevertheless, while a higher number of blades is preferred for low wind speed sites, the optimum number of blades is between two and five. The rotor rpm will be very low if the blade number is increased beyond five. As most of the Darrieus turbines are intended for electricity generation, a low rpm rotor demands high reduction ratio gearbox or larger permanent magnet generator if it is a direct drive power train. El-Samanoudy *et al.*[50] reveals that increasing the number of blades from two to four recorded significant power increases, but the difference between three and four blades is iota. More blades have significant aerodynamic torque loss due to the drag caused by supporting struts. Though two-bladed and three-bladed turbines do not significantly vary in the power performance, a two-bladed turbine is not preferred from the self-starting perspective as the self-starting capability depends on the initial starting orientation with respect to oncoming wind as described by Worasinchai *et al.*[51]

A two-bladed rotor is prone to vibrations from the unsteady loading of blades on a single plane. The cyclical peak torque for a two-bladed turbine occurs at an narrow azimuthal angle, whereas for a high solidity rotor with higher number of blades, the torque generation happens on a wider azimuthal angle greatly reducing the fatigue stress on the blades and support struts. Staelens *et al.*[52] found that the optimum solidity to extract maximum energy lies between $\sigma = 0.25$ and $\sigma = 0.4$. Another vital consideration is the noise, especially for the turbines intended for urban installations.

Number of blades has direct correlation on the noise generation. Mohamed[53] investigated noise from the Darrieus turbine with solidity $\sigma = 0.1$ and $\sigma = 0.25$ and revealed that the turbine with solidity of $\sigma = 0.25$ generate 7.6 dB more. The higher noise was attributed to the blade and wake interaction. A two-bladed turbine operating at a higher TSR can generate attention-grabbing swishing noise. Increasing the number of blades invariably increases the cost due to additional weight without significant increase in power, yet from the rotor dynamics and longevity of the structure, a more aesthetically pleasing three-bladed turbine is an apt choice. Hence it can be concluded that for low solidity rotors either two or three-bladed can be optimum for maximizing the performance, and for low wind speed operation high solidity rotors of four or higher number of blades are preferred.

2.4.4 *Helical Blades*

The helical turbine was proposed by Gorlov[54] to generate starting torque irrespective of the initial orientation of the rotor. The helical Darrieus turbine is constructed by constantly twisting the straight blades around the rotor axis as shown in Figure 2.14. The number of blades, size of the turbine and the airfoil characteristics dictate the swept angle. The starting torque generated is constant over 360^0 azimuthal position reducing the torque ripple during continuous operation. Poor self-starting capability

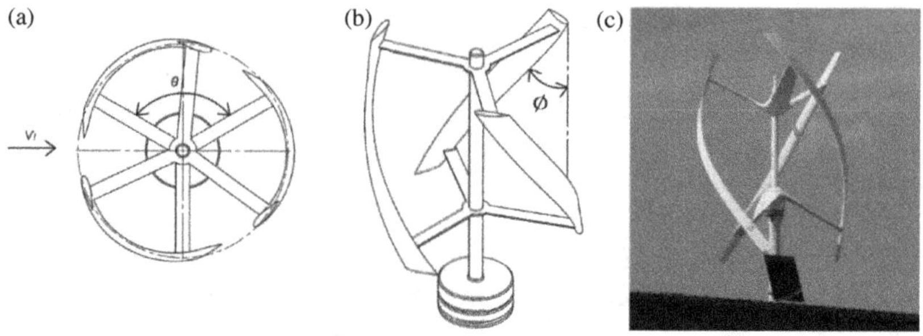

Figure 2.14 (a) Top view of helical rotor. (b) Front view of helical rotor. (c) Commercial helical turbine. Image (c) is reproduced with permission from http://www.visionairwind.com/visionair-5/.

of a Darrieus rotor is due to the lack of total energy generated by the blades to overcome the resistive torque to accelerate beyond the dead band. Baker[55] calculated the total energy that can be obtained for the Darrieus rotor, shown in Eq. (2.18) with respect to azimuthal position.

$$E = \varnothing rqC_t(\alpha)d\varnothing \tag{2.18}$$

To increase the energy yield either the blade tangential thrust should be increased or the blade span angle $(\alpha)d\varnothing$ has to be modified. The angle of inclination of the blade can be either outward or forward as given by Eq. (2.19).

$$\alpha = \sin^{-1}\left(\frac{V_f \sin\varnothing \cos\varnothing}{V}\right) \tag{2.19}$$

$$V = ((r\omega - V_f \cos\varnothing)^2 + (V_f \sin\varnothing \cos\varnothing)^2)^{\frac{1}{2}} \tag{2.20}$$

The forward lean of the blades influences the airfoil characteristics as the 2D flow assumption is no longer valid due to the introduction of spanwise flow converting it into complex 3D flow. Nevertheless, a reduction in lift coefficient was reported for such 3D flows, as the stall angle is increased due to the introduction of the span-wise flow component to the attached boundary layer.

For a static airfoil with an inclination of 30^0, an increase in the stall angle from 16^0 to 20^0 was reported. Purser and Spearman[56] reported that as the inclination angle was increased further, stall angle rises to 55^0 yielding a lift coefficient of 2.25. Shiono *et al.*[57] experimentally compared the helical and straight three-bladed water turbine and showed that the straight-bladed turbine has higher efficiency. A maximum Cp of 0.34 can be achieved with the straight blades, whereas a helical turbine with helix angle of 43.7° was able to achieve only a maximum Cp of 0.16 as shown in Figure 2.15 (right). The comparison of starting characteristics of the helical and the straight blade is shown in Figure 2.15 (left). Another important aspect of the helical blades is the wake pattern in the downstream and its interaction with the blades during the downstream travel. The blade wake is continuous instead of vortices as formed in the conventional straight-bladed Darrieus rotor. The blade vibrations are greatly reduced due to smooth interaction of wake. The tangential force loading of the blade is distributed along the blade span due to progressive shift in the AoA.

The helical turbine compared to straight-bladed turbines is aesthetically pleasing and well suited for the built-up environment as they are quiet in operation. On the downside Blackwell and Reis[58] claim that the cost

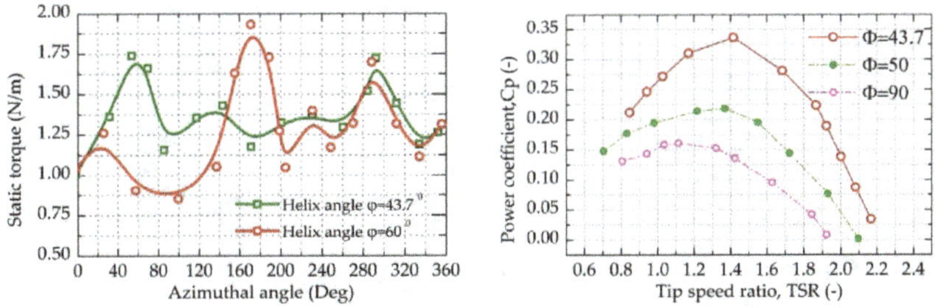

Figure 2.15 Effect of helical angle on starting torque (left) and Cp (right).

of manufacturing for the helical blade is higher as the manufacturing process is limited to molding and material is limited to composite fibers. The manufacturing cost of the helical blades is substantial out of total cost as most small wind turbines employ well-established aluminum extrusion process for manufacturing inexpensive blades.

2.4.5 *Blade Thickness*

Blade thickness has significant influence on both aerodynamic and structural characteristics. An optimum thickness should perform even in low Re for improved low wind speed operation. Top view of Darrieus rotors with thick and thin airfoils are shown in Figure 2.16. The airfoil thickness should be chosen cautiously, as the local relative velocity is different for the upstream and downstream halves. A thicker airfoil may perform better in the upwind half and suffer high separation bubble drag in the downwind half especially at low TSR and high AoA. A better overall performance was displayed by 18% thick airfoil for the Re range of $2 \times 10^5 - 3 \times 10^5$ as claimed by Healy[47] for low wind operation. Danao *et al.*[59] demonstrated that a thicker airfoil of 19% improved the starting capability than thinner airfoil. Batista *et al.*[60] suggested that the maximum thickness can be increased to 28.5% without any penalty on the Cp and to achieve better startup torque.

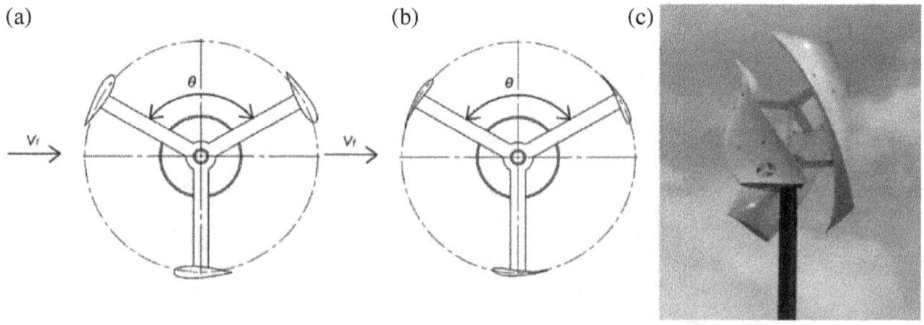

Figure 2.16 (a) H-Rotor with NACA 0021 (thick) airfoil. (b) H-Rotor with S1210 (thin) airfoil. (c) Commercial turbine with thicker blade. Image (c) is reproduced with permission from https://enerlice.fr/en/uge-v-air-wind-turbines/.

The turbine loses self-starting ability if the thickness is increased beyond this limit, due to increase in profile drag especially at low AoA substantially reducing the lift/drag ratio. The effect of airfoil thickness on the starting torque was investigated by Alam and Iqbal[61] at constant Re of 2×10^5. Varying thicknesses of symmetric NACA airfoils were tested and results showed that thicker airfoils perform better at low Re and early start.

From Figure 2.17 (left), the thinner airfoil NACA 0012 exhibits a sudden drop in the lift coefficient, whereas thicker airfoils such as NACA 0018 and NACA 0021 show smooth drop in lift. The sudden drop will induce vibrations in the rotor, which is more pronounced when it is rotating at relatively higher rpm. Also, it is evident from Figure 2.17 that the Cl variation between NACA 0012 and NACA 0018 is minimal, but the advantages that are gained structurally are remarkable. Figure 2.17 (right) shows the Cp comparison with TSR. The difference in the peak Cp between NACA 0012 and NACA 0018 is 16%, whereas the difference in peak Cp with NACA 0021 is 79%. The peak Cp is reduced as the thickness is increased but the turbine is able to operate at a wider range of TSR.

A thicker airfoil also facilitates the integration of blade load-alleviating devices such as actuators or flaps. As the turbine size grows, these devices play a critical role in reducing the blade load thereby extending blade life. A thicker airfoil will have more space to accommodate these devices.

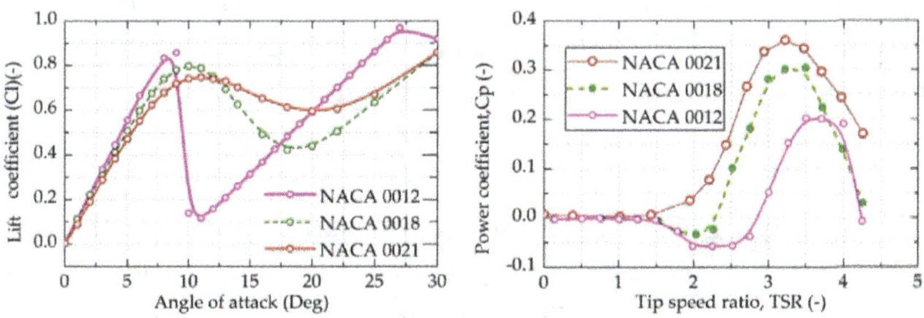

Figure 2.17 Variation of Cl (left) and Cp (right) with airfoil thickness.

Apart from the aerodynamic gains, a thicker airfoil provides higher bending stiffness, facilitating the addition of blade spar and enhancing the overall structural characteristics. The structurally improved blades are able to withstand high centrifugal forces allowing the rotor to rotate at higher rpm.

A high rpm rotor in turn reduces the generator and the drive train size with significant cost savings. Parchen *et al.*[62] found that the noise generated is inversely proportional to the thickness. The thicker airfoil will generate less noise compared to thinner airfoils especially at higher TSR. Hence, an optimum airfoil thickness is 18~20%c to achieve a better overall performance at low wind speed and for self-starting for the small wind turbine.

2.4.6 *Vortex Generators*

Vortex generators are passive devices commonly employed to delay the flow separation.[63] Though airfoil stall cannot be avoided, the onset can be delayed, maximizing the energy yield of a Darrieus turbine. Apart from delaying the stall, introducing the vortex generators on the Darrieus rotor blade is of interest to increase the lift at low Re during the low wind speed operation and starting as shown in Figure 2.18(a). The principle behind any vortex generator design is to induce stream-wise vortices that can exchange the high momentum fluid from freestream to relatively slower-moving boundary layer. This process energizes the boundary layer to overcome the adverse pressure gradient during high AoA, thus postponing the stall.[64] The dimensions of the vortex generator rely on the boundary layer thickness and may result in drag penalty if not optimized.

Though vortex generators can induce co-rotating or counter-rotating flows, the paired counter-rotating vortex generators arranged at 30° to the blade angle was most effective as suggested by Seshagiri *et al.*[65] An

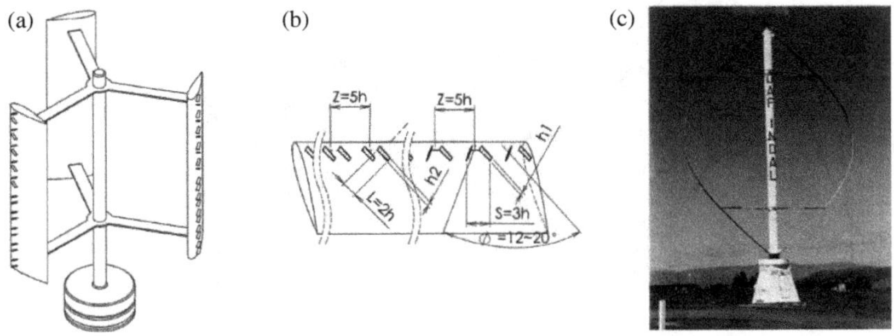

Figure 2.18 (a) H-Rotor with vortex generator. (b) Typical vortex generator details. (c) DAF INDAL turbine with vortex generator.[64]

optimal chord-wise position and span-wise spacing are crucial in determining the performance as shown in Figure 2.18(b). The chord-wise spacing has greater influence on the post-stall behavior with sudden decrease in Cl and optimal location is found to be 30%c rather than 10–15%c especially at higher AoA. The vortex generators, however, did not eliminate the laminar separation bubble but it reduced the size by splitting into segments with marked increase in the Cl of 25% at Re 8×10^4. The lift and drag characteristics of airfoil with and without vortex generator are shown in Figure 2.19 (left) and the influence on the power generated is shown in Figure 2.19 (right). The span-wise location of the vortex generator dictates the efficiency especially in an egg beater-shaped Darrieus turbine, since the root and equator regions operate at different tip speed. A counter-rotating vortex occupying 12% blade span reported a 5% increase in annual energy output if placed at root region[66] and 20% increase in power output at low TSR. When placed on the equator region of DAF 50 kW turbine, annual energy output fell from 7%. Though vortex generators are anticipated to improve the starting capability marginally, the peak power reduction is significant. Vortex generators are relatively simple in construction and are low cost, but associated drag at higher TSR is inevitable. Since the flow over VAWT blades is inconsistent, detailed turbine-specific

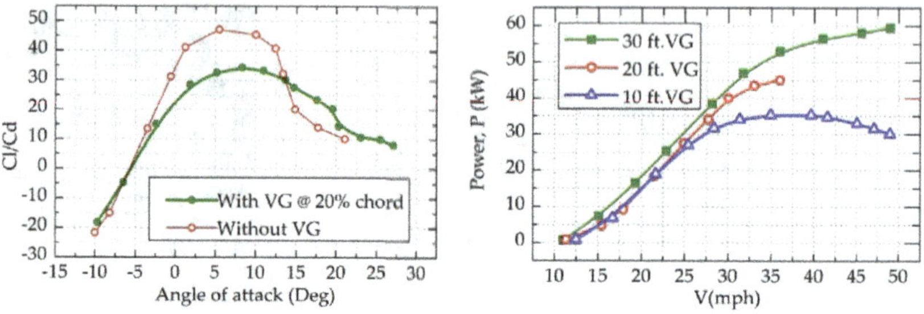

Figure 2.19 Cl/Cd vs AoA for NACA 63-618 with and without vortex generator (left) and power vs wind speed for various dimensions of vortex generators (right).

numerical and experimental optimization has to be carried out to ascertain the influence on low and high TSR. Noise has to be accounted for if it has to be installed in urban environments.

2.4.7 *Stepped Airfoils*

This is the family of airfoils with abrupt step or cavities introduced either on the suction or pressure side or both. The airfoils are first proposed by Kline and Fogleman[67] (KF) claiming high lift coefficient and higher stall angle than the conventional airfoils. Lumsdaine *et al.*'s[68] study on KF airfoils revealed that the higher lift coefficient was achieved at the expense of higher drag and the study concluded that the KF airfoil lift/drag ratio is similar to that of a flat plate.

Hence the applications of KF airfoils are limited to model planes in the past. The initial claim for high lift coefficient was attributed to the trapped vortex in the cavity and subsequently creation of a low-pressure zone. A typical airfoil with step integrated with the straight-bladed Darrieus rotor is shown in Figure 2.21. The variation of lift coefficient of NACA 2412 with step location and NACA 23012 with step depth is shown in Figure 2.20.

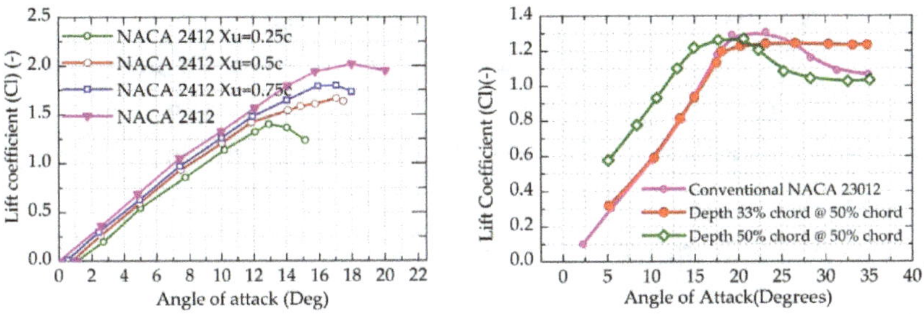

Figure 2.20 Variation of Cl with step location (Xu, left) and step depth (right).

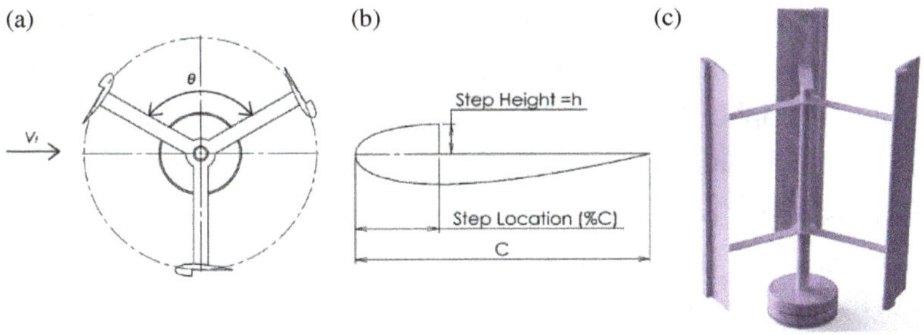

Figure 2.21 (a) H-rotor with stepped airfoil. (b) Details of an airfoil with step. (c) Illustration of a H-rotor with stepped airfoil.

There is a resurgence of KF airfoils after Finaish and Witherspoon's[69] investigation revealed that at low Re 1×10^5 the airfoils with the step on the suction side indeed have high lift for AoA up to 10^0 and are able to delay the stall compared to conventional airfoils. Most of the studies on KF airfoils are limited to pre-stall region expecting higher lift/drag ratio. Though it does not outperform conventional airfoils in the pre-stall region,[70] it demands attention for its aerodynamic behavior in the post-stall region as described by Cox *et al.*[71] Mishriky and Walsh[72] revealed that the stall angle can extend up to 40°, a desirable property for the Darrieus turbines operating at low TSR. An extended cavity on the suction side can increase the starting torque drastically by reverse drag.

A thorough literature search shows that KF airfoils have not been evaluated for Darrieus turbines so far and it is attractive for their simple design and low cost. The stepped airfoils seem to be a potential concept that can be investigated for straight-bladed Darrieus turbines.

2.4.8 Gurney Flaps

Gurney flaps are passive devices located at the trailing edge of an airfoil with the intention of enhancing lift as shown in Figure 2.22. Initially employed in racecar applications, these flaps are able to increase traction by altering the moment coefficient on an inverted airfoil with high nose down pitching moment.[73] Due to its simple features in construction and operation its applications are widespread in subsonic, supersonic aircraft, wind turbines[74] and helicopter horizontal stabilizers. Gurney flap is a small plate mounted perpendicular to the pressure side of the airfoil at the trailing edge. The typical height of the flap will be 1–2%c. Though lift coefficient continues to increase beyond 2%c, the associated drag increases significantly.[75] A systematic analysis on the gurney flap was performed by Liebeck.[76] Lift augmentation is achieved by increasing the pressure on the pressure side of the airfoil and reducing the pressure on the suction side to create a recirculating counter-rotating vortex at front and rear of the flap especially at high AoA. Though the drag increases,

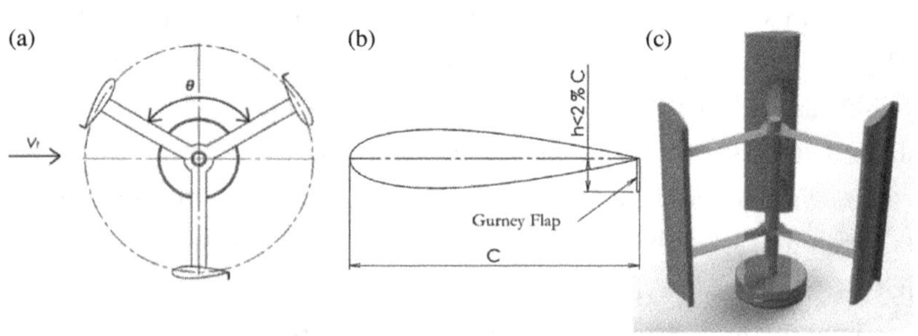

Figure 2.22 (a) H-rotor with gurney flap. (b) Details of an airfoil with gurney flap. (c) Illustration of a H-rotor with gurney flaps.

the increase in lift is much higher compared to drag. It should be noted that the airfoil characteristics may differ if the airfoil encounters unsteady wind and virtual camber effects when incorporated in Darrieus turbines compared to static airfoils. The notable positive aspect is the delay in dynamic stall but at maximum AoA, there is sudden drop in lift and moment coefficient which may lead to structural vibrations.[77] The gurney flap may generate more power especially in the upstream stroke compared to baseline turbine without gurney flap but may display a poor performance in the downstream half. Further experimental investigation is necessary to get an insight on starting behavior and low wind speed operation with gurney flap.

2.4.9 *Trailing Edge Flaps*

The trailing edge flaps are active high-lift devices hinged close to the trailing edge of an airfoil. The nominal length of the flap is 10c–20c. Initially these flaps are employed to reduce vibrations in the helicopter rotor by varying the flap angle in relation to the flight mode.[78] The Darrieus rotor integrated with a typical trailing edge flap is shown in Figure 2.23. Though a wide variety of flaps exists,[79] a plain hinged flap is simple in construction and can be intelligently manipulated to achieve the desired airfoil characteristics.

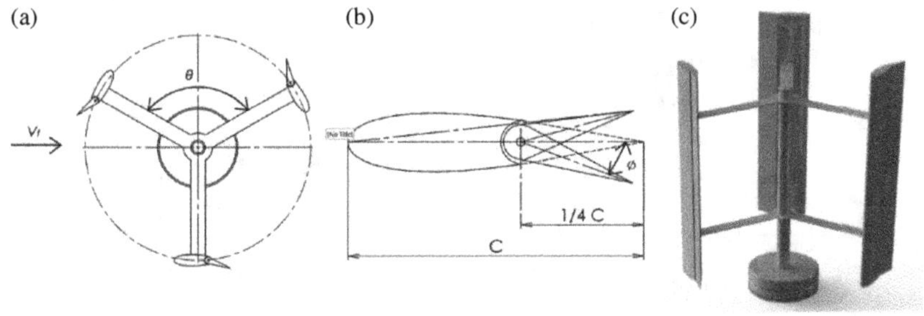

Figure 2.23 (a) H-rotor with trailing edge flap. (b) Details of an airfoil with trailing edge flap. (c) Illustration of a H-rotor with trailing edge flaps.

These flaps are successfully employed in HAWTs for load alleviation and power control.[80] The flaps are investigated for VAWTs to increase the lift at low Re and load regulation at high winds. Closed loop flaps[81] have the potential to increase the starting torque by sensing the wind direction and altering the flap angle accordingly. Overall turbine performance can be improved by varying these flaps with respect to azimuthal position to reduce the negative performance in the downstream half. The change in the AoA due to new chord line is given by Eq. (2.21):

$$\Delta\alpha = \frac{d\alpha}{d\theta} = \arctan\left(\frac{\sin\theta}{3+\cos\theta}\right) \tag{2.21}$$

The nominal AoA as a function of azimuthal angle is obtained by Eq. (2.22):

$$\alpha = \alpha_0 + \Delta\alpha = \arctan\left(\frac{\cos\Psi}{\lambda+\cos\psi}\right) + \arctan\left(\frac{\sin\theta}{3+\cos\theta}\right) \tag{2.22}$$

Yang *et al.*[82] conducted a 2D computational study on VAWTs with trailing edge flaps and the results are encouraging. The study compared the trailing edge flap at ¼c location from the trailing edge on NACA 0012 blade at 10 m/s with unmodified NACA 0012. The airfoil characteristics and the Cp variation with TSR are depicted in Figure 2.24. The average lift

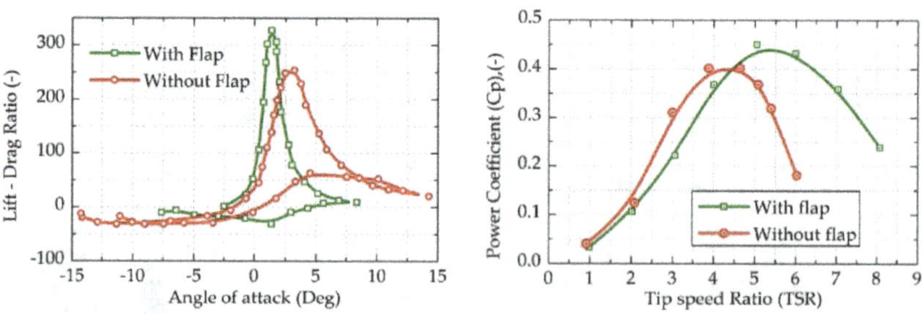

(a) (b)

Figure 2.24 (a) Cl and Cd of NACA 0012 with and without trailing edge flap. (b) Cp vs TSR for NACA 0012 with and without trailing edge flap.

coefficient of NACA 0012 with flap is increased by 0.15 and the maximum lift-to-drag ratio increased by 73.

With the flap control the peak Cp is enhanced by 10% at a higher TSR compared to conventional blade. The power improvement is attributed to the delay in the dynamic stall and by the minimization of blade vortex interaction. The implementation of trailing edge flaps to straight-bladed turbine is dictated by the size of the turbine. The trailing edge flaps are not possible to integrate in the helical blades as the hinge axis of the flap is twisted along the rotor axis and it is highly complicated due to actuation and blade loading.

Though it is widely employed in HAWTs, the flaps are mostly installed on the portion on the blade with minimum twist which is mostly at the tip of the blade, where the twist variation is minimal compared to the blade span. The use of trailing edge flaps along with individual blade pitching on HAWTs is expected to find its place especially for the turbines of 8 MW.

The initial computational study sheds light on the applicability of trailing edge flaps for Darrieus turbines and a possible means of achieving overall turbine performance at low wind speed assuring a good start for further study.

2.4.10 *J-Blade*

The J-profiled blades are formed by removing a portion of blade from the trailing edge either on the pressure or suction side of the conventional airfoil. The objective of introducing such an opening is to increase the starting torque and power generation at low wind speed in the range of 3–5 m/s especially for small wind turbines. The opening can be continuous for the entire blade span or intermittent as dictated by the blade thickness, startup wind speed and structural requirements. Figure 2.25 displays a H-Darrieus with incorporated J-profiled blade.

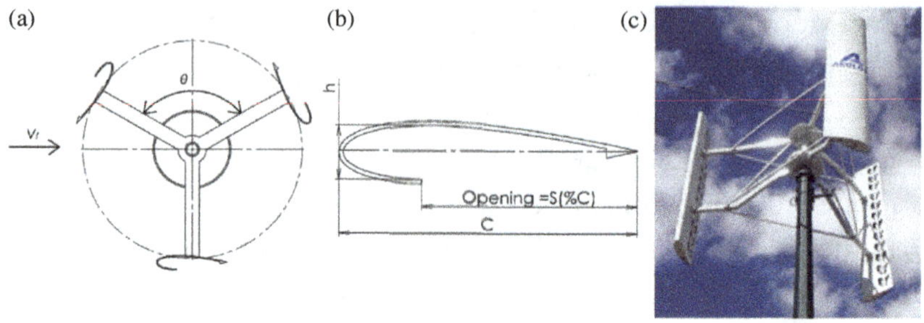

Figure 2.25 (a) H-Rotor with J-profile airfoil. (b) Details of J-profile airfoil. (c) Commercial turbine with J-profiled blade. Image (c) is reproduced with permission from https://www.windturbinestar.com/italy-5kw-vertical-axis-wind-turbine.html.

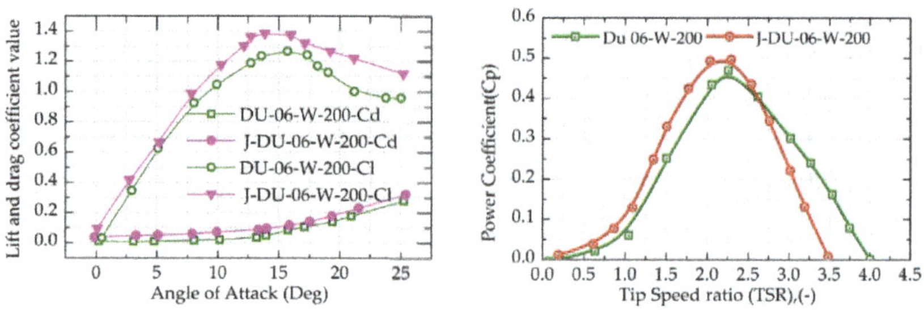

Figure 2.26 Cl and Cd of J-profile blade (left) and Cp vs TSR comparison with parent airfoil (right).

The J-blades operate both in lift and drag modes and exhibit lower lift/drag ratio compared to its parent conventional airfoil due to reduced lift and higher form drag. Zamani *et al.*[83] performed a computational study on DU-06-W-200 airfoil and the J-profile airfoil J-DU-06-W-200 derived from it. The drag coefficient for J-DU-06-W-200 is lower than DU-06-W-200 when compared for a 3 kW turbine at Re 5×10^5. The results indicate that the torque amplitude is widened for larger azimuthal angles, reducing the vibration and fatigue stress in the case of J-DU-06-W-200, but the operating TSR range is reduced from 4 to 3.5 as indicated in Figure 2.26. The performance of the turbine is greatly influenced by the location of the opening and the blade thickness. A thicker blade has wider drag bucket and tends to generate higher static torque. The study on NACA 0015 by Zamani *et al.*[84] with 250 mm chord concluded that the wake region on the rear of the rotor decreases due to vortices trapped in the opening. Chen *et al.*[85] studied the opening ration and suggested, from the self-starting perspective, that for the turbines operating at higher TSR an inner opening ratio, which is defined by the ratio of opening length (h) to the distance from the leading edge (s), should be 0.48 to 0.60.

For the turbines operating at low TSR and dominated by drag with the objective of replacing the Savonius turbine, an outer opening ratio of 0.72 to 0.84 is optimal. Though larger opening will generate higher torque, trade-off should be made with the Cp loss. Added advantage is from the inexpensive manufacturing process. Since the opening facilitates easy access, the blades can be manufactured by bending aluminum sheet over a pre-cut profile that performs the function of ribs to strengthen the blade. This manufacturing process can significantly curtail the manufacturing cost compared to traditional glass fiber molding process. The J-blade was implemented commercially and has been proven to be efficient especially for small wind turbines compared to the Hybrid Darrieus-Savonius turbine. Application of J-blades for higher kW capacity turbines requires further optimization study at high Re and TSR.

2.4.11 *Hybrid Savonius-Darrieus Rotor*

Savonius rotors generate higher starting torque as the blades are drag driven, hence these machines can be wisely integrated with Darrieus turbines to enhance the performance at low wind speeds. Past systematic studies by Rassoulinejad-Mousavi *et al.*[86] optimized the size of the Savonius rotor in a hybrid configuration. A troposkein Darrieus rotor (4 m) with Savonius rotor is available commercially for power capacity up to 3 kW[87] as shown in Figure 2.27(c). An experimental study by Gupta *et al.*[88] established the performance in comparison to conventional Darrieus, yet there is a lack of understanding on the complex aerodynamics and interaction between the rotors. A detailed study was performed by Wakui

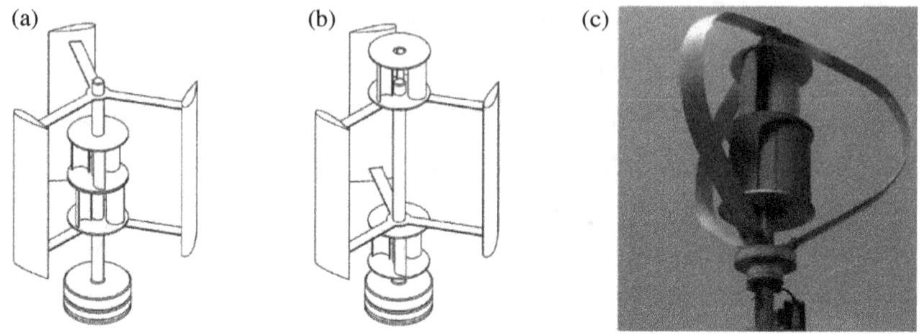

Figure 2.27 (a) Hybrid rotor with Savonius inside Darrieus. (b) Hybrid rotor with Savonius outside Darrieus. (c) Commercial hybrid turbine. Image (c) is reproduced with permission from http://www.hi-vawt.com.tw/en/ds1500w.html.

et al.[89] on two different configurations with two-stage Savonius offset at 90°. In the first configuration, the Savonius rotor was placed inside the Darrieus turbine as shown in Figure 2.27(a) and subjected to experimental test to compute the power and torque coefficients. The experimental results show that the starting torque of the combined machine is higher than that of the Darrieus rotor, but is not able to match in terms of Cp. A possible reason may be wake generated by the Savonius as described by Gupta and Sharma,[90] which severely affects the power generation of the Darrieus rotor in the downwind half. Another reason is that the Darrieus rotor reaches the peak Cp at higher TSR (>1) while the Savonius rotor generates resistive torque when the TSR is greater than 1.

Second configuration is to mount the Savonius rotor outside the Darrieus rotor as shown in Figure 2.27(b). Though the configuration eliminates the wake interaction of Savonius with Darrieus blades, still the Cp of the combined machine is less than that of the Darrieus rotor due to the above said resistive torque by the Savonius rotor. From the structural design perspective, mounting the Savonius rotor outside the Darrieus rotor demands longer shaft resulting in vibration.

Since both configurations achieve the same Cp, comparatively it is advisable to mount the Savonius rotor inside the Darrieus rotor to achieve peak Cp at their respective TSRs. Potential solutions have been proposed to address the above problems associated with Darrieus-Savonius rotor integration. Kyozuka[91] proposed to disengage the rotors when the Darrieus rotor accelerates beyond TSR 1. Though the solution is technically feasible the energy captured by the Savonius rotor after disengagement will not contribute to the total energy output and the wake generated by the Savonius rotor will still exist. Hence a strategy to eliminate the wake of the Savonius rotor will make this hybrid rotor a potential solution for low wind speed operation.

2.4.12 *Radial Arms with Drag Device*

Struts in Darrieus turbines refer to the structures that extend from the rotating tower or center shaft and connect the blades, through which the generated torque is transmitted and designed to withstand the aerodynamic loads and vibrations. Apart from these positive aspects, struts do cause flow disturbances and reduce the generated aerodynamic torque through parasitic drag. Johnston Jr.'s[92] study on the 17 m rotor revealed that there is a drop of 26% in power production due to the presence of six struts of NACA 0012 in the X-configuration compared to the turbine without struts. The effect of frontal drag on the power performance can be predicted by introducing an equivalent drag coefficient acting on the blade as proposed by Moran.[93] The Cp reduction by the frontal drag of the strut can be calculated as described by Li and Calisal.[94] Hence optimized struts are mandatory to increase the power production and that is where most of the studies are focused on. The struts are optimized for high bending stiffness and low drag which is substantial near the blade joint. The conventional struts of large turbines comprise welded pipes enclosed in NACA profiles of sufficient thickness. The profile can be constant along the entire length or it can be optimized to withstand more bending near the tower with higher thickness. To provide higher starting torque, the rear side of the struts is modified by hollowing out to create rear drag that will be added to the torque generated by the blades as shown in Figure 2.28(a). High bending

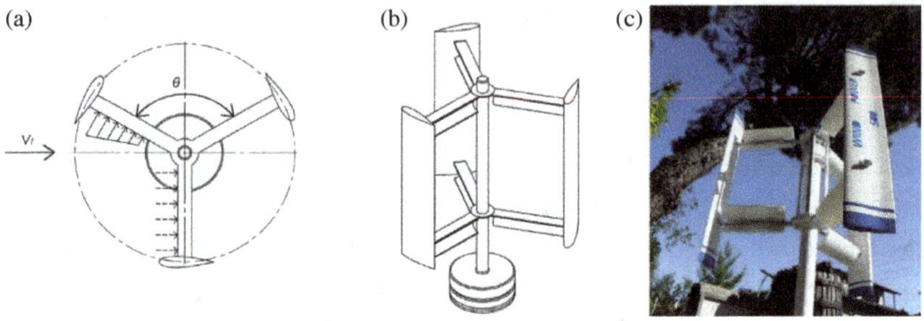

Figure 2.28 (a) Top view of H-rotor indicating the rear drag. (b) H-rotor with drag-inducing strut. (c) Commercial turbine with modified strut. Image (c) is reproduced with permission from https://www.makemu.it/prodotto/wind-generator-domus/?lang=en.

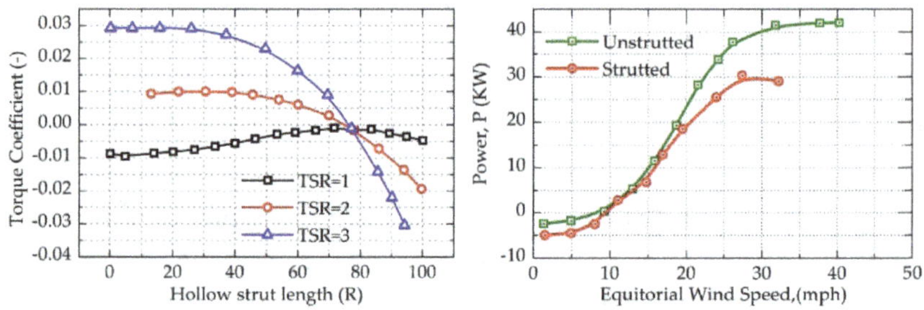

Figure 2.29 Drag torque of hollow strut at different TSR (left) and power output comparison of strut and unstrutted rotor (right).

stiffness, low frontal drag and high rear drag are the criteria for the optimization of the struts. A more cost-effective strategy than hollowing out, especially for small wind turbines of power capacity less than 3 kW, is to attach the semicircular blades similar to the Savonius turbine blades to the struts as shown in Figure 2.28(b). The opening on the rear part has to be optimized; otherwise the power loss will be significant when the turbine operates at high TSR as shown in Figure 2.29.[66] Another vital consideration in optimization is the strut wake and blade interaction. A larger opening with sharp edges can shed vortex, which causes flow separation of the blades and also induces vibration in turbines with low aspect ratio. Though commercially turbines are available with modified struts, literature on systematic optimization is not available and none has quantified the % increase in starting torque with modified strut. The radial extension of strut has greater influence on the starting torque and a trade-off should be reached with the power loss due to frontal drag.

2.4.13 *Double Darrieus Rotor*

Q. Y. Li[95] proposed the addition of secondary Darrieus rotor inside the primary Darrieus rotor. The addition of secondary rotor as shown in Figure 2.30(a, b) was claimed to increase the starting torque and persistent rotation at low wind speed rather than the power output of the primary rotor. Added claim was that the AoA can be reduced thereby delaying the stall on the primary blades. Though the study did not elaborate on the positive aspects, Double Darrieus rotor is anticipated to perform better even at higher TSR, as the resistive torque is minimal due to secondary blades and the starting torque also will be distributed to wider azimuthal angles rather than three peaks at 120° apart as in conventional Darrieus. Since a detailed aerodynamic study is lacking, a preliminary experimental study to check the feasibility is needed to ascertain the applicability to low wind speed operation in comparison to conventional Darrieus rotors.

(a) (b) (c)

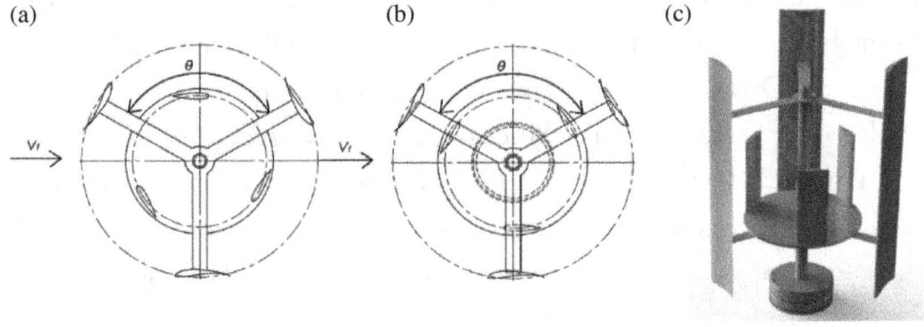

Figure 2.30 (a) Double Darrieus in offset configuration. (b) Double Darrieus in parallel configuration. (c) Illustration of a Double Darrieus rotor.

2.4.14 *Low Resistance Bearings*

The drive train architecture for small and large wind turbines are different. The small VAWT typically consists of a permanent magnet generator with integrated bearings. Whereas for the large VAWT, the drive train consists of bearings, disc brake, couplings, speed increaser and generator. The type of bearing and the arrangement are dictated by the type of support and type of generator used. The various types of tower support structures are Guy wired, cantilever type or a combination of both. The bearing arrangement will be decided largely based on these support systems. The primary function of the bearing system is to support the center shaft and the drive train with low frictional resistance from balls and seals, dampen the vibrations and make the turbine structurally stable at high winds. The drive train components have greater influence on the startup characteristics of Darrieus other than the cogging torque of the generator as given in Eq. (2.23):

$$T_R = T_F + T_G + (I_G + I_C + I_B + I_G) \times \alpha \tag{2.23}$$

where T_R, T_F, T_G are the resistive torque, frictional torque, and the cogging torque, respectively, in Nm. I_G, I_C, I_B, I_G are the inertias of gearbox, couplings, brake, and generator, respectively, in kg-mm^2. The bearing system along with the protective seals contribute substantially to the total resistive torque. A typical bearing arrangement is shown in Figure 2.31(a, b) for a small wind turbine of cantilevered type. The Darrieus rotor generates large moment load and significant axial load. Hence the bearings and the bearing housing should transfer these loads safely to the ground. A typical commercial Darrieus turbine employs a set of angular contact ball bearings and deep groove ball bearings. Since the angular contact bearings are able to sustain both axial and radial loads in back-to-back arrangement, they are placed on the top end of the generator shaft and the bottom end will be single- or double-row deep groove ball bearings based on the dynamic load rating. However, for large VAWT a combination of deep groove roller bearing and thrust bearing or taper roller bearing replacing the angular contact bearings for axial load is employed as shown in Figure 2.31(c).[96] Very few studies attempt to reduce the frictional

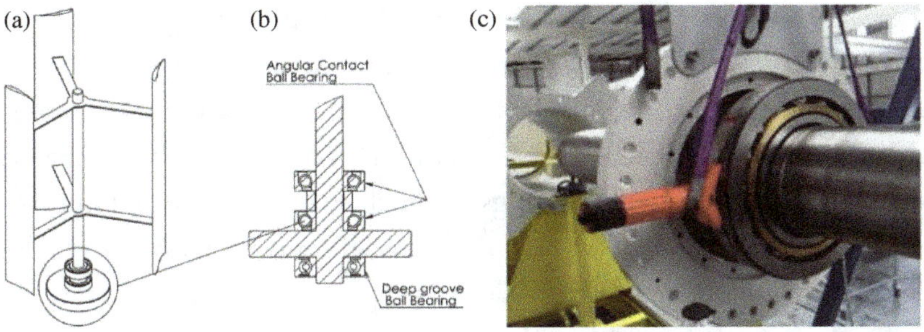

Figure 2.31 (a) Typical small VAWT with generator. (b) Bearing arrangement on small VAWT. (c) Angular contact ball bearing on large VAWT. Image (c) is reproduced with permission from https://www.schaeffler.de/en/products-and-solutions/industrial/industry_solutions/wind_sector_cluster/wind/generator/.

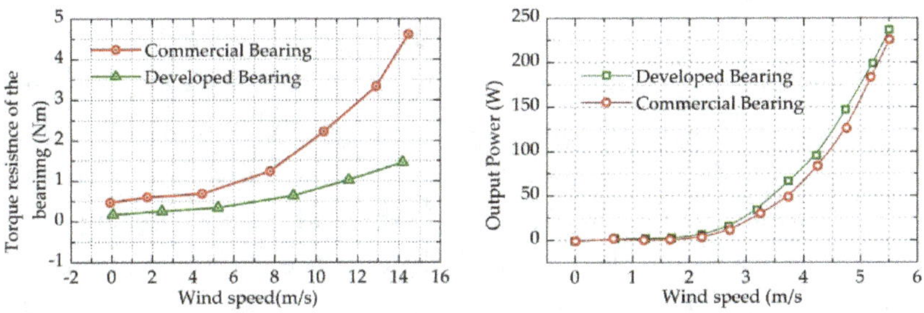

Figure 2.32 Torque resistance comparison of commercial and developed bearing (left) and power output vs wind speed of commercial and developed bearing (right).

resistance by developing bearings dedicated to the VAWTs. Aso *et al.*[97] reduces the rolling friction by forming an inner raceway on the main shaft, without affecting the service life of the bearings. The study optimized the degree of conformance of the raceway[98] and the ball-diameter ratio for the dynamic load rating computed through IEC 61400-2. The developed bearing is incorporated on a 3.2 kW Darrieus turbine and subjected to field test to compare with commercially available bearings in terms of starting wind speed. The results show that the turbine with commercial bearings starts to rotate at 2.9 m/s, whereas the optimized bearing starts at 1.8 m/s, reducing the starting speed by 62% and starting torque by 38% as shown in Figure 2.32. From the energy output perspective, with average wind speed of 4.7 m/s, the developed shaft and the bearing unit can generate 148 W, while the turbine with commercial turbines are able to generate only 128 W as shown in Figure 2.32. The study concludes by improving the efficiency of the bearing system by 25~28% compared to commercially available bearings. Though there is a notable improvement in the efficiency of the turbine, special bearings for VAWTs may incur high cost of manufacturing and replacement as they are non-standard and not manufactured in high volume. In case of raceway wear after prolonged operation, the entire shaft has to be replaced. Hence reliability is a question for the current concept.

2.4.15 *Multi-Stage Rotor*

The concept of multi-stage Darrieus turbine was proposed by Gorelov and Krivospitsky[99] to address the self-starting problem without compromising the efficiency of the rotor. The proposed turbine termed as Gorelov-Masgrowe turbine has two tiers, with each tier comprising two or more blades as shown in Figure 2.33. The blades in each tier are offset by 360/2*n*, where *n* is the number of blades. The blades are connected either by individual struts or by a common ring. The individual strut connections are prone to blade tip loss as calculated by Shen *et al.*,[100] whereas the common ring acts as blade end plates preventing the blade tip vortices. The study on the startup performance and Cp with varying

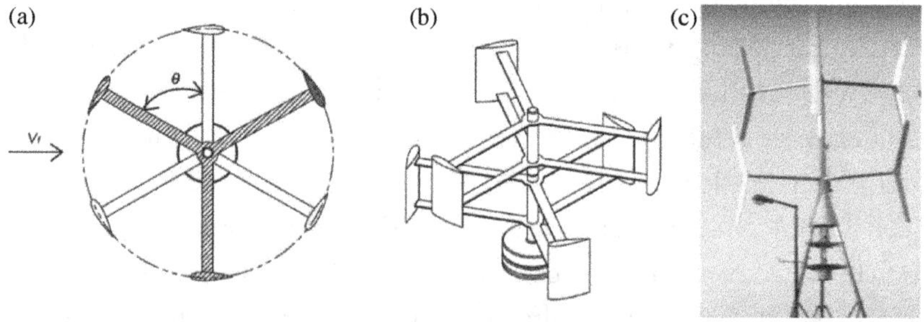

Figure 2.33 (a) Top view of two-tier Gorelov-Musgrove turbine. (b) Isometric view of a two-tier Gorelov-Masgrowe turbine. (c) Commercial two-tier turbine. Image (c) is reproduced with permission from https://twitter.com/ANewWindTurbine/status/411490790112641024/photo/1.

chord length for two- and three-bladed configuration shows that for a two-bladed configuration the maximum performance is achieved when the solidity is 0.28. Increasing the solidity beyond that improves the startup, but the value of Cp deteriorates. A full-scale 1 kW was constructed with varying blade thickness of 16, 18, 20, and 22% and subjected to field test. Higher efficiency of 34–36% was achieved for 20–22% thick airfoils starting at 2.4–2.6 m/s. A 3 kW two-tier turbine with three blades on each tier demonstrated a startup wind speed of 1.6–2 m/s with best efficiency of 39–40%. The improvement in startup performance is not documented, which may shed additional light on the low wind performance. The concept is lucrative especially for small wind turbines and it can be a potential alternative to complex blade pitching, hybrid Savonius-Darrieus turbine. However, the concept seems too simple and passive without any actuating mechanisms. For larger utility-scale turbines, the proposed concept has to be assessed for structural stability. A well-designed center shaft can eliminate the vibrations induced by the blade offset. Additional struts may be required to keep the blade in position, in which case it is inevitable to eliminate the power loss associated with the struts. Compared to conventional Darrieus, two-tier configuration may incur additional cost in manufacturing struts and blade support structures.

2.4.16 *Diffuser-Augmented Turbines*

Augmentation systems are intended to increase the wind speed locally over the rotor. Since the power generated is proportional to the cube of the wind speed, a small increase in the wind speed can significantly increase the power output exceeding Betz limit for the given rotor size.[101] A notable progress has been made in the diffuser-augmented tidal Darrieus turbine. The diffuser is much more suitable for the tidal turbines due to their smaller size and as the housing for the generator and the drive train.[102] The diffuser for the tidal turbines adds overall rigidity as the hydrodynamic loads are much higher compared to the wind turbines of the same size. Some of the commonly employed flow augmentation devices are guide vanes, diffuser, stator, and duct as shown in Figure 2.34.

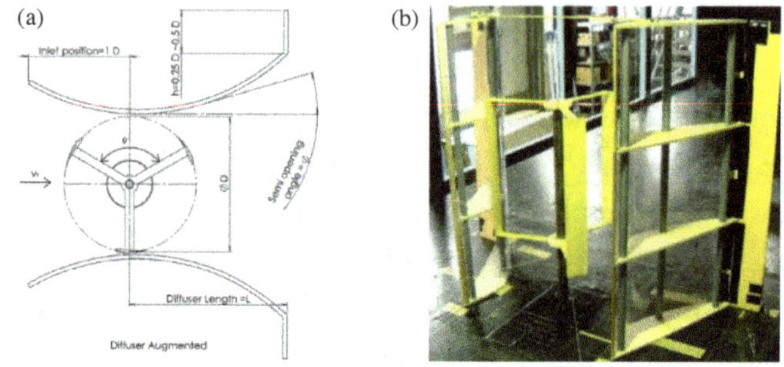

Figure 2.34 (a) Top view of a diffuser-augmented turbine. (b) Experimental setup of a diffuser turbine.[104]

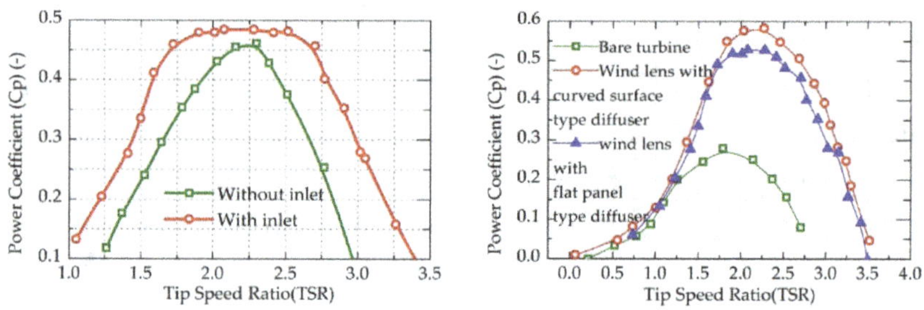

Figure 2.35 Effect of diffuser inlet on Cp (left) and Cp comparison with and without inlet (right).

Flow augmentation techniques for Savonius-type VAWTs are well studied[103] as apart from accelerating the flow, reducing the drag on the advancing blade is a notable gain. Another reason for much of the studies being concentrated on the Savonius turbines is that they can be easily fabricated and tested. Though studies on the flow augmentation system of Darrieus wind turbines are limited, it is sufficient to draw a sensible conclusion. A two-bladed H-Rotor with NACA 0018 airfoil of chord length 0.15 m was experimentally investigated with diffuser and flanges (wind lens) by Watanabe *et al.*[104] and results are indicated in Figure 2.35. Compared to the bare turbine, the wind lens turbine output increases by a factor of 2 with optimized diffuser angle and the flange height. An effective diffuser angle for Darrieus turbine is 20° and the length of the diffuser can be half the throat width. Though flat panel diffusers are cost effective in manufacturing, a curved diffuser has better output by a factor of 2.1 compared to bare turbine. Influence of presence of inlet was compared by testing with and without inlet for the diffuser length of 0.57 D and flange width of 0.5 Dm. The results as displayed in Figure 2.35 shows no significant power augmentation with inlet, except the turbine can operate at a wider TSR.

Geurts *et al.*[105] focused on spacing between the diffuser and the rotor. Closer spacing between the rotor and the diffuser wing can cause turbulence and lead to a large variation in AoA. The optimized spacing between the diffuser and the rotor is 1.8 R to 2 R, where R is the rotor diameter considering the vortex shedding. Added advantage of incorporating the diffuser is the reduction of torque ripple by a factor of 4.15[105] and extending the peak torque for a wider azimuthal angle. With optimized diffuser design the Cp can be increased up to 1.5 times. The study also concludes that rotor should be located between 0.9 and 1.3 m from the duct inlet for maximum power extraction. On the downside the distinctive advantage of omnidirectionality is lost due to the addition of nozzle, though the diffuser can align to the wind direction. The yaw error substantially affects the efficiency and induced vibrations on the structure. The wind loading on the diffuser or nozzle is significant and the cost incurred in manufacturing a rigid structure, tower and foundation outweighs the turbine cost.

2.4.17 *Generator Starting*

For maximum energy output, it is necessary for the turbine to accelerate to the rated rpm in the shortest possible time. Wind turbines are either connected to the electrical grid or an energy storage system such as a battery and in either case it is possible to use the energy to accelerate the rotor to the rated rpm. Since most turbines use direct drive permanent magnet generator (PMG),[106] which can operate efficiently in both generator and motor modes, the overall system is less complex compared to other mechanical torque-inducing devices. Kjellin and Bernhoff[107] conducted a study on a 12 kW PMG for H-rotor-type turbine by introducing an auxiliary winding along with the main winding in the generator to provide starting torque. The auxiliary winding is dedicated for the rotor acceleration and has 1/10th of the length of the main winding and can be operated at low voltage. Precise control on acceleration can be achieved by adjusting the voltage on the windings. The proposed solution can reduce the cost involved in high-voltage components as the windings demand only low-voltage components. The concept of active motoring was simulated on a 5 kW HAWT by Aner *et al.*[108] with two blades accelerated typically in 12 s to the rated rpm at 10 m/s wind speed. The system includes a PMG, maximum power point actuation with very spare directional converter and field-oriented controller as shown in Figure 2.36(a). The PMG is switched to generate the electromagnetic torque if the controller detects any movement of rotor. The generated motor torque will be varied by comparing the aerodynamic torque measured by the MPPT controller with the reference torque. The simulation results indicate that the acceleration time can be reduced from 13.29 s with pure aerodynamic torque to 1.9 s with motoring torque at 10 m/s wind speed as shown in Figure 2.37. Though the motoring start has

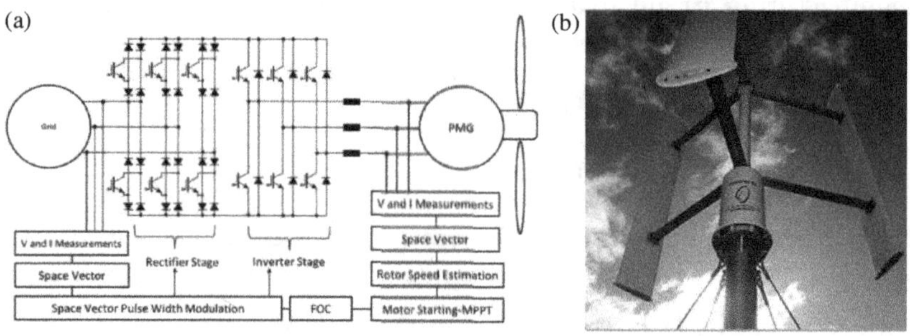

Figure 2.36 (a) Electrical circuit of a motor startup and (b) Commercial turbine incorporated with electrical startup.[108]

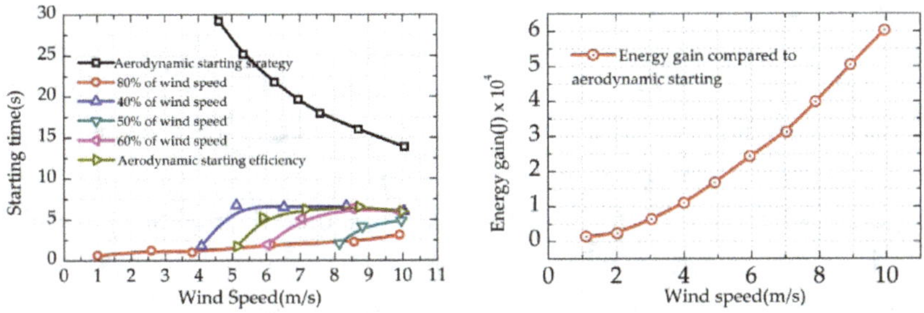

Figure 2.37 Starting time for different ramp-up speed (left) and energy gain compared to aerodynamic start (right).

negligible impact on the peak Cp of a turbine, the significant startup time reduction aids in maximizing the energy extraction. Cogging torque is an inherent characteristic of any PMG and arises from cogging between magnets and the armature windings and can be measured by the torque required to rotate the shaft under no load. The self-starting of Darrieus turbines occurs when the aerodynamic torque exceeds the cogging torque added with other resistive torques. Cogging torque of a commercially available PMG of 1 kW is <0.45 Nm and for 12 kW it is <14 Nm.[109] Several studies in the past proposed multiple solutions such as pole shifting and uneven distribution of stator slots to reduce the cogging torque and hence the starting wind speed of Darrieus turbines. 38% reduction in the cogging torque was achieved by Hwang *et al.*[110] by maintaining a non-uniform gap between the magnet and the stator via modifying the magnetic pole shape. The electric startup can be a potential alternative compared to the complex mechanical startup method in terms of cost and reliability. Though the auxiliary winding or bidirectional converter claims to be simple, there is a considerable power loss to accelerate the rotor and especially during inconsistent wind that oscillates around the startup wind speed. The electric start can provide an initial acceleration, but may not able to sustain the low wind speed rotation and hence will not be suitable for Darrieus turbines operating at low wind speed sites.

2.4.18 *Aspect Ratio*

The power extracted by the Darrieus rotor is proportional to the swept area, i.e., blade height multiplied by the diameter of the rotor, and the ratio between them is termed as aspect ratio as illustrated in Figure 2.38(a, b). The same swept area can be achieved either by increasing the blade height with smaller diameter or vice versa. Though the frontal area is the same, the power performance of the rotor is different both aerodynamically and structurally. Very few studies had focused on the finite aspect ratio effects, which has a strong correlation with blade tip vortices and tip loss.[111]

A straight-bladed Darrieus turbine has finite blades with blade tips that induce two velocity components: a span-wise component along the blade chord and downwash along the blade span. The downwash results from the wind flow around the tip instead of flowing over the blade. DMST predicted that the streamtubes adjacent to the blade tips are less efficient due to less transfer of momentum from the air, leading to power loss. The aspect ratio influences AoA and lift and drag coefficients significantly. The finite aspect ratio effects on Darrieus rotors were analyzed through a DMST model by Paraschivoiu *et al.*[112]

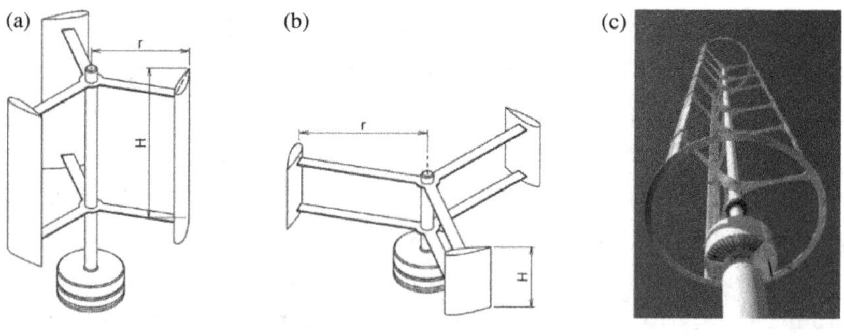

Figure 2.38 (a) Low aspect ratio. (b) High aspect ratio H-rotor. (c) Commercial high aspect ratio turbine. Image (c) is reproduced with permission from https://www.wind-spireenergy.com/3kW-wind-turbine.htm.

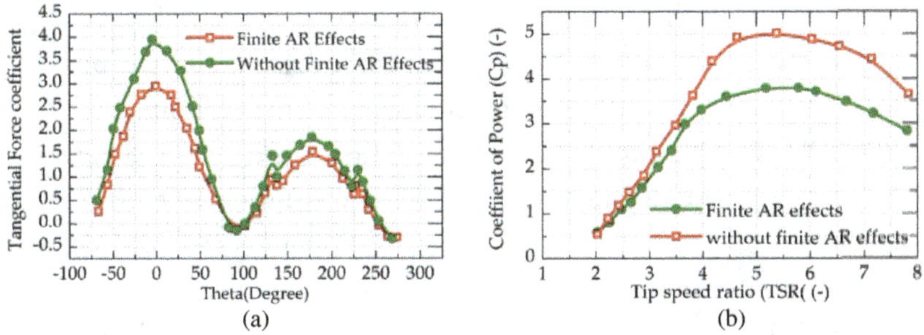

Figure 2.39 (a) Finite aspect ratio effect on tangential load coefficient vs azimuthal angle. (b) Finite aspect ratio effect on Cp vs TSR.

DMST simulation during the wind turbine design phase on the annual energy yield calculation reveals a power loss of 7% for the aspect ratio of 40, 10% for 30, and 26% for 10 as depicted in Figure 2.39(a, b). Brusca *et al.*[113] contradicts that the turbines with low aspect ratio have an advantage of increase in Re due to increase in rotor diameter, thereby resulting in higher performance.

The reason can be attributed to the assumption of constant rpm rotor irrespective of the rotor diameter and not accounting for the aerodynamic loss. The effect of aspect ratio and the blade tip loss is proportional to number of blades and a possible solution to reduce the losses is to incorporate wing tips. A turbine with high aspect ratio is structurally sound with reduced centrifugal force and is more dynamically stable. Turbines with low aspect ratio are not favored from the self-starting perspective due to their increased inertial resistance, which delays the startup and slows acceleration. Hence a high aspect ratio of above 8 is preferred for self-starting and low wind speed performance.

2.4.19 *Circulation-Controlled Blades*

Circulation flow control (CC) is one of the aerodynamic flow control techniques that had been put forward in the 1970s and successfully demonstrated on short take-off aircraft such as C-17.[114] The concept gained momentum after the advancement of micro-pneumatic devices, which enables this technology to be implemented in medium- and large-sized HAWTs for load alleviation at high winds. The dynamic circulation technique can be implemented as a potential solution to enhance the performance of VAWTs in low wind speeds by lift augmentation without any moving device exterior of blades as shown in Figure 2.40(a), and (c) shows VAWT with CC blades under wind tunnel test.[115] The strategy is to increase the torque by maximizing the tangential force coefficient (C_t) via increasing the lift coefficient (C_l) and reducing the drag coefficient (C_d) as given by Eq. (2.24).

$$C_T = C_l \sin\alpha - C_D \cos\alpha \tag{2.24}$$

CC is implemented by blowing air on a curved surface, mostly on the large radius-rounded trailing edge[116] as shown in Figure 2.40(b). This phenomenon reenergizes the boundary layer to remain attached to the curved surface due to the Coanda effect and pushes the rear stagnation point further down. A high value of lift can be achieved up to seven times compared to conventional airfoils.[117] Figure 2.41(a) illustrates the comparison of NACA 0015 and generic CC airfoil. The aerodynamic property of the CC technique relies on blowing coefficient C_μ, a non-dimensional

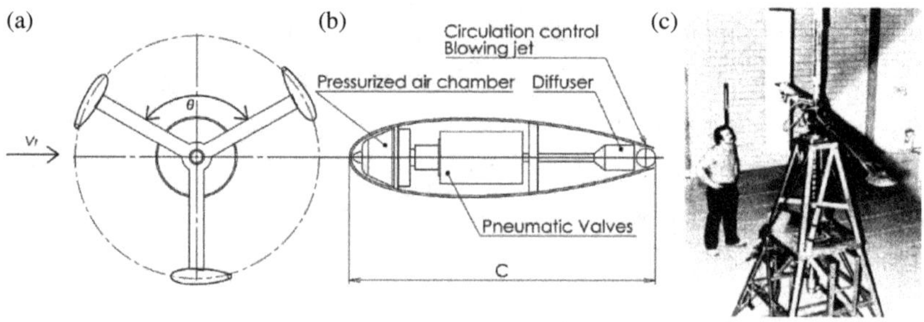

Figure 2.40 (a) Top view of H-rotor with CC airfoil. (b) Generic CC airfoil with internal components. (c) H-rotor with CC airfoil under wind tunnel test.[115]

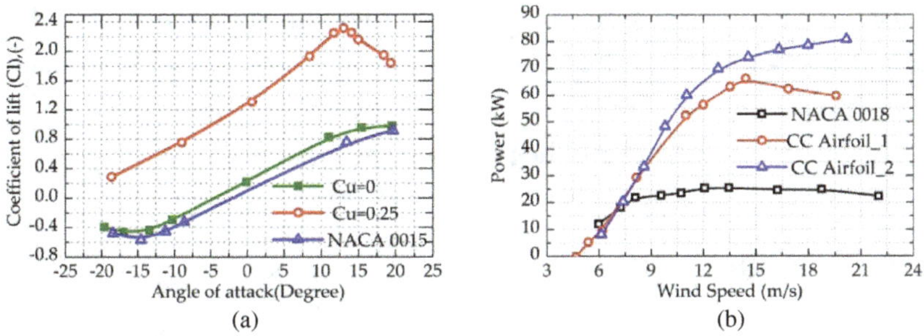

Figure 2.41 (a) Cl of NACA 0015 and CC airfoils at Re = 270,000 and (b) power vs windspeed comparison of NACA 0018 and CC airfoils.

parameter, which is the ratio of jet momentum to the freestream momentum.

$$C\propto = \frac{\dot{m}Uj}{Qc} \tag{2.25}$$

The choice of design parameters is a compromise between lift augmentation, additional base drag as well as power required to pump air jet. The CC technique was reported to increase turbine performance by 5–50% by carefully manipulating the blowing coefficient, TSR, wind distribution, height of the blown jet and span of the jet as shown in the Figure 2.41(b).[118] Some of the key factors that attract researchers to pursue the CC technique for application in VAWTs are the ability to generate high torque at low wind speed, reduction of torque ripples thereby extending the longevity of the turbine structures, and alleviating the loads by decreasing the lift at high winds to act as an aerodynamic braking. The CC technique can be easily tweaked to operate at its maximum performance for a wider TSR. However, CC blade sections require a higher trailing edge thickness than conventional sections, giving rise to additional base drag. Apart from these positive aspects, the CC technique does have shortcomings such as increase in noise compared to conventional airfoils, significant loss of power to generate the jet stream, and sensitivity of aerodynamic coefficients to jet control parameters. The complexity and the requirement of sophisticated control systems make it suitable for aircrafts rather than VAWTs, where economics and reliability over a long period are the primary concerns.

2.4.20 *Morphing Blades*

The objective of the morphing blade is to obtain certain characteristics that are intractable from rigid blades, rather than increasing the lift coefficient. A conventional morphing blade incorporated in VAWT with details on the blade is illustrated in Figure 2.42(a, b). Currently this technique has been implemented in large offshore HAWTs for blade load alleviation.[119] Wind tunnel test on trailing edge morphing blades are shown to reduce the blade root bending moment by 95%.[120] The primary components of morphing blades are the skins that allow large deformation and actuation. The vital reasons for employing morphing blades in VAWTs are enhancement of starting torque, delaying the dynamic gradient and regulating the rpm by decreasing the lift coefficient at high winds.

Early endeavors on flexible sail wings and morphable sheet metal blades on VAWTs are not encouraging as their performances are always

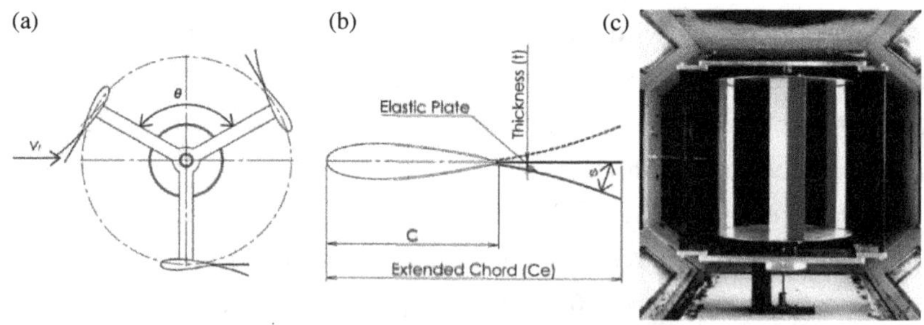

Figure 2.42 (a) Top view of H-rotor with morphing blade. (b) Trailing edge elastic blade. (c) Wind tunnel test of an elastic blade.[120]

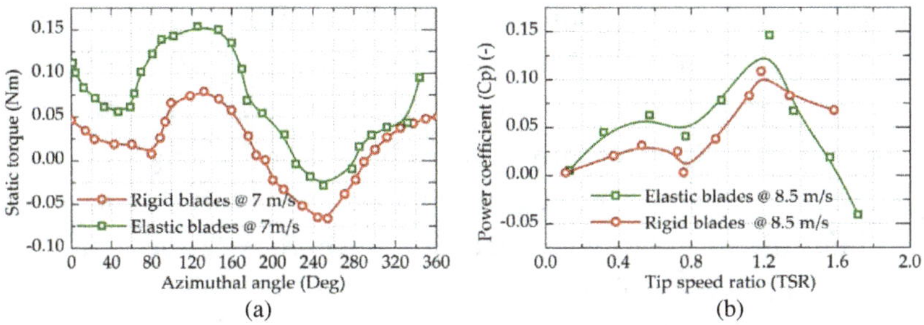

Figure 2.43 (a) Static torque vs azimuthal angle for elastic and rigid blades. (b) Cp vs TSR for elastic and rigid blades.

diminished compared to rigid airfoils.[121] From the starting perspective of VAWTs, cambered airfoils are preferred for their higher starting torque, though it results in higher drag on the downwind half of rotation. The morphing blades are capable of regulating the camber in the upwind and downwind portions of rotation leading to enhanced performance at low TSR and comparatively higher starting torque than rigid blades as demonstrated by Nakajima *et al.*[122] The starting torque comparison depicted a lower negative torque value as portrayed in Figure 2.43(a), and the Cp comparison with rigid blades as shown in Figure 2.43(b) shows that the turbine is able to operate with increased Cp for a wide range of TSR.

The development of morphing blades for VAWTs is hindered by fatigue failure in early attempts due to unavailability of fatigue-resistant polymer materials. The desirable characteristics for morphing material are large skin deformation, low weight, high load-carrying capacity, high out-of-plane stiffness for high load-carrying capacity and low in-plane stiffness to keep the actuating forces to a minimum. Environmental resistance and recovery rate are added advantages. Elastomeric skins made of Tecoflex 80A or Hexcel flex core are suitable prospects for VAWT application. The actuating force to deform the plate or skin can be obtained passively or actively. Active actuation involves the use of piezoelectric or pneumatic source, whereas the passive system can be implemented by converting the centrifugal force to a controlled actuation source such as hydraulic or pneumatic energy. The key advantages of this concept are low actuating force, high aerodynamic efficiency at low wind speed, better starting characteristics and reduction of blade loads. Though the morphing blade technology is still evolving, some of the current drawbacks that limit this technology are reduction in chord-wise and span-wise dynamic instability, flutter instability and noise at high rate of actuation.[123]

2.4.21 *Blade Pitching*

Blade pitching was proposed as a pragmatic solution to avert dynamic stall and to improve self-starting capabilities[124] as shown in Figure 2.44(a). The AoA of a straight-bladed Darrieus rotor blade continuously varies during each rotation and the airfoils undergo stall in specific azimuthal position in relation to TSR due to large AoA. The blades can be pitched actively by cams and servomotors or passively by using blade center of pressure to reduce the AoA. The passive pitch systems are suited for smaller VAWTs as they are less complex and rely on blade mass to create centrifugal force. Springs and stoppers are used by Kirke and Lazauskas[125] to restore the blade to its home position as shown in Figure 2.44(b). The

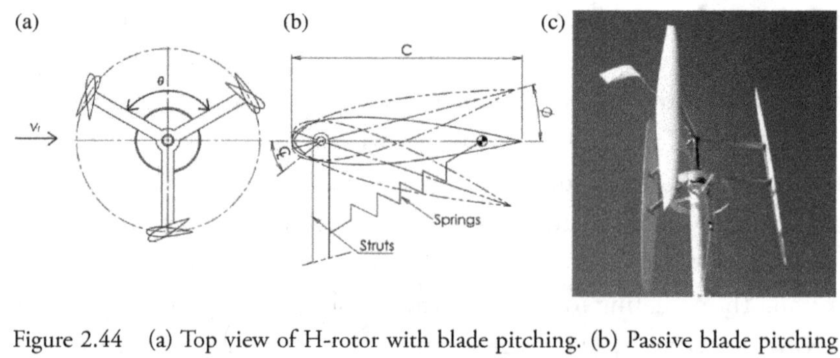

Figure 2.44 (a) Top view of H-rotor with blade pitching. (b) Passive blade pitching with centrifugal force and springs. (c) Commercial turbine with blade pitching. Image (c) is reproduced with permission from www.envergate.com.

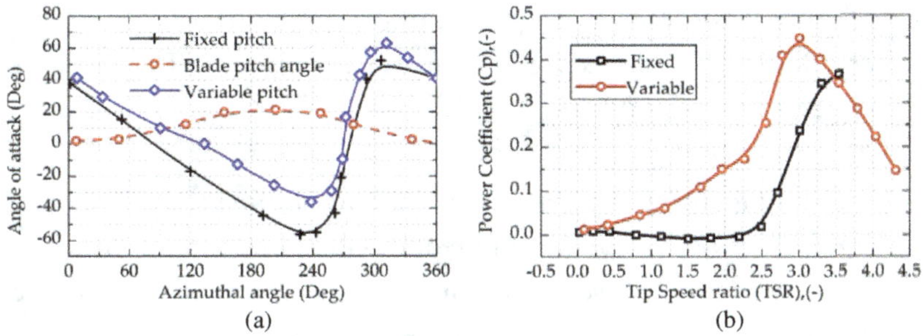

Figure 2.45 (a) Variation of AoA vs TSR. (b) Effect of passive blade pitching on Cp.

passive pitch system allows the blade to swing around a pivot in a controlled manner to achieve its favorable pitch. Variation of AoA with TSR is illustrated in Figure 2.45(a) and the influence of blade pitching on power performance is shown in Figure 2.45(b).

The primary drawback of passive pitching compared to active pitching is that both upwind and downwind will pitch in the same direction and still the rotor is prone to negative power production at downwind. Dynamic and active pitch control are not economically viable to incorporate in smaller VAWTs, but can be readily integrated in larger VAWTs due to low rotor rpm. High rate of pitching is not reliable for smaller VAWTs from economical and structural aspects.[126] Continuously varying the blade pitch for every azimuthal position is referred to as cycloidal turbine. The blade pitching is electronically controlled with a closed loop system including wind vane, wind speed and rpm sensors. The pitch of each wing is actively positioned to the optimum angle during the 360° rotations, so that the efficiency and subsequent power production are maximized.

Achieving an optimal angle of blade pitch is in fact a technical challenge. The variable pitch has to follow an optimized algorithm to extract maximum energy. A genetic algorithm based on variable pitch system can increase the annual energy output by 30% as predicted by the DMST model.[126] Pitch amplitude has significant influence on the performance related to the TSR. Higher magnitude of the pitch shows better performance at lower TSR, but reduced performance at higher TSR as the blade encounters large AoA at lower TSR.[127]

Apart from enhancing the performance of the turbine the active pitch can regulate power production and reduce vibrations and torque loads. From an overall perspective, the starting characteristics will be improved significantly, but a sustained rotation in low winds is not feasible with blade pitching. Since significant increase in energy output was reported, the active blade pitching systems are implemented in commercial turbines of capacity 50 kW and above.

2.4.22 *Omnidirectional Guide Vanes*

The diffuser-augmenting devices demand the yawing system to align with the wind direction. Though they are not active yawing systems, they do incur significant cost to enable the rotation of the diffuser. The Omnidirectional Augmenting Guide Vane (ODGV) systems are proposed as an alternative for diffuser systems by eliminating the yawing. A typical ODGV consists of circumferentially arranged guide vanes projecting radially from the turbine axis as shown in Figure 2.46(a). The stator or guide vanes enclose the turbine providing the required

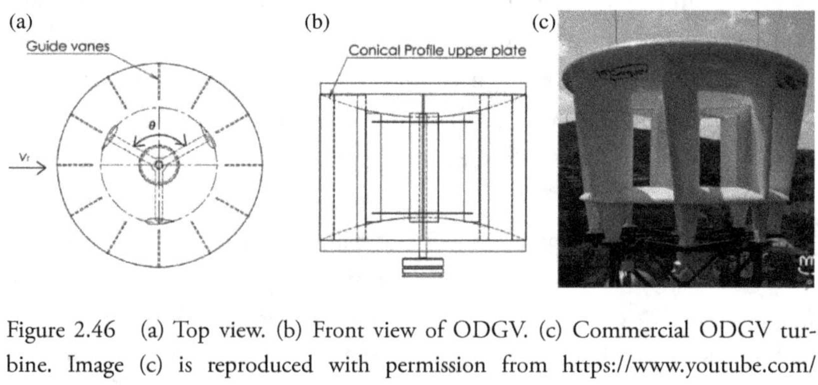

Figure 2.46 (a) Top view. (b) Front view of ODGV. (c) Commercial ODGV turbine. Image (c) is reproduced with permission from https://www.youtube.com/watch?v=0M-w5y5l8yk&ab_channel=McCamleyLtd.

structural support and housing for the bearings. Apart from increasing the wind speed, the guide vanes reduce turbulence and noise. The ODGV enhances the overall structural rigidity of the system by distributing the load and provides the capability to withstand high winds. Other than the guide vanes, ODGV consists of upper and lower plates with either flat or conical profile (Figure 2.46(b)). Five-bladed Giromill was put under test by Chong *et al.*[128] with conical upper and lower plates along with four pairs of guide vanes. The study claims that the wind is accelerated to the optimum angle of AoA, thus enhancing the performance in weak wind flows.

The wind tunnel study of the above system shows that the rotational speed was increased by 182% leading to an increase of 3.48 times in power performance. A similar system of two segmented guide vanes with the included angle of 10° was analyzed by computational studies[129] and it was concluded that the Cp increases by 31.65% compared to straight guide vanes and 147.1% increase compared to the bare turbine.

The computational study on the conical surface with guide vanes made of NACA 0018 profile reports an increase in the mass flow rate through the rotor and great reduction in the negative torque by decelerating the flow at specific regions. Nobile *et al.*[130] found that the guide vanes with conical surfaces increase the turbulent mixing, boosting the Cp and Ct up to 30~35% compared to bare turbines. In addition, the top surface aids in collecting rainwater and is tapped with a hollow shaft rotor. The top plates can also be designed to support solar panels[131] to increase the net output of the system.

The effect of guide vane geometry such as setting angle and the effect of gap between the vanes and the rotor blades are investigated with NACA 4518 blades of chord length 0.1 m and rotor diameter of 0.6 m.[132] The influence of these parameters on Cp are displayed in Figure 2.47(a, b). As suggested by Pope *et al.*,[133] apart from increasing the turbine

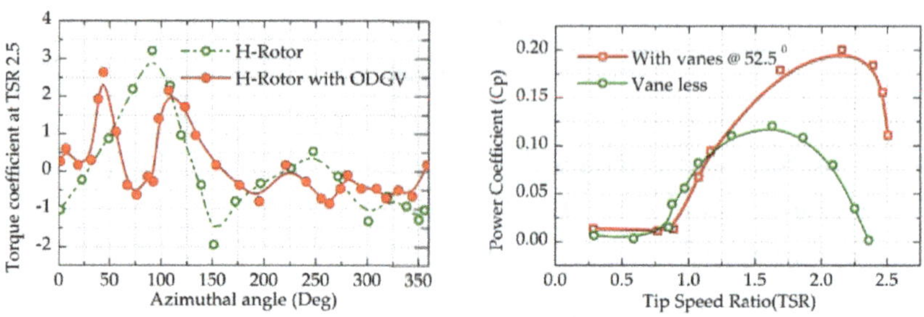

Figure 2.47 Effect of guide vanes on Ct (left) and Cp (right).

performance the ODGV can be a feasible solution for the issues that haunt urban wind turbines such as poor aesthetics, bird strike, safety and vibration, and increased net energy output for the given footprint. Nevertheless, ODGVs incur high initial capital cost and demand a strong foundation due to high wind loading and strong vortex shedding in the downstream. Though they accelerate wind flow and enhance the low wind speed performance, heir economics need to be evaluated for commercialization.

2.4.23 *Trailing Edge Cavity Airfoil*

The truncated airfoil with a cavity is proposed as a potential solution to sustain the rotation at low wind speeds. Baker *et al.*[134] demonstrated experimentally the advantages of truncated airfoils optimized for maximum aerodynamics characteristics[135] with excellent structural properties[136] due to high thickness, and they are preferred for the root of HAWT blades.[137] The cavity on the trailing edge is believed to induce the reverse drag that acts in the same direction as the aerodynamic torque to increase the net drag as shown in Figure 2.48. Essentially the rotors with truncated airfoil work with both drag and lift forces. At low wind speed, the drag forces dominate similar to the Savonius blades with minimal resistive torque from the incoming blades due to airfoil profile. As the Re rises, the drag force diminishes, the lift force dominates and the machine

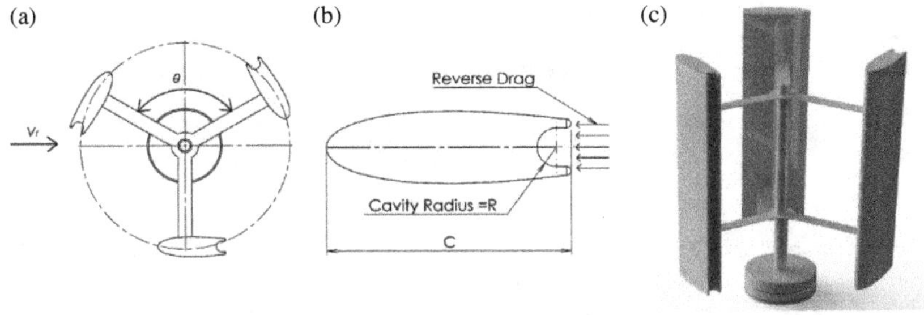

Figure 2.48 (a) H-rotor with trailing edge cavity airfoil. (b) Reverse drag on trailing edge cavity airfoil. (c) Illustration of a H-rotor with trailing edge cavity airfoil.

Table 2.1 Comparison of existing strategies on low wind speed operation.

Strategy	Merits	Demerits
Cambered airfoil	Better starting capability, delayed stall, less sensitive to surface roughness, high pitching moment	Increased drag on the down half, reduction in Cp
Solidity	Increased starting torque, low centrifugal forces due to low rpm	Large AoA, large drive train size and additional cost due to more number of blades
Helical blades	Improved aesthetics, smooth torque pulsation leading to reduced vibration	High manufacturing cost of blades, low torque for every azimuthal position
Blade thickness	Better performance at low Re, delayed stall and structurally sound	Increased profile drag at high Re and noise
Vortex generators	Improved performance at low Re, extended dynamic stall and marginal improvement in starting torque	Increased drag at high TSR and increases noise due to vortex shedding
Steeped airfoils	Feasibility check	
Gurney flaps	Delayed dynamic stall with increase in starting torque compared to conventional airfoil	Vibration after stall and noise due to vortex shedding

(*Continued*)

Table 2.1 (*Continued*)

Strategy	Merits	Demerits
Trailing edge flaps	Better performance in high and low TSR, able to regulate the rotor rpm aerodynamically	Power has to be expended to operate flaps, requires sophisticated control system
J-blade	Excellent startup torque, able to sustain low wind speed rotation, ease of blade manufacturing	Degraded performance at high TSR due to high form drag at high Re and increased fatigue failure of blades
Hybrid Darrieus-Savonius rotor	Excellent performance at low wind speed and reduced torque pulsation	Degraded performance at high TSR due to resistive torque from Savonius rotor
Drag-inducing struts	Higher starting torque, able to sustain low wind speed rotation, integrated in the strut	Higher form drag at high TSR and reduction in structural properties of strut.
Double Darrieus rotor	Feasibility check	
Low resistance bearings	Marginal decrease in the startup torque, improved structural stability	No notable influence in the low wind speed operation
Multi-stage rotor	Reduced peak torque pulsation, improved directional starting and minimized blade wake interaction	Power loss due to additional struts, blades are prone to vibrations
Diffuser-augmented turbines	Power output at low wind speed, reduced noise and birds safe	High structural loading at high winds, additional structures for yawing
Power-assisted startup	Low wind startup, minimum acceleration time and does not require additional mechanical structures	Additional sensor to monitor wind speed, not able to sustain low wind speed rotation
High aspect ratio	Low inertial resistance, reduced centrifugal force and strut parasitic drag	Power loss due to blade wake interaction, heavier and longer blades for same swept area

Table 2.1 (*Continued*)

Strategy	Merits	Demerits
Circulation-controlled blades	Marginal increase in startup torque, able to regulate rotor rpm without additional mechanical structures	Complex flow control devices, not much reward for low wind speed operation
Morphing blades	Able to regulate rpm and load alleviation at high rpm, increased annual energy yield	Require precise mechanical actuators and less incentive for low wind speed operation
Blade pitching	Favorable AoA can be maintained for maximum output and able to regulate rpm	Complex moving parts and increase in cost for actuation systems
Omnidirectional guide vanes	Reduced vibration and birds safe, capable of mounting photovoltaic panels on top, best suited for building top installation	High wind loading and high capital cost for manufacturing of guide vanes
Trailing edge cavity airfoil	Feasibility check	
Hybrid Darrieus-Savonius Telescopic rotor	Feasibility check	

operates by pure lift. The above hypothesis should be verified experimentally on an optimized truncated airfoil and performance has to be compared with conventional rotors to conclude on their applicability for low wind speed operation.

The current chapter critically reviewed the existing strategies that are commercially implemented, as well as those that have the potential to be implemented but are yet to be optimized for the suitability of Darrieus turbines for low wind speed operation. Following are the valuable conclusions that can be drawn out of this extensive literature search:

1. Very few airfoils have been developed for VAWTs and conventional symmetric NACA series are still widely preferred for higher kW capacity turbines, whereas cambered airfoils are adopted for smaller capacity turbines to generate higher starting torque compared to symmetrical airfoils. Desirable properties for Darrieus turbine airfoils are generation of sufficient lift at high AoA, reduced drag at negative AoA and structural soundness.
2. Generating substantial power for wind speed below 4 m/s is highly impossible by pure lift force due to higher AoA and static stall.
3. Integrating some form of drag-inducing device significantly outputs higher power at low wind speed. The performance is not curtailed by high turbulence when the Darrieus turbines are installed in rooftop environments.
4. Integrating drag features on airfoil such as stepped airfoil, trailing edge cavity airfoil, and J-blade has notable gain in low wind speed, but Cp deteriorates as Re increases.
5. The peak Cp reduction due to drag-inducing devices is inevitable. Hence averting the impact of drag devices at high TSR remains an open research question.
6. Another haunting issue is the lack of aerodynamic over speed regulation and load alleviation similar to blade pitching in HAWTs. A pragmatic solution that is economically feasible should improve low wind performance and also be capable of regulating over speed.
7. Flow-augmenting and flow-diverting devices such as diffuser and guide vanes have diminutive effect on the performance and are practically not feasible for higher capacity turbines.
8. Potential concepts that are capable of enhancing low wind performance and over speed regulation are morphing blades, blade pitching, and circulation-controlled blades. In order to implement

these concepts commercially they have to designed at low cost with improved reliability.

9. The power performance has to be compromised if a single solution is to cater to the above said problems at low cost.
10. The success strategy rests clearly on the economics, reliability, operational cost, low noise and simple design. The Darrieus community is still in search of such a solution.

Hence it is indispensable to integrate a drag device with sufficient frontal area to sustain the rotation and generate optimal power from the Darrieus rotor at weak wind flows. The drawbacks in the existing solutions that integrate the drag device should be identified and optimized. Based on the above summary, the following four concepts are shortlisted for feasibility check and the rest of this book will be dedicated to the optimization of these concepts through computation and experimentation.

1. **Rotor with stepped airfoils**
2. **Dual Darrieus rotor**
3. **Rotor with modified trailing edge cavity airfoil**
4. **Hybrid Darrieus-Savonius rotor**

Chapter Nomenclature

α	Angle of attack of blade element (°)
E	Net energy produced by the blade element in one revolution
q	Dynamic pressure
C_p	Power coefficient
C_t	Thrust coefficient
$\varnothing$	Angle between span line of the blade and the vertical
ω	Turbine rotational speed (rad/s)
V_f	Wind speed through plane of the rotor (m/s)
V	Wind velocity experienced by the blade (m/s)

r Blade radius (m)
U_J Jet nozzle velocity
Q Freestream dynamic pressure
c Airfoil chord
$\dot{m}$ Jet mass flow rate

References

1. Bos, R. *Self-Starting of a Small Urban Darrieus Rotor* (Master's Thesis). Delft University of Technology **2012**.
2. Wang, Z.; Zhuang, M. Leading-edge serrations for performance improvement on a vertical-axis wind turbine at low tip-speed-ratios. *Applied Energy* **2017**, *208*, 1184–1197.
3. Fujisawa, N.; Shibuya, S. Observations of dynamic stall on Darrieus wind turbine blades. *Journal of Wind Engineering and Industrial Aerodynamics* **2001**, *89*, 201–214.
4. Wilson, R. E.; Lissaman, P. *Applied Aerodynamics of Wind Power Machines*. Oregon State University **1974**.
5. Templin, R. Aerodynamic performance theory for the NRC vertical-axis wind turbine. *NASA STI/Recon Technical Report N* **1974**, *76*.
6. Strickland, J. A performance prediction model for the darrieus turbine. *Proceedings of the International Symposium on Wind Energy Systems*. Cambridge **1977**.
7. Holme, O. A contribution to the aerodynamic theory of the vertical-axis wind turbine. *Proceedings of the International Symposium on Wind Energy Systems*. Cambridge **1977**.
8. Paraschivoiu, I. Double-multiple streamtube model for Darrieus in turbines. **1981**.
9. Hirsch, I. H.; Mandal, A. A cascade theory for the aerodynamic performance of Darrieus wind turbines. *Wind Engineering* **1987**, 164–175.
10. Mandal, A.; Burton, J. The effects of dynamic stall and flow curvature on the aerodynamics of darrieus turbines applying the cascade model. *Wind Engineering* **1994**, 267–282.
11. Currie, I. G. *Fundamental Mechanics of Fluids*. CRC Press **2012**.

12. Strickland, J. H.; Webster, B.; Nguyen, T. A vortex model of the Darrieus turbine: an analytical and experimental study. *Journal of Fluids Engineering* **1979**, *101*, 500–505.
13. Migliore, P.; Wolfe, W.; Fanucci, J. Flow curvature effects on Darrieus turbine blade aerodynamics. *Journal of Energy* **1980**, *4*, 49–55.
14. Muraca, R.; Stephens, M. V.; Dagenhart, J. R. Theoretical performance of cross-wind axis turbines with results for a catenary vertical axis configuration. **1975**.
15. Sharpe, D. *A Theoretical and Experimental Study of the Darrieus Vertical Axis Wind Turbine.* Kingston Polytechnic **1977**.
16. Alexander, A.; Holownia, B. Wind tunnel tests on a Savonius rotor. *Journal of Wind Engineering and Industrial Aerodynamics* **1978**, *3*, 343–351.
17. Kamoji, M.; Kedare, S.; Prabhu, S. Performance tests on helical Savonius rotors. *Renewable Energy* **2009**, *34*, 521–529.
18. Akwa, J. V.; Vielmo, H. A.; Petry, A. P. A review on the performance of Savonius wind turbines. *Renewable and Sustainable Energy Reviews* **2012**, *16*, 3054–3064.
19. Sivasegaram, S. Secondary parameters affecting the performance of resistance-type vertical-axis wind rotors. *Wind Engineering* **1978**, 49–58.
20. Jeon, K. S.; Jeong, J. I.; Pan, J.-K.; Ryu, K.-W. Effects of end plates with various shapes and sizes on helical Savonius wind turbines. *Renewable Energy* **2015**, *79*, 167–176.
21. Mabrouki, I.; Driss, Z.; Abid, M. S. Performance analysis of a water Savonius rotor: effect of the internal overlap. *Sustainable Energy* **2014**, *2*, 121–125.
22. Mojola, O. On the aerodynamic design of the Savonius windmill rotor. *Journal of Wind Engineering and Industrial Aerodynamics* **1985**, *21*, 223–231.
23. Mahmoud, N.; El-Haroun, A.; Wahba, E.; Nasef, M. An experimental study on improvement of Savonius rotor performance. *Alexandria Engineering Journal* **2012**, *51*, 19–25.
24. Blackwell; Sheldahl, R. E.; Feltz, L. V. Wind tunnel performance data for two-and three-bucket Savonius rotors. *Journal of Energy* **1978**, *2*, 160–164.

25. Ogawa, T.; Yoshida, H. The effects of a deflecting plate and rotor end plates on performances of Savonius-type wind turbine. *Bulletin of JSME* **1986**, *29*, 2115–2121.
26. Saha, U.; Thotla, S.; Maity, D. Optimum design configuration of Savonius rotor through wind tunnel experiments. *Journal of Wind Engineering and Industrial Aerodynamics* **2008**, *96*, 1359–1375.
27. Damak, A.; Driss, Z.; Abid, M. Experimental investigation of helical Savonius rotor with a twist of 180. *Renewable Energy* **2013**, *52*, 136–142.
28. Blackwell, B. F.; Sheldahl, R. E.; Feltz, L. V. Wind tunnel performance data for two-and three-bucket Savonius rotors. *Journal of Energy* **1978**, *2*, 160–164.
29. Islam, M.; Ting, D. S.; Fartaj, A. Desirable airfoil features for smaller-capacity straight-bladed VAWT. *Wind Engineering* **2007**, *31*, 165–196.
30. Laneville, A.; Vittecoq, P. Dynamic stall: the case of the vertical axis wind turbine. *Journal of Solar Energy Engineering* **1986**, *108*, 140–145.
31. Timmer, W.; Van Rooij, R. Summary of the Delft University wind turbine dedicated airfoils. *Journal of Solar Energy Engineering* **2003**, *125*, 488–496.
32. Jasinski, W. J.; Noe, S. C.; Selig, M. S.; Bragg, M. B. Wind turbine performance under icing conditions. *Journal of Solar Energy Engineering* **1998**, *120*, 60–65.
33. Shepherd, D.; McBride, D.; Welch, D.; Dirks, K. N.; Hill, E. M. Evaluating the impact of wind turbine noise on health-related quality of life. *Noise and Health* **2011**, *13*, 333.
34. Kato, Y.; Seki, K. Vertical axis wind turbine designed aerodynamically at Tokai University. *Periodica Polytechnica Mechanical Engineering* **1981**, *25*, 47–56.
35. Snyder, M. H.; Furukawa, N. *Comparison of Performance of Darrieus Wind Turbines Having 12% And 21% Thick Sections*. Wichita State University **1979**.
36. Sheldahl, R. E.; Klimas, P. C. *Aerodynamic Characteristics of Seven Symmetrical Airfoil Sections Through 180-Degree Angle of Attack For Use in Aerodynamic Analysis of Vertical Axis Wind Turbines*. Sandia National Laboratories **1981**.

37. Sutherland, H. J.; Berg, D. E.; Ashwill, T. D. A retrospective of VAWT technology. *Sandia Report No. SAND2012.0304* **2012**.
38. Migliore, P. Comparison of NACA 6-series and 4-digit airfoils for Darrieus wind turbines. *Journal of Energy* **1983**, *7*, 291–292.
39. Zervos, A. Aerodynamic evaluation of blade profiles for vertical axis wind turbines. *Proceedings of the European Community Wind Energy Conference*. Denmark **1988**.
40. Claessens, M. *The Design and Testing of Airfoils for Application in Small Vertical Axis Wind Turbines* (Master's Thesis). Delft University of Technology **2006**.
41. McGhee, R. J.; Beasley, W. D. Wind-tunnel results for a modified 17-percent-thick low-speed airfoil section. **1981**.
42. Singh, M.; Biswas, A.; Misra, R. Investigation of self-starting and high rotor solidity on the performance of a three S1210 blade H-type Darrieus rotor. *Renewable Energy* **2015**, *76*, 381–387.
43. Kumar, M.; Surya, M. M. R.; Sin, N. P.; Srikanth, N. Design and experimental investigation of airfoil for extruded blades. *International Journal of Advances in Agricultural and Environmental Engineering* **2017**, *3*, 2349–1523.
44. Rainbird, J. M.; Bianchini, A.; Balduzzi, F.; Peiró, J.; Graham, J. M. R.; Ferrara, G.; Ferrari, L. On the influence of virtual camber effect on airfoil polars for use in simulations of Darrieus wind turbines. *Energy Conversion and Management* **2015**, *106*, 373–384.
45. Islam, M.; Ting, D. S.; Fartaj, A. Design of a special-purpose airfoil for smaller-capacity straight-bladed VAWT. *Wind Engineering* **2007**, *31*, 401–424.
46. Kirke, B.K. *Evaluation of Self-Starting Vertical Axis Wind Turbines for Stand-Alone Applications*. Griffith University **1998**.
47. Healy, J. The influence of blade camber on the output of vertical-axis wind turbines. *Wind Engineering* **1978**, 146–155.
48. Ågren, O.; Berg, M.; Leijon, M. A time-dependent potential flow theory for the aerodynamics of vertical axis wind turbines. *Journal of Applied Physics* **2005**, *97*, 104913.
49. Mohamed, M. Impacts of solidity and hybrid system in small wind turbines performance. *Energy* **2013**, *57*, 495–504.

50. El-Samanoudy, M.; Ghorab, A.; Youssef, S. Z. Effect of some design parameters on the performance of a Giromill vertical axis wind turbine. *Ain Shams Engineering Journal* **2010**, *1*, 85–95.
51. Worasinchai, S.; Ingram, G. L.; Dominy, R. G. The physics of H-Darrieus turbine starting behavior. *Journal of Engineering for Gas Turbines and Power* **2015**, *138*, 062605.
52. Staelens, Y.; Saeed, F.; Paraschivoiu, I. A straight-bladed variable-pitch VAWT concept for improved power generation. *ASME 2003 Wind Energy Symposium.* Nevada **2003**.
53. Mohamed, M. Aero-acoustics noise evaluation of H-rotor Darrieus wind turbines. *Energy* **2014**, *65*, 596–604.
54. Gorlov, A. *Development of the Helical Reaction Hydraulic Turbine. Final Technical Report.* Northeastern University **1998**.
55. Baker, J. Features to aid or enable self starting of fixed pitch low solidity vertical axis wind turbines. *Journal of Wind Engineering and Industrial Aerodynamics* **1983**, *15*, 369–380.
56. Purser, P. E.; Spearman, M. L. *Wind-Tunnel Tests at Low Speed of Swept and Yawed Wings Having Various Plan Forms.* NASA **1951**.
57. Shiono, M.; Suzuki, K.; Kiho, S. Output characteristics of Darrieus water turbine with helical blades for tidal current generations. *Proceedings of the The Twelfth International Offshore and Polar Engineering Conference.* Kitakyushu **2002**.
58. Blackwell, B.; Reis, G. *Blade Shape for a Troposkien Type of Vertical-Axis Wind Turbine.* Sandia National Laboratories **1974**.
59. Danao, L.A.; Qin, N.; Howell, R. A numerical study of blade thickness and camber effects on vertical axis wind turbines. *Proceedings of the Institution of Mechanical Engineers, Part A: Journal of Power and Energy* **2012**, *226*, 867–881.
60. Batista, N.; Melício, R.; Marias, J.; Catalão, J. Self-start evaluation in lift-type vertical axis wind turbines: Methodology and computational tool applied to asymmetrical airfoils. *Proceedings of the Power Engineering, Energy and Electrical Drives (POWERENG).* Malaga **2011**.
61. Alam, M.; Iqbal, M. A low cut-in speed marine current turbine. *The Journal of Ocean Technology* **2010**, *5*, 49–61.
62. Parchen, R.; Bruggeman, J.; Dassen, A. The effect of the blade thickness of wind turbine blades on the noise due to inflow turbulence. *Acustica* **1996**, *82*, S82.S82.

63. Mueller-Vahl, H.; Pechlivanoglou, G.; Nayeri, C.; Paschereit, C. Vortex generators for wind turbine blades: A combined wind tunnel and wind turbine parametric study. *ASME Paper No. GT2012.69197* **2012**.
64. Quinlan, P.; Scheffler, R.; Wehrey, M. Performance test results of vortex generators attached to a vertical axis wind turbine. *Proceedings of the 7th ASME Wind Energy Symposium*. Louisiana **1987**.
65. Seshagiri, A.; Cooper, E.; Traub, L. W. Effects of vortex generators on an airfoil at low Reynolds numbers. *Journal of Aircraft* **2009**, *46*, 116.
66. Paraschivoiu, I. *Wind Turbine Design: With Emphasis on Darrieus Concept*. Presses inter Polytechnique **2002**.
67. Kline, R. L.; Fogleman, F. F. Airfoil for aircraft having improved lift generating device. **1977**.
68. Lumsdaine, E.; Johnson, W.; Fletcher, L.; Peach, J. Investigation of the Kline-Fogleman airfoil section for rotor blade applications. **1974**.
69. Finaish, F.; Witherspoon, S. Aerodynamic performance of an airfoil with step-induced vortex for lift augmentation. *Journal of Aerospace Engineering* **1998**, *11*, 9–16.
70. Voona, R. Enhancing the aerodynamic performance of stepped airfoils. **2012**.
71. Cox, M.; Avakian, V.; Huynh, B. Performance of a stepped airfoil at low Reynolds numbers. *Proceedings of the Proceedings of the 19th Australasian Fluid Mechanics Conference*. Melbourne **2014**.
72. Mishriky, F.; Walsh, P. Effect of step depth and angle in Kline-Fogleman (KFm-2) airfoil. *Global Journal of Research In Engineering* **2016**, *16*, 1–9.
73. Zerihan, J.; Zhang, X. Aerodynamics of Gurney flaps on a wing in ground effect. *AIAA Journal* **2001**, *39*, 772.780.
74. Zayas, J.; Van Dam, C.; Chow, R.; Baker, J.; Mayda, E. Active aerodynamic load control for wind turbine blades. *Proceedings of the European Wind Energy Conference*. Athens **2006**.
75. Liu, T.; Montefort, J. Thin-airfoil theoretical interpretation for Gurney flap lift enhancement. *Journal of Aircraft* **2007**, *44*, 667–671.
76. Liebeck, R. H. Design of subsonic airfoils for high lift. *Journal of Aircraft* **1978**, *15*, 547–561.
77. Xu, Z.; Wang, Q.; Dai, G.; Tan, H.; Zhong, Y. Study on improvement in aerodynamic performance of vertical axis wind turbine using Gurney flap. *2011 Second International Conference on Mechanic Automation and Control Engineering*. Inner Mongolia **2011**.

78. Viswamurthy, S. R.; Ganguli, R. An optimization approach to vibration reduction in helicopter rotors with multiple active trailing edge flaps. *Aerospace Science and Technology* **2004**, *8*, 185–194.
79. Joncas, S.; Bergsma, O.; Beukers, A. Power regulation and optimization of offshore wind turbines through trailing edge flap control. *Proceedings of the 43th AIAA Aerospace Science Meeting and Exhibit*. Nevada **2005**.
80. Castaignet, D.; Barlas, T.; Buhl, T.; Poulsen, N. K.; Wedel-Heinen, J. J.; Olesen, N. A.; Bak, C.; Kim, T. Full-scale test of trailing edge flaps on a Vestas V27 wind turbine: active load reduction and system identification. *Wind Energy* **2014**, *17*, 549–564.
81. Carla, A.; Amne, E.; Joelle, N.; Michel, E. Performance enhancement of a vertical axis wind turbine using a slotted deflective flap at the trailing edge. *Energy Conversion and Management* **2022**, *273*, 116388.
82. Yang, Y.; Li, C.; Zhang, W.; Guo, X.; Yuan, Q. Investigation on aerodynamics and active flow control of a vertical axis wind turbine with flapped airfoil. *Journal of Mechanical Science and Technology* **2017**, *31*, 1645–1655.
83. Zamani, M.; Maghrebi, M. J.; Varedi, S. R. Starting torque improvement using J-shaped straight-bladed Darrieus vertical axis wind turbine by means of numerical simulation. *Renewable Energy* **2016**, *95*, 109–126.
84. Zamani, M.; Nazari, S.; Moshizi, S. A.; Maghrebi, M. J. Three dimensional simulation of J-shaped Darrieus vertical axis wind turbine. *Energy* **2016**, *116*, 1243–1255.
85. Chen, J.; Yang, H.; Yang, M.; Xu, H. The effect of the opening ratio and location on the performance of a novel vertical axis Darrieus turbine. *Energy* **2015**, *89*, 819–834.
86. Rassoulinejad-Mousavi, S.; Jamil, M.; Layeghi, M. Experimental study of a combined three bucket H-rotor with savonius wind turbine. *World Applied Sciences Journal* **2013**, *28*, 205–211.
87. http://www.hi-vawt.com.tw/en/ds3000w.html.
88. Gupta, R.; Das, R.; Sharma, K. Experimental study of a Savonius-Darrieus wind machine. *Proceedings of the International Conference on Renewable Energy for Developing Countries*. University of Columbia **2006**.
89. Wakui, T.; Tanzawa, Y.; Hashizume, T.; Nagao, T. Hybrid configuration of Darrieus and Savonius rotors for stand-alone wind turbine-generator systems. *Electrical Engineering in Japan* **2005**, *150*, 13–22.

90. Gupta, R.; Sharma, K. Flow physics of a combined Darrieus-Savonius rotor using computational fluid dynamics (CFD). *International Research Journal of Engineering Science, Technology and Innovation* **2012**, *1*, 1–13.
91. Kyozuka, Y. An experimental study on the Darrieus-Savonius turbine for the tidal current power generation. *Journal of Fluid Science and Technology* **2008**, *3*, 439–449.
92. Johnston Jr, S. F. *Proceedings of the Vertical Axis Wind Turbine (VAWT) Design Technology Seminar for Industry*; Sandia National Laboratories **1980**.
93. Moran, W. A. *Giromill Wind Tunnel Test and Analysis*. U.S. Department of Energy **1977**.
94. Li, Y.; Calisal, S. M. Three-dimensional effects and arm effects on modeling a vertical axis tidal current turbine. *Renewable Energy* **2010**, *35*, 2325–2334.
95. Li, Q. Y.; Wu, S. L.; Peng, F. B.; Li, G. J. A study on aerodynamic of vertical-axis wind-mill with double blades. *Solar World Congress* **1983**, 14–19.
96. https://www.4navitas.com/vertical-axis-wind-turbines/.
97. Aso, T.; Seki, K.; Ushiyama, I. Study on the bearing resistance in a wind turbine generator system. *Wind Engineering* **2015**, *39*, 113–128.
98. Kakuta, K. Friction moment of radial ball bearing under thrust load. *Transactions of the Japan Society of Mechanical Engineers* **1961**, *27*, 945–956.
99. Gorelov, D.; Krivospitsky, V. Prospects for development of wind turbines with orthogonal rotor. *Thermophysics and Aeromechanics* **2008**, *15*, 153–157.
100. Shen, W. Z.; Mikkelsen, R. F.; Sørensen, J. N.; Bak, C. Evaluation of the Prandtl tip correction for wind turbine computations. *Proceedings of the 2002 Global Windpower Conference and Exhibition*. Paris **2002**.
101. Igra, O. Research and development for shrouded wind turbines. *Energy Conversion and Management* **1981**, *21*, 13–48.
102. Gaden, D. L.; Bibeau, E. L. A numerical investigation into the effect of diffusers on the performance of hydro kinetic turbines using a validated momentum source turbine model. *Renewable Energy* **2010**, *35*, 1152.1158.
103. Tartuferi, M.; D'Alessandro, V.; Montelpare, S.; Ricci, R. Enhancement of Savonius wind rotor aerodynamic performance: A computational study of new blade shapes and curtain systems. *Energy* **2015**, *79*, 371–384.

104. Watanabe, K.; Takahashi, S.; Ohya, Y. Application of a diffuser structure to vertical-axis wind turbines. *Energies* **2016**, *9*, 406.
105. Geurts, B.; Simao Ferreira, C.; Van Bussel, G. *Aerodynamic Analysis of a Vertical Axis Wind Turbine in a Diffuser*. European Wind Energy Association **2010**.
106. Liu, S.; Li, S.; He, L. Direct-driven permanent magnet synchronous wind-power generating system with two three-level converters based on SVPWM control. *Procedia Engineering* **2012**, *29*, 1191–1195.
107. Kjellin, J.; Bernhoff, H. Electrical starter system for an H-rotor type VAWT with PM-generator and auxiliary winding. *Wind Engineering* **2011**, *35*, 85–92.
108. Aner, M.; Nowicki, E.; Wood, D. Accelerated starting by motoring a grid-connected small wind turbine generator. *2011 International Conference and Utility Exhibition on Power and Energy Systems: Issues & Prospects for Asia (ICUE)*. Pattaya **2011**.
109. http://www.ginlong.com/en/Wind_Products/.
110. Hwang, S.-M.; Eom, J.-B.; Jung, Y.-H.; Lee, D.-W.; Kang, B.-S. Various design techniques to reduce cogging torque by controlling energy variation in permanent magnet motors. *IEEE Transactions on Magnetics* **2001**, *37*, 2806–2809.
111. Branlard, E.; Gaunaa, M. Development of new tip-loss corrections based on vortex theory and vortex methods. *Journal of Physics: Conference Series* **2014**, *555*, 012012.
112. Paraschivoiu, I.; Desy, P.; Masson, C. Blade tip, finite aspect ratio, and dynamic stall effects on the Darrieus rotor. *Journal of Propulsion and Power* **1988**, *4*, 73–80.
113. Brusca, S.; Lanzafame, R.; Messina, M. Design of a vertical-axis wind turbine: how the aspect ratio affects the turbine's performance. *International Journal of Energy and Environmental Engineering* **2014**, *5*, 333–340.
114. Wood, N. J.; Nielsen, J. N. Circulation control airfoils-Past, present, future. **1985**.
115. Migliore, P.; Wolfe, W.; Walters, R. Aerodynamic tests of Darrieus turbine blades. *Proceedings of the Proceedings of the AIAA/SERI Wind Energy Conference*. Boulder, **1980**.
116. Jones, G. S.; Englar, R. J. Advances in pneumatic-controlled high-lift systems through pulsed blowing. **2003**.

117. Englar, R. J. Two-dimensional subsonic wind tunnel tests of two 15-percent thick circulation control airfoils. **1971**.
118. Shires, A.; Kourkoulis, V. Application of circulation controlled blades for vertical axis wind turbines. *Energies* **2013**, *6*, 3744–3763.
119. Lachenal, X.; Daynes, S.; Weaver, P. M. Review of morphing concepts and materials for wind turbine blade applications. *Wind Energy* **2013**, *16*, 283–307.
120. Van Wingerden, J.; Hulskamp, A.; Barlas, T.; Marrant, B.; Van Kuik, G.; Molenaar, D. P.; Verhaegen, M. On the proof of concept of a 'smart'wind turbine rotor blade for load alleviation. *Wind Energy* **2008**, *11*, 265–280.
121. Liu, W.; Xiao, Q. Investigation on Darrieus type straight blade vertical axis wind turbine with flexible blade. *Ocean Engineering* **2015**, *110*, 339–356.
122. Nakajima, T.; Tsutahara, M.; Yamauchi, M.; Kuroda, K. Performance of vertical-axis type straight-bladed Darrieus turbine with elastic blades (Performance improvement by deformation of blades). *Journal of Fluid Science and Technology* **2015**, *10*, JFST0009.
123. Ai, Q.; Azarpeyvand, M.; Lachenal, X.; Weaver, P. M. Aerodynamic and aeroacoustic performance of airfoils with morphing structures. *Wind Energy* **2016**, *19*, 1325–1339.
124. Hwang, I. S.; Min, S. Y.; Jeong, I. O.; Lee, Y. H.; Kim, S. J. Efficiency improvement of a new vertical axis wind turbine by individual active control of blade motion. *Proceedings of the SPIE* **2006**, *6173*, 617311.
125. Kirke, B.; Lazauskas, L. Enhancing the performance of vertical axis wind turbine using a simple variable pitch system. *Wind Engineering* **1991**, 187–195.
126. Paraschivoiu, I.; Trifu, O.; Saeed, F. H-Darrieus wind turbine with blade pitch control. *International Journal of Rotating Machinery* **2009**, *2009*, 505343.
127. Chougule, P.; Nielsen, S. R.; Basu, B. Active blade pitch control for straight-bladed darrieus vertical axis wind turbine of new design. In *Proceedings of the Key Engineering Materials*, **2013**; pp. 668–675.
128. Chong, W.; Pan, K.; Poh, S.; Fazlizan, A.; Oon, C.; Badarudin, A.; Nik-Ghazali, N. Performance investigation of a power augmented vertical axis wind turbine for urban high-rise application. *Renewable Energy* **2013**, *51*, 388–397.

129. Chong, W.; Fazlizan, A.; Poh, S.; Pan, K.; Hew, W.; Hsiao, F. The design, simulation and testing of an urban vertical axis wind turbine with the omni-direction-guide-vane. *Applied Energy* **2013**, *112*, 601–609.
130. Nobile, R.; Vahdati, M.; Barlow, J. F.; Mewburn-Crook, A. Unsteady flow simulation of a vertical axis augmented wind turbine: A two-dimensional study. *Journal of Wind Engineering and Industrial Aerodynamics* **2014**, *125*, 168–179.
131. Bhandari, B.; Lee, K.-T.; Lee, G.-Y.; Cho, Y.-M.; Ahn, S.-H. Optimization of hybrid renewable energy power systems: A review. *International Journal of Precision Engineering And Manufacturing-Green Technology* **2015**, *2*, 99–112.
132. Takao, M.; Kuma, H.; Maeda, T.; Kamada, Y.; Oki, M.; Minoda, A. A straight-bladed vertical axis wind turbine with a directed guide vane row — Effect of guide vane geometry on the performance. *Journal of Thermal Science* **2009**, *18*, 54–57.
133. Pope, K.; Rodrigues, V.; Doyle, R.; Tsopelas, A.; Gravelsins, R.; Naterer, G.; Tsang, E. Effects of stator vanes on power coefficients of a zephyr vertical axis wind turbine. *Renewable Energy* **2010**, *35*, 1043–1051.
134. Baker, J.; Mayda, E.; Van Dam, C. Experimental analysis of thick blunt trailing-edge wind turbine airfoils. *Journal of Solar Energy Engineering* **2006**, *128*, 422.431.
135. Bangalore Gangadharacharya, K. Vortex shedding and aerodynamic drag on truncated trailing edge airfoil. **2015**.
136. TPI Composites, Inc. *Parametric Study for Large Wind Turbine Blades.* Sandia National Laboratories **2002**.
137. Kahn, D. L.; van Dam, C.; Berg, D. E. *Trailing Edge Modifications for Flatback Airfoils*. Sandia National Laboratories **2008**.

Chapter 3

Feasibility Check on Potential Concepts

3.1 ERIAN Subsonic Wind Tunnel

The experimental investigation on the wind tunnel for the rest of this book is conducted on ERIAN subsonic wind tunnels. Open-loop suction-type wind tunnel was designed and constructed for testing HAWT and VAWT micro wind turbines. The wind tunnel spans a length of 4.8 m with the test section extending for 1.4 m. The bell mouth inlet of 1.28 m × 1.28 m is reduced to the test section size of 0.7 m × 0.7 m. Square cross-section flow straightener is placed in the upstream and downstream flows in the convergent and diffuser sections, respectively. The downstream flow is in line with the upstream flow straightener to prevent the rotational effect induced by the axial flow fan, since the diffuser is comparatively short. The wind tunnel speed can be varied from 0–15 m/s with maximum recorded turbulence of 0.2% and mean variation of 2%. Uniform wind flow was generated by 0.9 m diameter axial flow fan (multi-wing). The variable frequency drive regulates the wind speed and can be programmed to simulate unsteady wind conditions. The axial flow fan is powered by Permanent Magnet Motor (PMM) with a power capacity of 5 kW. The smaller size of the PMM minimizes the blockage compared to the induction motor. The rotor shaft is supported by deep groove ball bearing on bottom and spherical ball bearing on top to accommodate any shaft misalignment. The schematic of the wind tunnel is shown in Figure 3.1.

3.1.1 *Instrumentation*

The instrumentation on the wind tunnel is shown in Figure 3.2. The wind speed was measured with omnidirectional hot wire anemometer probe (Kanomax-0965) with an accuracy of 0.1 m/s mounted at 300 mm from the downstream of the settling chamber exit. Measured air velocity is converted into 0–5 V DC by airflow transducer (Kanomax-6332 D) with a sampling rate of 0.1 s. The rotational speed was measured as rpm by proximity sensor (Omron-E2B-S08LN02-WP-B1 2M) that can

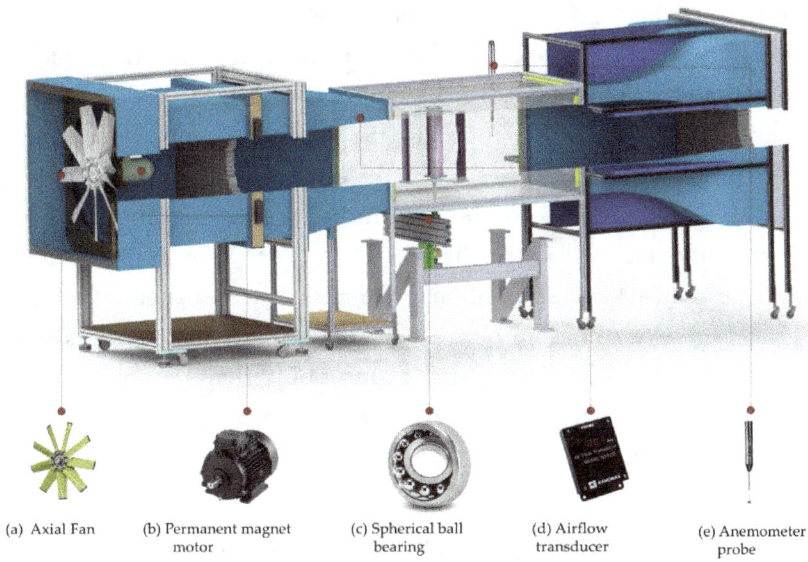

Figure 3.1 ERIAN open loop subsonic wind tunnel.

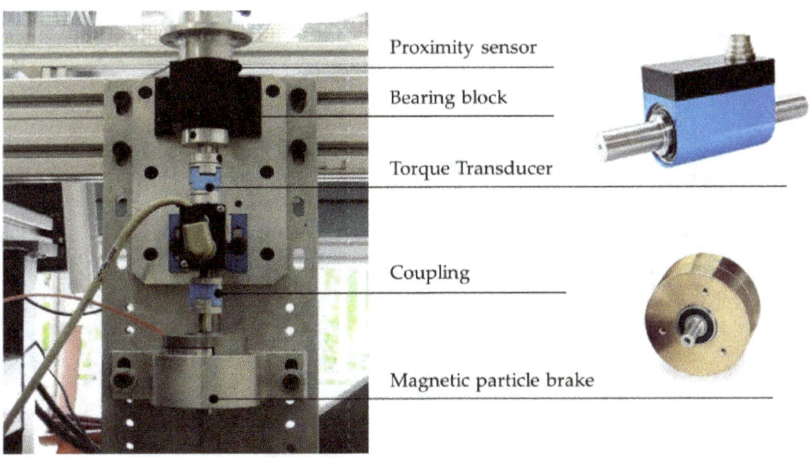

Figure 3.2 Instrumentation details of the ERIAN wind tunnel.

output 4–20 mA pulse for each rotation. The braking torque was applied through magnetic particle brake (Placid Industries-B2). The braking torque was adjusted by varying the input current to the brake through variable DC power supply, curtailing the turbine to operate at required TSR. The difference between the generated rotor torque and the braking torque was measured by torque transducer (Lorenz-DR-2112-R) with repeatability of ±0.02%. The thrust load on the rotor was measured by S-type resistance load cell fixed on both ends of the rotor shaft (Omega-LCEA-1kg) with an output of ±0.5 mV DC. All the above sensors were connected to the DEWE 43V data acquisition system for synchronous data logging. The data is averaged for 5 s before being recorded at the sampling rate of 5 s.

3.1.2 *Data Reduction and Blockage Correction*

The performance characteristics of the investigated models are evaluated by the following equations. The required equations other than the ones listed below will be mentioned in the respective chapters.

$$C_p = \frac{P_{rotor}}{P_{wind}} \tag{3.1}$$

$$P_{wind} = 0.5 \times \rho \times A_s \times V_f^3 \tag{3.2}$$

$$P_{rotor} = T \times \omega \tag{3.3}$$

$$C_t = \frac{T}{0.5 \times R \times \rho \times V_f^2 \times A_S} \tag{3.4}$$

$$C_{TS} = \frac{T_S}{0.5 \times R \times \rho \times V_f^2 \times A_S} \tag{3.5}$$

$$C_H = \frac{T_H}{0.5 \times R \times \rho \times V_f^2 \times A_S} \tag{3.6}$$

The test rotor for most of the investigated models have frontal swept area of at least 420 mm × 420 mm and occupies 30~40% of the test section area, which is more than the allowable blockage. High solid blockage ε_s will induce errors by increasing the upstream velocity locally, which is more pronounced for the closed than the open section. The deflected flow interacts with the wind tunnel wall surfaces influencing the accuracy of the testing as shown in Figure 3.3. Wake blockage ε_w is cumbersome to model compared to blockage modeling for a stationary body. The wake patterns of Darrieus and Savonius rotors vary to a greater extent with the wake blockage more pronounced on the Savonius rotor as described by Fujisawa.[1] Since the Savonius rotor has high solid blockage producing large wake disturbances, it becomes more difficult to predict the degradable effect on the flow and the wake generated has a lower mean velocity than the freestream.

The resulting flow is highly turbulent, asymmetric, periodic, unsteady and separated.

Since the current study involves both Darrieus and Savonius rotors a generic correction method proposed by Pope and Harper[2] was adapted to offset the blockage effects. If the tested model is of a regular shape,

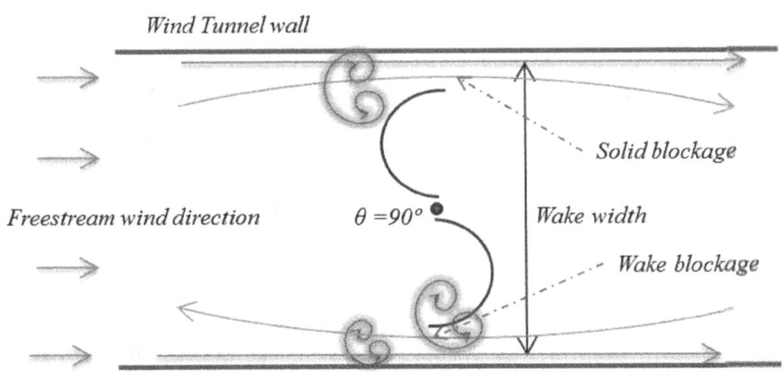

Figure 3.3 Solid and wake blockage of a Savonius rotor.

the total blockage is the summation of solid and wake blockages. It is given by

$$B_T = \varepsilon_s + \varepsilon_w \tag{3.7}$$

The total blockage including solid and wake blockages for the non-standard shapes is

$$B_T = \frac{Model\ \ Frontal\ \ Area}{Test\ \ Section\ \ Area} \tag{3.8}$$

$$v_c = v_f(1 + B_T) \tag{3.9}$$

$$Re_c = Re_f(1 + B_T) \tag{3.10}$$

$$q_c = q_u\,(1 + 2B_T) \tag{3.11}$$

3.2 Airfoil with Step (KF-N-21)*

3.2.1 *Design of KF-N-21 Airfoil*

An airfoil that is capable of maintaining the lift over high AoA without significant increase in drag will be an apt choice for Darrieus turbines. Kline and Fogleman[3] (KF) conceived airfoils with abrupt discontinuity claiming to have enhanced lift. The initial profile designed by the inventors has a distinctive abrupt step on the suction side of the airfoil. The airfoil was claimed to perform better than the conventional airfoil by trapping vortex bubbles in the step. Extensive experimental investigations Cox *et al.*[4] and Krishnamurthi[5] revealed that gain in lift was achieved at the expense of increased drag with lower lift-to-drag ratio. Though the airfoil does not outperform in the pre-stall region, it demands attention for its aerodynamic behaviors in the post-stall region when the step is

* Parts of this section are adapted with permission from Kumar, P. M.; Surya, M. M. R.; Kethala, R.; Srikanth, N. Experimental investigation of the performance of Darrieus wind turbine with trapped vortex airfoil. *2017 3rd International Conference on Power Generation Systems and Renewable Energy Technologies (PGSRET)*. Johor Bahru 2017.

located on the suction side of the airfoil.[6] The ability of KF airfoil to extend the stall even at high AoA (>40°) is a desirable property for Darrieus turbines. The stall extension was attributed to flow separation due to adverse pressure gradient and subsequent reattachment at the end of the turbulent circulation zone in the cavity. The reattachment length, which has significant impact on lift-to-drag ratio, depends on the skin friction coefficient and depth of step. The vital parameters for the KF airfoil are the location on chord, depth of the step, number of steps, and surface on which the step exists (suction side or pressure side). These geometrical parameters can be intelligently varied to obtain the required aerodynamic characteristics. The impact of these parameters on the lift and drag coefficients of the airfoil along with AoA, camber, and Re are well investigated in previous studies. The new airfoil KF-N-21 was created by introducing the step discontinuity borrowed from the suction side of the NACA 0021 airfoil as shown in Figure 3.4. In order to create maximum drag during the reverse flow to aid in startup, the depth of step is located at 18%c, while the maximum thickness of airfoil is 21%c. The step length is 50%c and originates at 30%c from the leading edge, though location near to the trailing edge is favorable for high lift coefficient. The step terminates at 5%c from trailing edge to prevent it from flapping.

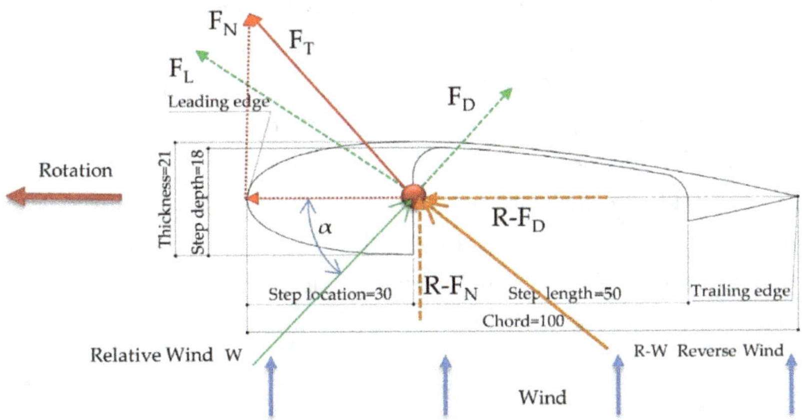

Figure 3.4 Flow velocities and force diagram of KF-N-21 airfoil.

3.2.2 *Computational Optimization of KF-N-21 Airfoil*

A 2D numerical simulation was carried out on a 3-straight-bladed Darrieus turbine with KF-N-21 airfoil and NACA 0021 airfoil. The results are compared in terms of torque coefficient Ct and power coefficient Cp. The current arrangement of blade support with end plates eliminates the parasitic drag due to support arms and blade tip vortices. With these enhancements, 2D simulation will be appropriate for the current study. The inlet side of the domain is assigned to inlet boundary condition with the wind speed varying from 4 m/s to 9 m/s. The rear side of the domain was assigned to outlet boundary condition of atmospheric pressure. Rest of the sides of the domain are considered as walls. A structured mesh has been generated in the vicinity of the blade, with $y^{+} < 1$ and edge sizing of 0.1 mm to capture the vortex bubble formation. A gradual cell growth is achieved from the blade boundary to adopt to the unstructured triangular elements of domain. Mesh independent study is performed to determine the optimum number of cells. The cell count is varied from 0.3 million to 0.6 million and their influence on the performance coefficients is less than 1.1%. Hence an optimum cell count of 0.5 ± 0.05 million was adopted for the entire study. Second-order spatial discretization was employed to improve the numerical accuracy of the sliding mesh technique. Five complete revolutions are computed with constant time steps. The performance coefficients are computed for every degree of revolution, averaging the values for the last three revolutions, until a stabilized flow field is established. The TSR varies from 1 to 3 by 0.2 increment and the azimuthal angle varies from 0° to 360°. The rotor under investigation and the computational domain dimensions are shown in Figure 3.5(a) and mesh details in Figure 3.5(b, c). Shear stress transport (SST) $k-\omega$ turbulence model along with the Reynolds-averaged Navier-Stokes (RANS) model are employed.

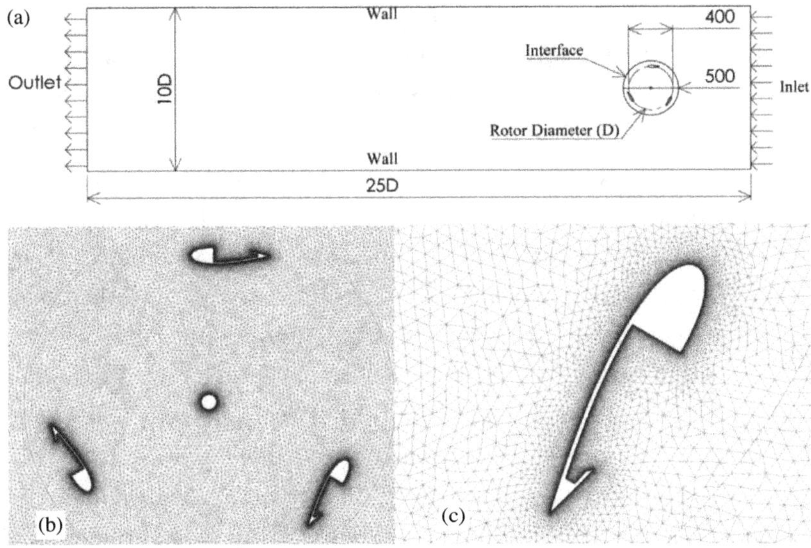

Figure 3.5 (a) Domain characteristics. (b, c) Mesh details of KF-N-21 airfoil.

3.2.3 *Discussion on the KF-N-21 Airfoil Computational Results*

The performance of the turbine is measured in terms of Ct and Cp. The Ct is obtained after solving the RANS equation by ANSYS and computing the moment coefficient after averaging the last three revolutions. The Cp is calculated from the Ct. Ct values are recorded for every degree of rotation of the turbine. The average Ct is obtained for the last three revolutions and this average value is used to compute Cp. It can be noted that both the results have minor variations as long as the TSR is less than 2. But as the TSR increases the deviation between the results also widens. A possible reason is that the 2D assumption does not include the blade tip vortices, which have significant effect on the performance of the turbine at higher TSR. Cp vs TSR curves for NACA 0021 and KF-N-21 airfoil are plotted for a range of Re as shown in Figure 3.9. Results from CFD show that the peak Cp of KF-N-21 is 21% higher than that of NACA 0021 airfoil for Re of 1.2×10^5 and for Re of 1.2×10^5 to 1.8×10^5, the peak Cp of KF-N-21 is 14% higher than that of NACA 0021 airfoil. At higher Re from 2.1 to 2.7×10^5, the Cp curve of KF-N-21 matches that of NACA 0021 as shown in Figure 3.9. In Figure 3.10, the variation of static Ct with respect to the azimuthal angle at different Re is shown. It can be noted that, for rotation angles between 30° and 60°, the torque coefficient of KF-N-21 airfoil is higher than that of NACA 0021. This behavior repeats after every 90° rotation of the turbine. At high AoA stall is encountered and the drag force is high in NACA 0021 airfoil blades. In KF-N-21, the lift force generated increases at higher AoA due to the trapped vortex bubble formation in the suction side of the airfoil where a step is located. This gain in lift also increases drag but the lift-to-drag ratio obtained is less. The trapped vortex bubble begins its regime to form instantaneously right from 45° azimuthal angle and a complete vortex is generated at about 60° as shown in Figure 3.6(b). Figure 3.8(a), (b) and (c) show the velocity, vorticity and the coefficient of pressure contours, respectively, at 15°, 20°, 45° and 90°. The vortex is trapped by the cavity present in the suction side of the airfoil. The two edges of the cavity serve as a barrier to the wind flow forming a separation bubble inside the cavity. The pressure contour comparison is shown in Figure 3.7.

Front and rear edges act as reflection planes, while edges located near the nose of the airfoil are responsible for the formation of separated flow region and a shear layer. The shear layer has vorticity in it which concentrates into a vortex when the fluid detaches from the suction orifice. As the vortex strength increases with time, the separated flow region is dominated and the generation of shear layer from the tip of the cavity edge will eventually get terminated. The vortex strength stabilizes and

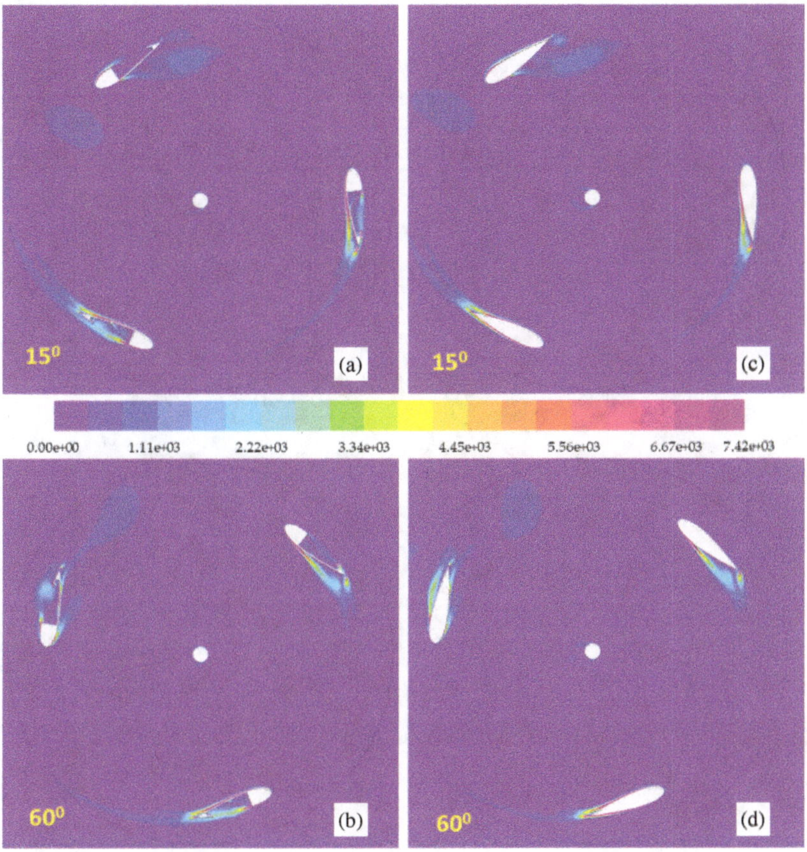

Figure 3.6 Pressure contours (in Pascal) of (a, b) KF-N-21 and (c, d) NACA 0021 airfoil.

attains an equilibrium point. By this vortex-trapping technique, high lift coefficients in the range of $C_l = 6$ can be achieved. With vortex strength and mass flow rate, the lift and drag due to the vortex trapping process can be obtained. The increase in the lift directly influences the tangential force component, which is the primary reason for the increase in Ct and thereby Cp. The mechanism of the vortex bubble formation has an impact on the Ct curves. Between the azimuthal angles 30° and 60°, there exists

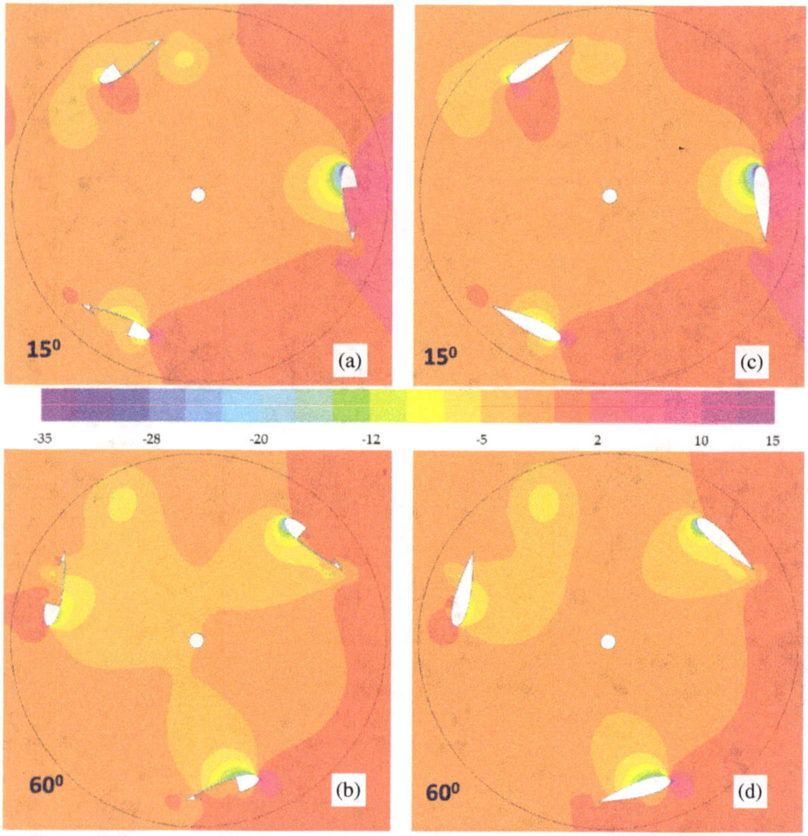

Figure 3.7 Vorticity (1/s) contours of (a, b) KF-N-21 and (c, d) NACA 0021.

a double peak formation. This is due to the generated vortices which cause the Ct values to be high, and it is maintained at higher rotation angles compared to NACA 0021 airfoil.

3.2.4 *Experimental Setup and Procedure for KF-N-21 Airfoil*

The airfoils are prepared as explained in the previous section. The mounted test rotors in the wind tunnel are shown in Figure 3.11 and the design parameters are shown in Table 3.1.

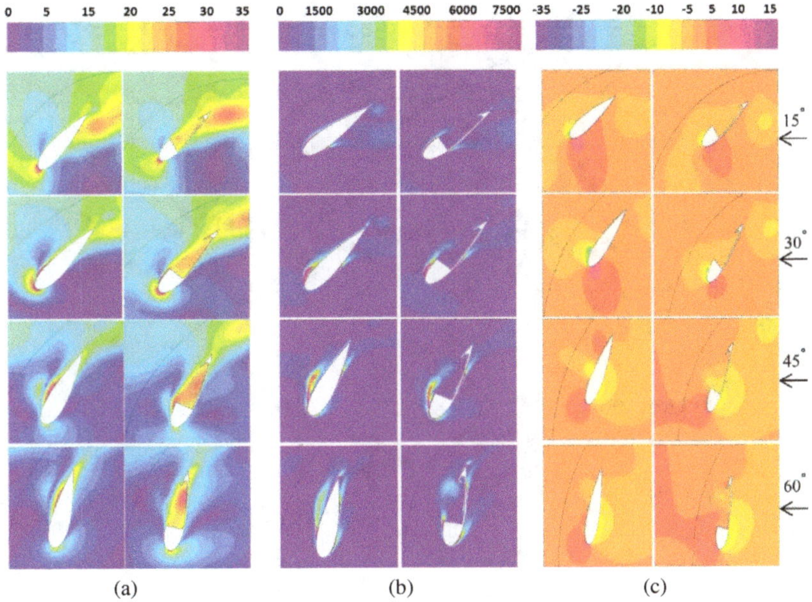

Figure 3.8 Contours of (a) velocity, (b) vorticity, and (c) pressure.

3.2.4.1 *Power Coefficient*

The Cp curves as a function of TSR are presented in Figure 3.13. At Re 1.5×10^5, the maximum Cp for the KF-N-21 rotor was achieved at TSR 1.1, which implies that the rotor torque is largely generated by the drag force rather than the lift force. NACA 0021 rotor was able to achieve a maximum Cp of 0.09 at TSR value of 1.3 as enough lift is not generated due to premature flow separation. As Re increases the Cp curves move closer and one possible reason might be the behavior of KF-N-21. Significant lift is generated compared to NACA 0021 due to trapped vortex as the flow has sufficient strength to recirculate inside the cavity before it sheds off. Same pattern was exhibited with increasing Re. The maximum Cp obtained for the tested wind speed of 8 m/s was 0.19 for TSR 1.3. The Cp curve of KF-N-21 rotor was expected to display a smooth stall rather than a sharp decline at maximum Cp. But the experimental results does not reflect a similar pattern for high Re. For the Re

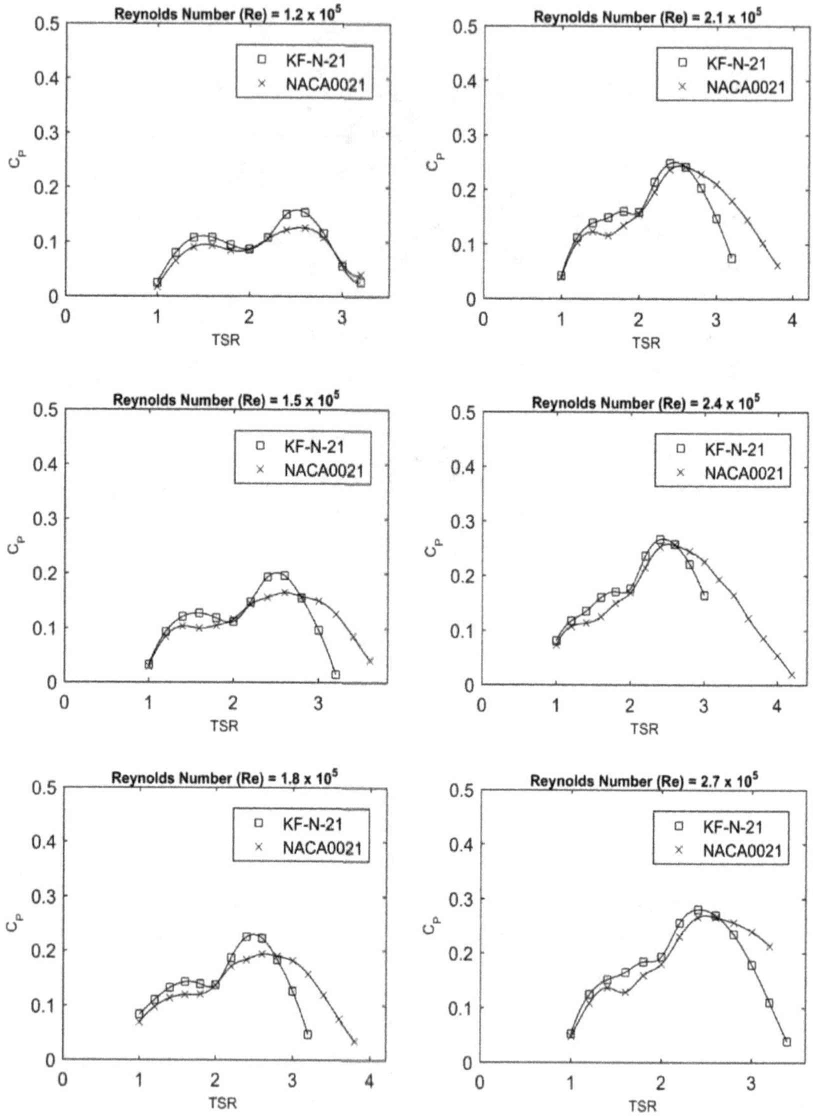

Figure 3.9 CFD prediction of Cp vs TSR for KF-N-21 airfoil.

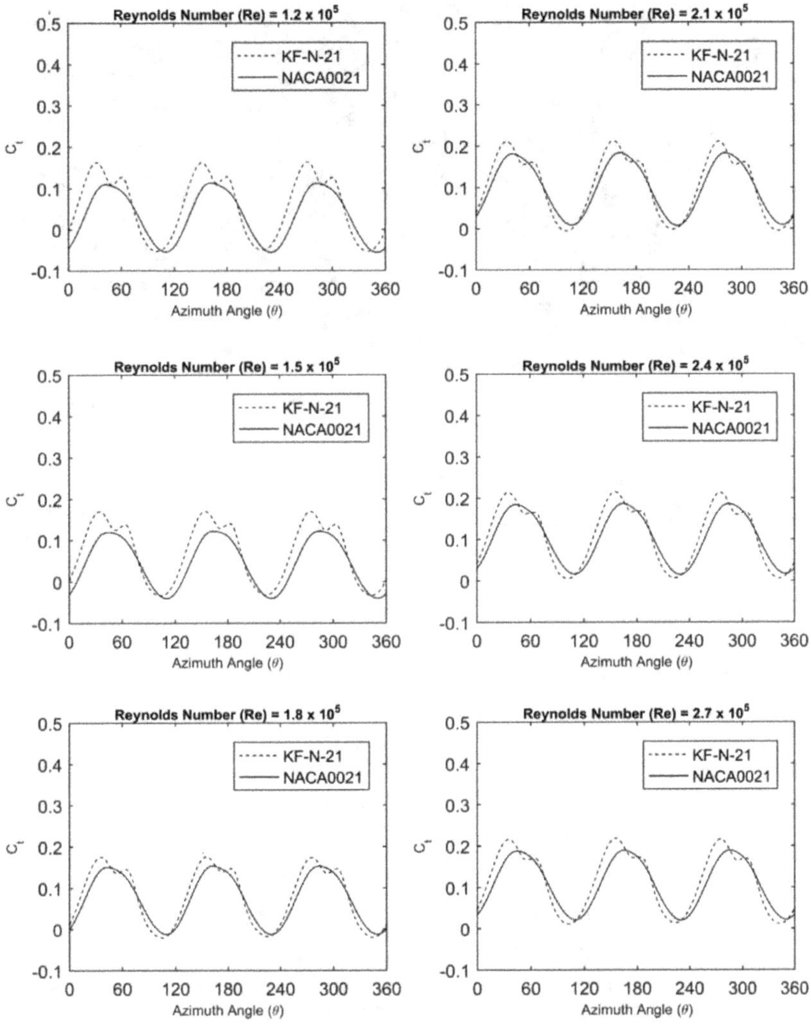

Figure 3.10 CFD prediction of C_{ts} vs TSR for KF-N-21 airfoil.

above 2.1×10^5 the Cp curves of both the rotors follow the same trajectory. Further investigations through computational analysis will reveal the flow behavior pertaining to the sharp drop in Cp. The overall Cp values for both the rotors are reduced due to significant frictional torque in the measurement train such as bearings and couplings.

Figure 3.11 KF-N-21 rotor (left) and NACA 0021 rotor (right) on wind tunnel setup.

Table 3.1 Design parameters of KF-N-21 and NACA 0021 rotors.

Design Parameters	Values
Rotor diameter	400 mm
Rotor height	300 mm
Solidity	0.43
Center shaft diameter	20 mm
Operating Re range	100,000–280,000
Airfoil chord length	100 mm
Airfoil designation	NACA 0021, KF-21
Number of blades	3
End plate diameter	420 mm
End plate thickness	8 mm
Support	Both end supported

3.2.4.2 *Static Torque Coefficient*

It can be noted from Figure 3.12 that for NACA 0021 there exists peaks of higher Ct. In contrast KF-N-21 rotor displays positive torque for most of the azimuthal angles but of lower magnitude. It is also interesting to note that KF-N-21 generates negative torque for the same azimuthal angle while the NACA 0021 rotor generates peak positive torque. For Re 2.1×10^5, the instantaneous peak Ct is at azimuthal angle of 245°

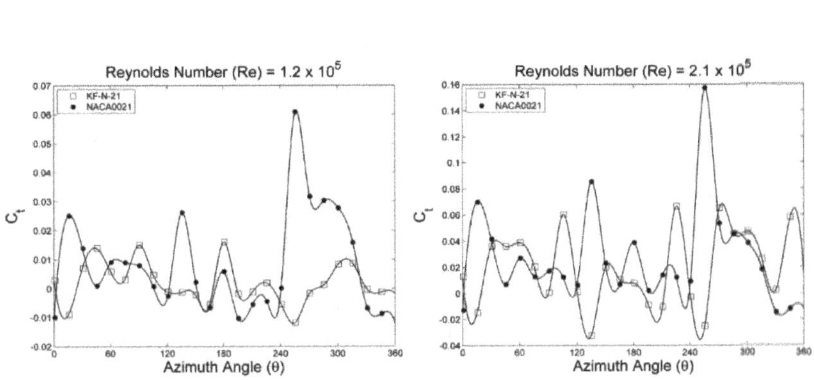

Figure 3.12 Static Ct variation with Re.

and sudden drop will push the rotor into oscillation mode, whereas with KF-N-21, the Ct drops smoothly, which accelerates the rotor to pass through the stall angle. In overall perspective, KF-N-21 rotor has better starting characteristics than NACA 0021 rotor.

3.3 Dual Darrieus Rotor*

3.3.1 *Description of Dual Darrieus Rotor*

The proposed Dual Darrieus (DD) rotor comprises two sets of blades sharing a common shaft. The outer rotor, hereafter referred to as the primary rotor, has three straight blades of NACA 0018, and inner rotor, hereafter referred to as the secondary rotor, has three blades of NACA 0018. NACA 0018 airfoil profile was chosen, as the performance and airfoil characteristics are well documented. Combination of airfoils, either symmetric or cambered, thick or thin, for the rotors greatly influences the characteristics of the DD turbine. Primary rotor is similar to a conventional Darrieus rotor, but the secondary rotor has significant modifications. The modifications incorporated in the secondary rotor are to minimize the downstream wake thereby maintaining the Cp of the

---

* Parts of this section are adapted with permission from Kumar, P. M.; Ajit, K. R.; Surya, M. R.; Srikanth, N.; Lim, T.-C. On the self-starting of Darrieus turbine: An experimental investigation with secondary rotor. *2017 Asian Conference on Energy, Power and Transportation Electrification (ACEPT)*. Singapore 2017.

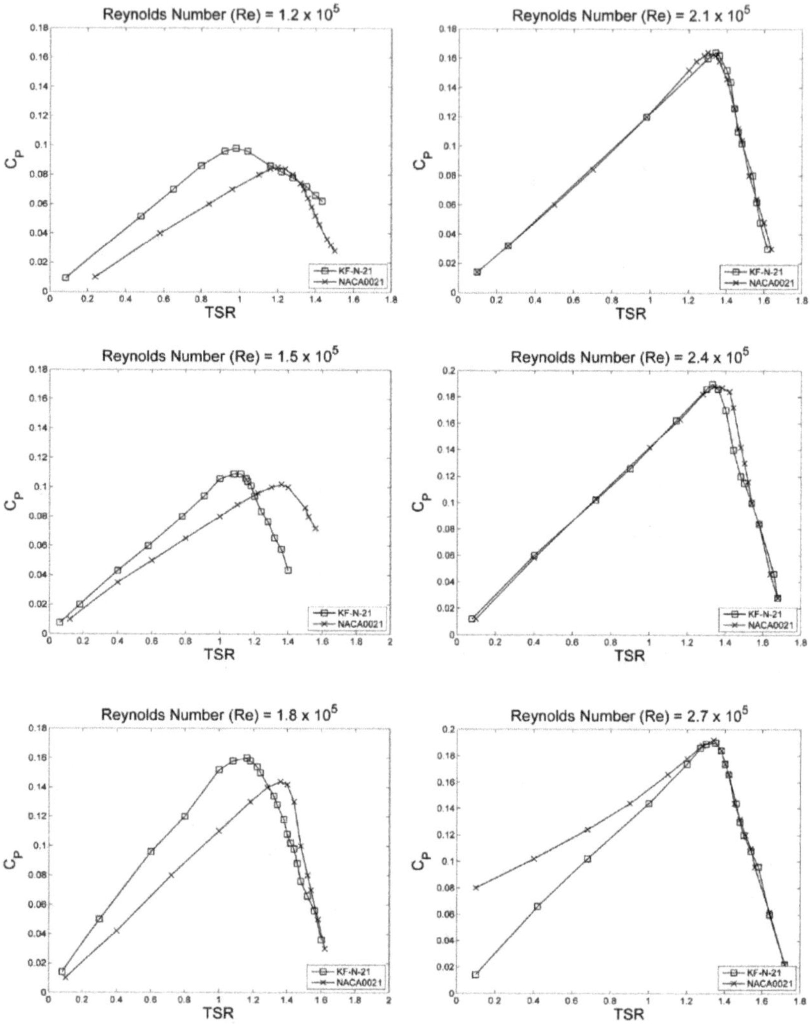

Figure 3.13 Experimental results on Cp vs TSR for various Re.

primary rotor at an acceptable level. The design parameters of the rotors are tabulated in Table 3.2. The configurations under investigation are shown in Figure 3.15. The primary rotor configuration is a conventional H-rotor. The parallel configuration is similar to a biplane wing, with their aerodynamic centers aligned. In the azimuthal offset configuration,

Table 3.2 Design parameters of Dual Darrieus rotor.

Design Parameters	Primary Rotor	Secondary Rotor
Rotor diameter (mm)	400	200
Rotor height (mm)	300	150
Solidity	1.5	1.5
Center shaft diameter (mm)	20	20
Operating Re range	100,000–280,000	100,000–240,000
Airfoil chord (mm)	100	50
Airfoil	NACA 0018	NACA 0018
Number of blades	3	3
End plate diameter (mm)	420	220
End plate thickness (mm)	8	5

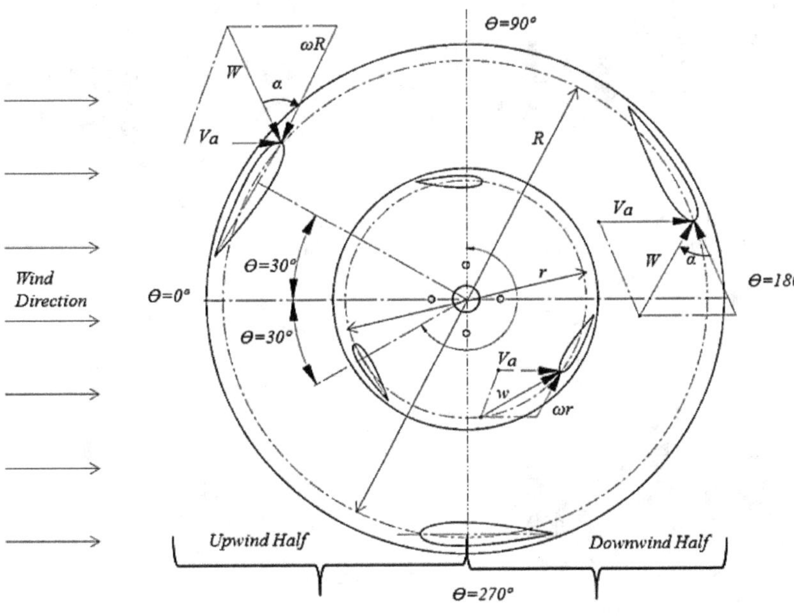

Figure 3.14 Velocity diagram of Dual Darrieus rotor.

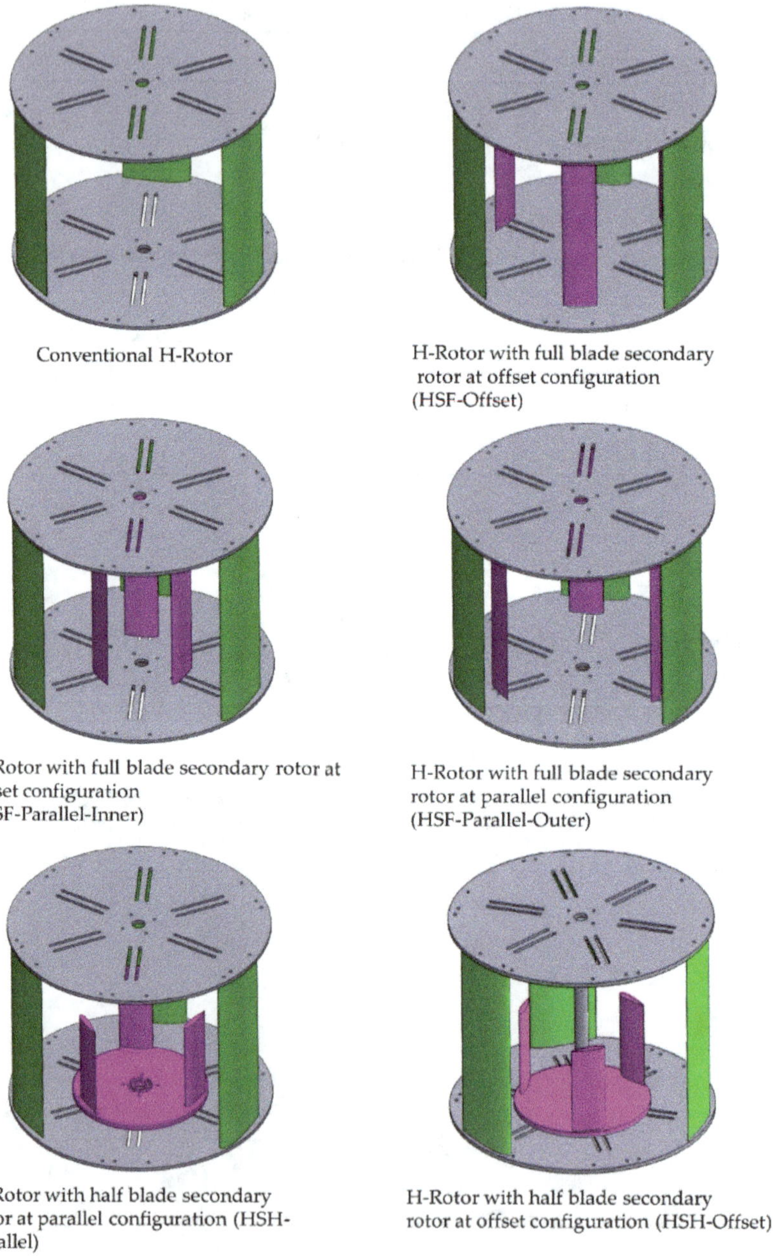

Conventional H-Rotor

H-Rotor with full blade secondary rotor at offset configuration (HSF-Offset)

H-Rotor with full blade secondary rotor at offset configuration (HSF-Parallel-Inner)

H-Rotor with full blade secondary rotor at parallel configuration (HSF-Parallel-Outer)

H-Rotor with half blade secondary rotor at parallel configuration (HSH-Parallel)

H-Rotor with half blade secondary rotor at offset configuration (HSH-Offset)

Figure 3.15 Dual Darrieus rotor configurations under investigation.

the secondary rotor is at an angular offset to the primary rotor and is anticipated to have significant influence on the starting capability and low wind speed performance. The ratio between primary and secondary rotor diameters is crucial in determining the operating performance of the turbine as a whole and has been arrived at by considering the solidity of the secondary rotor. As the current section aims to check the feasibility of the dual rotor configuration and its influence on the starting characteristics, it is out of scope to optimize the rotor radii, selection of airfoils or determining the optimum design parameters. The aerodynamics of conventional Darrieus turbines is complex by itself, due to inconsistent AoA, dynamic stall effects, adverse blade fluid interaction, and wind shear. The preliminary velocity diagram is shown in Figure 3.14. The complexity is compounded by the addition of the secondary rotor. The DMST model will be able to predict the DD rotor performance and has to be modified to include the secondary rotor. The DMST model for DD rotor will have four actuator disks in contrast to two for the conventional Darrieus rotor. Each actuator disk will have its own induction factors to compute the induced velocity.

3.3.2 *Experimental Setup and Procedure for Dual Darrieus Rotor*

The dimensions of the investigated rotors are given in Table 3.2. The mounted rotors in the wind tunnel are shown in Figure 3.16.

Figure 3.16 HSF-Offset and HSH-Offset configuration of DD on wind tunnel setup.

3.3.3 *Coefficient of Power*

The Cp curves as a function of TSR are presented in Figures 3.17 and 3.18. For Re = 193,827, the H-rotor in parallel configuration displays the highest Cp of 0.15, whereas the H-rotor is able to achieve a Cp of 0.11. Wider operating TSR is a more favorable property than the peak Cp, as the turbine will be able to generate power for a wider wind speed range. An interesting finding is that the peak Cp of H-rotor with secondary rotor is achieved at TSR 1, irrespective of the Re, whereas the H-rotor alone reaches the peak Cp at 1.2 TSR. As the Re increases to 268,376, the H-rotor displays highest Cp of 0.17 and with secondary rotor can reach Cp of 0.15. By poring over the experimental data, the power loss can be attributed to the aerodynamics losses associated with two sets of blades. A considerable amount of energy in the freestream wind is not converted to useful torque due to the formation of tip vortices, as the length of the secondary rotor is half that of the primary rotor and does not have any wing tip to prevent such losses. The second notable reason could be that the primary and secondary rotors will not reach the peak Cp for the same rotor rpm, i.e., when primary rotor is at design TSR, the secondary rotor will not be at optimum and blades tend to generate more drag due to large variation in AoA and vice versa. The primary and secondary rotor radii optimization is vital for the blades to operate in the range of AoA, without significant loss in lift. A maximum Cp of 0.09 is obtained for the H-rotor and 0.13 for the H-rotor in parallel configuration corresponding to Re = 193,827 leading to 22% increase in efficiency. As the Re increases to 268,376 the Cp value of the conventional H-rotor increases to 0.17, whereas the H-rotor in parallel configuration is 0.14. A possible reason for the reduction in power is interaction between the wake of the primary blade leaving the trailing edge and immediately encountering the secondary blade generating significant drag and reducing the overall torque generation. At specific azimuthal angles combined with AoA, the secondary blade operates only with apparent wind rather than freestream wind due to the blockage of primary rotor blades. Of all the

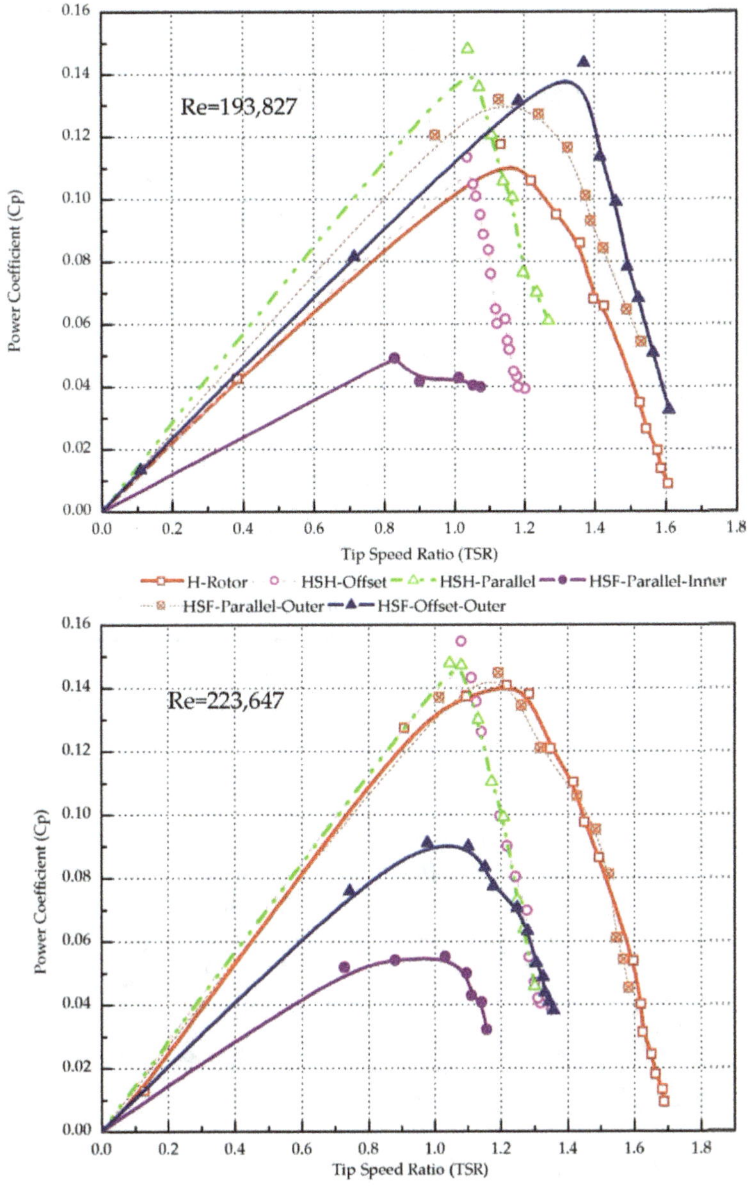

Figure 3.17 Cp of Dual Darrieus configurations for Re 1.9×10^5 (top) and 2.2×10^5 (bottom).

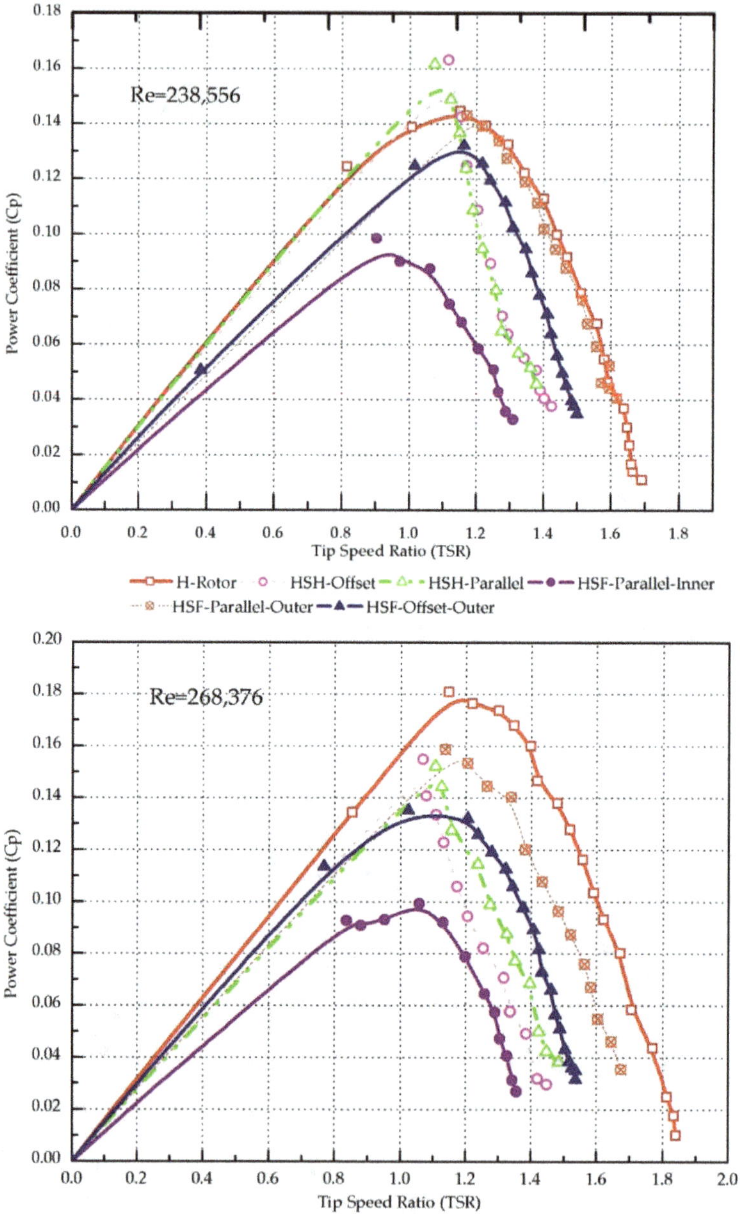

Figure 3.18 Cp of Dual Darrieus configurations for Re 2.3×10^5 (top) and 2.6×10^5 (bottom).

configurations, the H-rotor in offset configuration is the poorest performer in terms of Cp for the investigated Re. The performance of H-rotor in parallel configuration degrades with increase in the Re. Except for the conventional H-rotor, the other configurations operate in a smaller TSR range, similar to the rotor with higher solidity.

3.4 Modified Trailing Edge Airfoil*

3.4.1 *Design of* NACA 0018TC-39 *Airfoil*

The truncated airfoil finds its application in wind turbines, especially in the large HAWTs at the blade root region. The transition region from airfoil to cylindrical root requires thicker airfoil up to 40% to withstand the high bending moment as the structural properties are more crucial than the aerodynamic properties, where the truncated trailing edge (TTE) airfoils are indispensable. The structurally superior blunt airfoil has some excellent aerodynamic features which make it superior to conventional airfoils. Research from NASA along with Ohio State University suggested that blunting the 40% thick airfoil increases the maximum lift coefficient by almost 90%. Purpose of the truncation process is to eliminate the sharp and steep curvature at the trailing edge portion which will in turn reduce the adverse pressure gradient caused in that area due to full pressure recovery. The reduced adverse pressure gradient helps in delaying the flow separation, which is a prime concern in low speed flows.[7]

Typically, blunt trailing edge airfoils are created either by truncation or flatback. Truncating the airfoil involves cutting off some portion of the airfoil from the trailing edge and making it blunt, which may modify the camber. An alternate solution is flatback airfoils, which essentially involves adding sufficient thickness on both pressure and suction sides

---

* Parts of this section are adapted with permission from Kumar, P. M.; Kulkarni, R.; Srikanth, N.; Lim, T.-C. Performance assessment of Darrieus turbine with modified trailing edge airfoil for low wind speeds. *Smart Grid and Renewable Energy* **2017**, *8*, 425–439.

symmetrically, preserving its camber. Since this study involves the symmetrical airfoil, the methodology used is simple airfoil truncation. Standish and Van Dam[8] showed that for symmetric Go¨-490 airfoil at Re 5×10^5 the lift coefficient increases with the truncation percentage up to a certain limit. The maximum lift coefficient increased by almost 45% when the airfoil was truncated at 35%. According to Tangier and Somers,[9] sharp trailing edge airfoils with thickness/chord ratio (t/c) < 0.25 can be modified to have lifting properties that are vastly susceptible to surface disturbances.

Hence a simple yet effective guideline for blunt trailing edge airfoils is to restrict the difference between the maximum thickness and trailing edge thickness to ~20% of chord length. In doing so, the adverse pressure gradient on later portion of the suction surface is controlled, which in turn recovers the expected lifting properties under disturbed conditions. The airfoil has been designed in this research based on this rule of thumb. As Baker *et al.*[10] described, the flatback[11] series of airfoils was discovered by mixing the suction side shape drawn from the thick, high lift inboard NREL airfoils, and a structurally superior pressure side derived from the LS-1 series airfoils. The family of FB airfoils was developed by Sandia National Laboratories in collaboration with University of California–Davis as part of the Blade System Design Study.

The strategy behind the new airfoil design is to dramatically increase the output torque at low wind speed and low TSR. A non-cumbersome way is to intelligently use the airfoil drag if generated in the direction of rotor rotation. To make it possible, a drag-creating feature should be introduced at the trailing edge of the airfoil, as the flow is towards the leading edge from trailing edge when the blade retreats from the incoming wind. A thick trailing edge will create more base drag compared to sharp trailing edge due to vortex shedding. Gangadharacharya[12] suggests that the TTE airfoil generates higher lift at high AoA, where the conventional airfoil could have stalled. A major drawback of the TTE airfoil is the abrupt stall

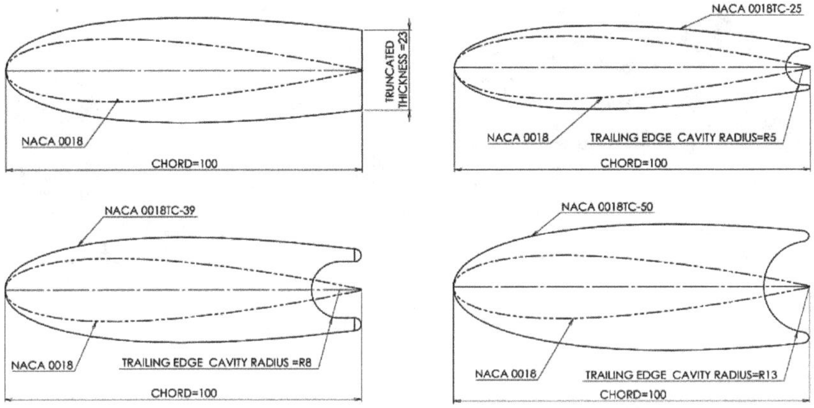

Figure 3.19 Truncated airfoils with the conventional NACA 0018 airfoil.

which may induce instability and vibrations if incorporated in Darrieus turbines. For smooth stall and to reduce the base drag, Baker[13] introduced splitter plates mounted perpendicular to the trailing edge. 25% reduction in drag was reported for single splitter plate. The study extends with double splitter plate similar to the cavity and the results indicated a smooth stall. For the current study, the blunt trailing edge is created from conventional NACA 0018 airfoil and a cavity is excavated inside the blunt trailing edge as shown in Figure 3.19. To incorporate a cavity with sufficiently large radius to maximize the Darrieus turbine performance, the trailing edge should be relatively thick and this necessitates the airfoil to be truncated. The truncation % from the trailing edge is a question for optimization that has to be answered computationally.

3.4.2 *Computational Optimization of Truncation Parameters*

The study is performed in a two-dimensional domain. The computational domain consists of a C-type grid. The distances from the airfoil leading edge to the domain inlet and outlet are 10c and 25c, with a domain width of 20c, whereas c is the airfoil (blockage ratio 5%). The computational

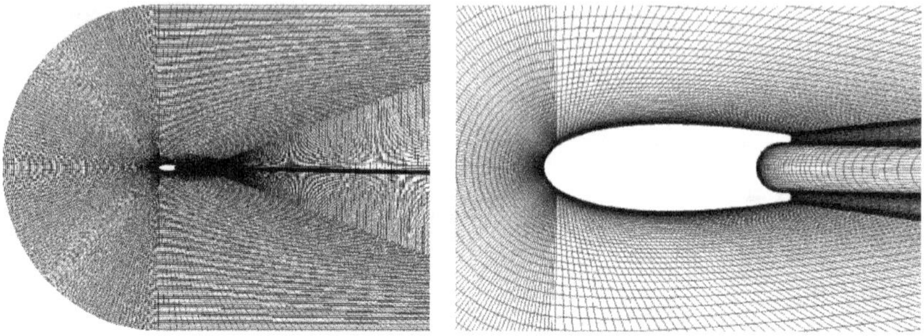

Figure 3.20 Mesh details of truncated airfoils.

grid consists of quadrilateral cells as illustrated in Figure 3.20, where a boundary layer grid restrains the y^+ value to below 1 on the walls. The two-dimensional grid has approximately 80,000 cells. The grid resolution resulted from a grid convergence study using three grids generated based on an overall linear factor $\sqrt{2}$ for coarsening and refining.

The boundary conditions are set as constant inlet velocity of 44 m/s in order to validate with results of NACA 0018 by matching the Re, zero static gauge pressure at the outlet, symmetry on the sides and a no-slip condition on the airfoil. The approach-flow uniform total turbulence intensity at the domain inlet is 5%. RANS calculations are performed using the commercial computational software package ANSYS/Fluent 16. Both spatial and temporal discretization are second order and pressure-velocity coupling is done with coupled scheme. Turbulence is modeled using the two-equation k-ω SST model. As we can observe from Figure 3.22(a–d), at lower AoA, the flow is getting curled inside the baseline cavity essentially creating a pair of counter-rotating vortices trapped inside the cavity. Due to trapping of these shaded vortices the vortex street formed behind the thick trailing edge is restricted in one manner. This phenomenon assists in reducing the wake region width behind the airfoil up to a certain extent. As the AoA increases, the rotating vortex gradually disappears and the flow separation moves towards the leading edge resulting in higher drag.

3.4.2.1 *Lift and Drag Characteristics*

From the obtained curves, it is evident that the lifting performance of 25% truncated airfoil is superior to the rest but will result in smaller baseline cavity. The 50% truncated airfoil should produce maximum lift according to truncated airfoil theory but may lead to early separation at high AoA. The separation can be attributed to high leading-edge thickness which presents itself as bluff body. All these factors considered, the overall aerodynamic performance of the 39% truncated airfoil is much more effective than the rest of the configurations for most of the AoA range. Also, larger baseline cavity will allow the airfoil to produce larger reverse drag in low TSR.

To quantify the reverse drag on the airfoil, a computational study has been done on the optimized airfoil. Similar grid size has been adapted for this study with y^+ value less than 1. The results are shown in Figure 3.22(e–h). For Re 300,000 the drag coefficient for the airfoil in reverse manner is found to be 0.08, which is exceptionally high. As the AoA increases to 10, 20 and 30, the drag coefficient continues to increase to 0.17, 0.22, and 0.26, respectively, as shown in Figure 3.21. These high drag coefficients in backward direction provide high moments which add up to the torque produced by lifting blades together, yielding better starting properties.

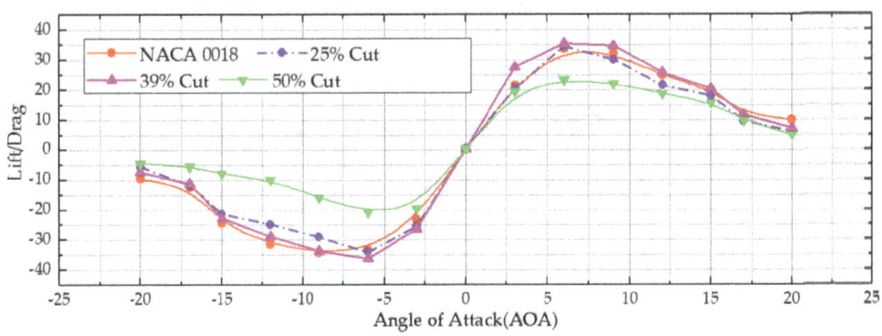

Figure 3.21 Cl/Cd vs AoA for different truncated airfoils.

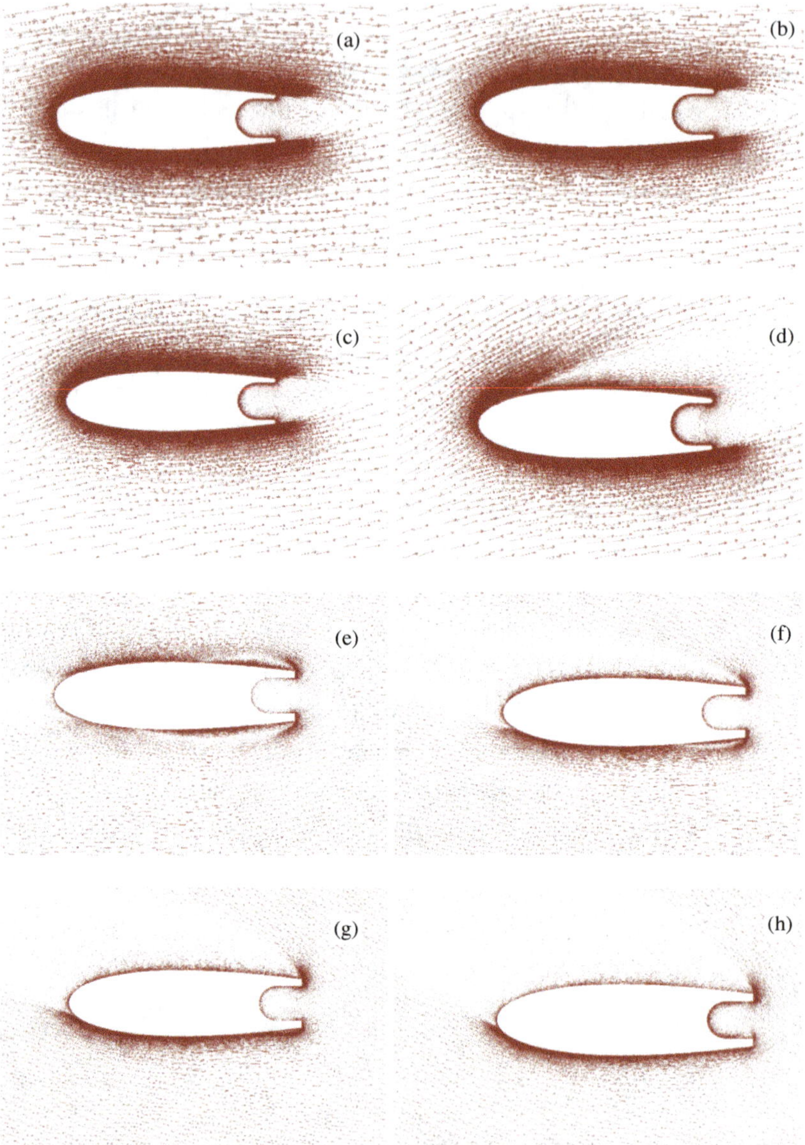

Figure 3.22 Vector plots of NACA 0018TC-39 airfoil in forward and reverse flows.

3.4.3 *Experimental Setup and Procedure for* NACA 0018TC-39 *Airfoil*

The blades and the experimental setup are similar to the previously carried out wind tunnel tests. The dimensions of the rotors under investigation are presented in Table 3.3. The aerofoils are interchanged on the same end plates to minimize any errors due to shaft misalignment. Figure 3.23 shows the mounted rotors in the wind tunnel.

Table 3.3 Design parameters of NACA 0018 and NACA 0018TC-39 rotors.

Design Parameters	NACA 0018	NACA 0018TC-39	Units
Rotor diameter	400	400	mm
Rotor height	300	300	mm
Solidity	1.5	1.5	—
Center shaft diameter	20	20	mm
Operating Re range	100,000–280,000	100,000–280,000	—
Airfoil chord	100	100	mm
Number of blades	NACA 0018	NACA 0018TC-39	—
End plate diameter	3	3	mm
End plate thickness	420	420	mm

Figure 3.23 NACA 0018 and NACA 0018TC-39 rotors on wind tunnel setup.

3.4.3.1 *Power Coefficient*

Dynamical wind turbine performances are obtained by computing the instantaneous Ct. Figure 3.24 shows the dynamic Cp for both NACA

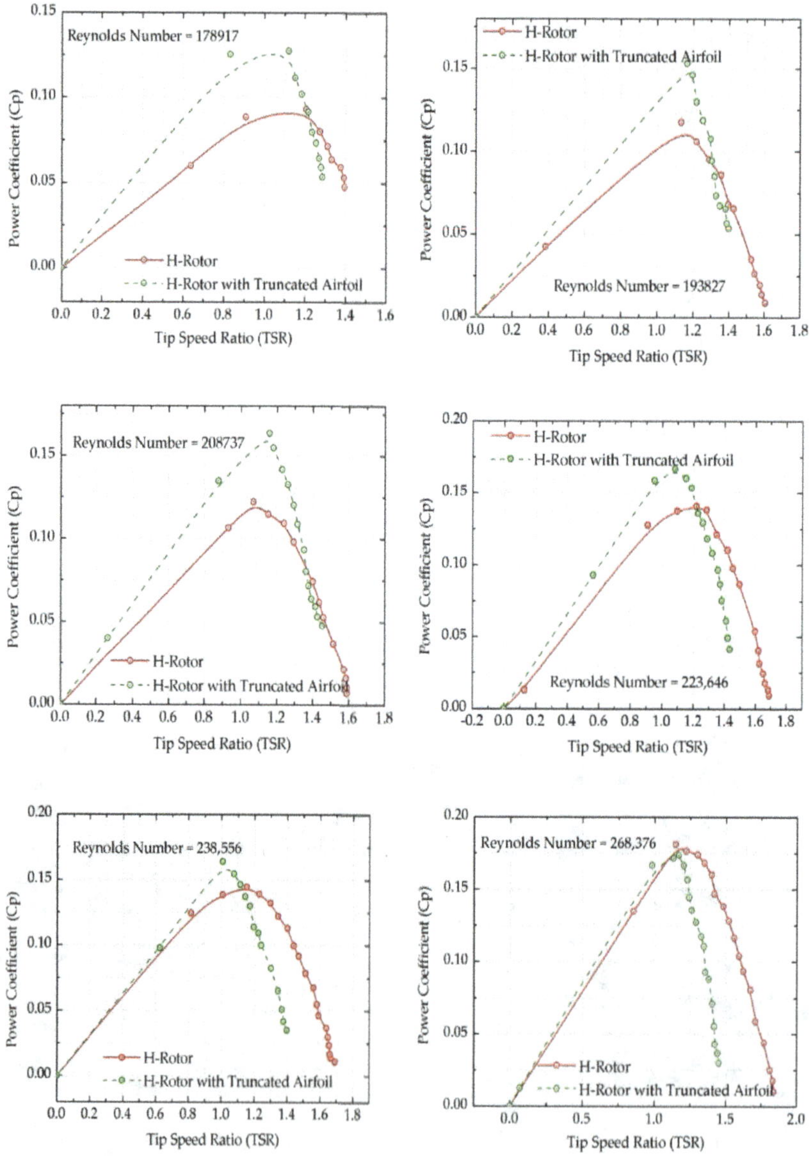

Figure 3.24 Cp of NACA 0018 and NACA 0018TC-39 rotors at different Re.

0018TC-39 and NACA 0018 airfoils in comparison. For a wide range of Re, the high Cp is achieved by NACA 0018TC-39 at lower TSR values signifying prominent increment in starting torque. For Re 178,917, for conventional turbine, highest Cp of 0.09 is achieved at TSR of 1.1 while for the same Re, for the NACA 0018TC-39 bladed configuration, the highest Cp of 0.13 was achieved at TSR of 1.07. For the lower values of TSR, NACA 0018TC-39 bladed rotor continues to perform significantly better than the conventional configuration in terms of Cp. For Re 268,376 the performance of NACA 0018TC-39 bladed turbine is still superior but the increment is not very high. A possible reason could be that the wake produced by the oscillating vortex street behind blunt trailing edge yields fluctuating lifting characteristics, causing the turbine to deviate from the design point. However, the thicker airfoil continues to provide the high lifting coefficients, making up for loss created due to vortex shedding. At lower TSR values the NACA 0018TC-39 bladed rotor provides exceptionally good characteristics due to additional reverse drag component obtained due to the baseline cavity, which adds up to the lift force eventually providing additional useful starting torque. The overall Cp value for both the configurations is reduced to some extent due to significant frictional torque in the measurement trains such as bearing and couplings. The Cp of NACA 0018TC-39 is 12.6% at 178,917, and at 238,566 is 1.6% higher than the conventional NACA 0018 airfoil. The difference in Cp between the investigated airfoils progressively diminishes as the Re increases. This is evidence that the truncated airfoil manifests itself as a conventional airfoil at higher Re with only a mild influence on lift.

3.4.3.2 *Static Torque Coefficient*

Static Ct is a vital parameter to judge the starting characteristics of the Darrieus wind turbine, which is characterized by the net torque generated by the blades at rest. At very low TSR the airfoils output a negligible torque and the cavity introduction is anticipated to boost the torque to significant values. Both the configurations are subjected to static Ct measurement for 360° with 15° increments computed as per

EOLO-3000.[14] The results reinforce the hypothesis that the trailing edge cavity will promote self-starting and increase the low rpm at low wind speed. For the tested Re, the NACA 0018TC-39 rotor demonstrates a higher torque for most of the azimuthal positions than the NACA 0018 rotor. Another notable behavior exhibited by NACA 0018TC-39 rotor is that static torque is distributed from 0° to 360°, rather than three peaks at an interval of 120° by conventional Darrieus which may lead to static stall if AoA is large at the renewed position of blades. A smooth positive torque distribution aids the rotor to accelerate to the rated rpm as illustrated in Figure 3.25. Comparing the Ct values, the NACA 0018TC-39 bladed turbine accomplishes 21% at 178,917, 23% at 193,827 and 5% at 266,896 higher static torque than the NACA 0018 rotor. In the case of NACA 0018TC-39, for Re 208,737, the minimum static Ct of 0.02 accelerates the rotor beyond stall. For the investigated Re, the NACA 0018TC-39 has better starting and low wind speed performance than the conventional configuration.

3.5 Hybrid Darrieus Telescopic Savonius Turbine*

3.5.1 *Description of Hybrid Darrieus Telescopic Savonius Turbine*

As pointed out earlier, the Hybrid Darrieus-Savonius rotor displays excellent performance at low wind speed, yet the resistive drag above TSR 1 is the limiting factor hindering the rotor performance as a whole. Although solutions such as decoupling the Savonius rotor from the main shaft or the use of relatively smaller Savonius rotors have been proposed, the turbulent wake severely degrades the performance of the Darrieus

* Parts of this section are adapted with permission from the following references: Mohan Kumar, P.; Surya, M. M. R.; Narasimalu, S.; Lim, T.-C. Experimental and numerical investigation of novel Savonius wind turbine. *Wind Engineering* **2019**, *43*, 247–262; Kumar, P. M.; Purimitla, S. R.; Shubhra, S.; Srikanth, N. Numerical and analytical study on telescopic Savonius turbine blade. *2017 3rd International Conference on Power Generation Systems and Renewable Energy Technologies (PGSRET)*. Johor Bahru **2017**.

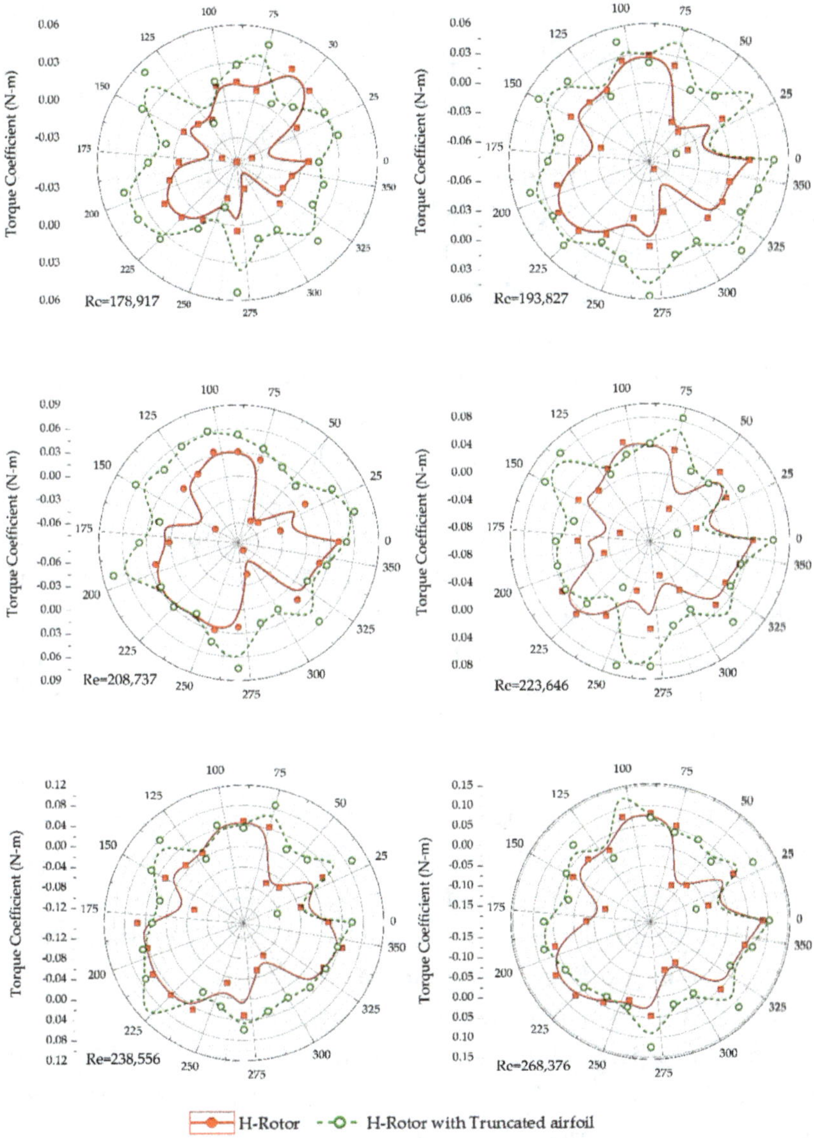

Figure 3.25 Static Ct vs azimuthal angle for different Re.

rotor. Employing a smaller-sized rotor will curtail the advantage as a low wind speed device and hence considerable size of the Savonius rotor is indispensable. A novel solution of Telescopic Savonius Turbine (TST) is proposed in which the Savonius buckets can slide within each other to reduce the frontal area and hence the wake. The bottom stage that houses the rest is enclosed in a cylinder for minimum turbulent wake when integrated into Darrieus rotors. The proposed concept will not limit the Darrieus rotor to accelerate beyond TSR 1 in contrast to the conventional Hybrid Darrieus-Savonius rotor. Added advantage is the over speed regulation of the Darrieus rotor and power regulation by dynamically varying the height of the Savonius buckets via retracting and extending. The operation sequence is, during standstill or in low wind speed the buckets are extended to operate as a Conventional Savonius Turbine (CST) in drag mode with negligible contribution from the Darrieus rotor. The buckets will continue in the extended configuration until the TSR reaches 1~1.2, after that the buckets are retracted into a cylinder and TST will not contribute to the power generation thereafter. During high winds the TST is extended to generate resistive torque for the required rate of deceleration.

The TST rotor under investigation consists of three blades, equally spaced at 120° apart. Each blade has three buckets (semicircular segment) of equal height and varying thickness as shown in Figure 3.26. The three stages are arranged concentrically and are able to slide within each other. The number of stages for the current study is limited to three based on the wind tunnel size and the blockage effect. The number of stages will be dictated by power requirement, expected thrust load reduction at high wind, available height for extension and retraction, rotational dynamics and structural aspects. Nevertheless, a compact turbine will be beneficial in transportation and for minimal thrust load, while increase in number of stages will impact the aerodynamic properties inversely due to increase in bucket thickness. The torque on the rotor shaft is generated by the difference in drag between concave and convex sides of the buckets. A bucket with higher thickness that has smaller concave radius and larger convex radius will have reduced

Figure 3.26 Telescopic Savonius at one, two and three stages.

drag coefficient difference resulting in lower torque generation. TSTs are meant to be high aspect ratio turbines reaching 6–10 compared to CST of 1–5. Hence, it occupies minimum footprint, a desirable property for Darrieus integration. The TST has to be optimized and experimentally compared with CST for their performance before integrating into Darrieus rotors. Hence a simple analytical model is developed followed by the computational optimization of TST buckets. A scaled TST is tested in the wind tunnel and performance is recorded along with thrust load. A Hybrid Darrieus Telescopic Savonius (HDTS) is experimented and compared with the conventional Darrieus rotor.

3.5.2 *Analytical Modeling of Telescopic Savonius Turbine*

To mathematically formulate the power generated by the TST, a simple configuration of two blades with three stages of bucket is considered. Figure 3.27 shows the critical notations pertaining to this mathematical formulation. The three stages are of equal height h, but radially offset at a distance f on each side with respect to the first stage. The first bucket is located at a distance from the shaft center.

With the aforementioned notations, the swept area A can be written as

$$A = 2h\{(d_1 + e) + (d_1 + e - 2f) + (d_1 + e - 4f)\} \tag{3.12}$$

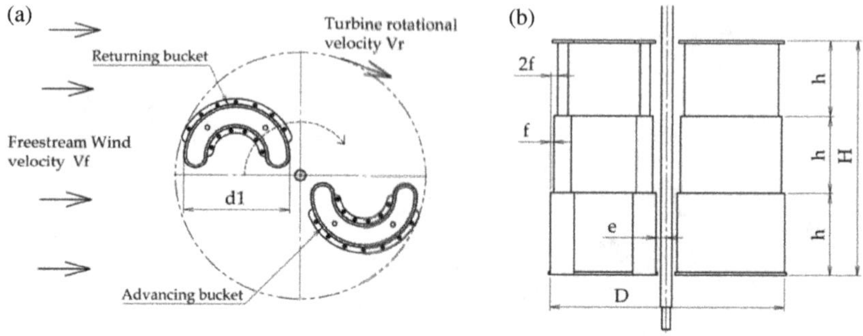

Figure 3.27 Analytical model parameters of TST.

Since each bucket has different thickness, the total drag coefficient C_c on the concave side is the sum of individual drag coefficients of first, second and third stages written as

$$C_c = C_{c1} + C_{c2} + C_{c3} \tag{3.13}$$

where C_{c1}, C_{c2}, C_{c3} represent the concave side drag coefficients of each bucket and C_{v1}, C_{v2}, C_{v3} represent convex side drag coefficients. Similarly for the convex side, the total drag coefficient C_v can be written as

$$C_v = C_{v1} + C_{v2} + C_{v3} \tag{3.14}$$

The force developed on the concave side F_c and convex side F_v is given by

$$F_c = \frac{\rho A V_f^2 C_c}{2} \tag{3.15}$$

$$F_v = \frac{\rho A V_f^2 C_v}{2} \tag{3.16}$$

in which ρ stands for density of air and V_f is the freestream wind velocity. The power generated on the turbine shaft is

$$P = F \times V_r \tag{3.17}$$

Here V_r stands for the rotational velocity, hence the power P_c generated only by the concave side of the bucket when it moves along with the wind is

$$P_c = \frac{1}{2}\rho A\left(V_f - V_r\right)^2 \times V_r \tag{3.18}$$

and the power P_v generated by the convex side is

$$P_v = \frac{1}{2}\rho A\left(V_f + V_r\right)^2 \times V_r \tag{3.19}$$

Net power available on the shaft P_s is given by

$$P_s = P_c - P_v \tag{3.20}$$

In order to condense the first derivative, it is assumed that

$$m = C_c - C_v \tag{3.21}$$

$$n = C_c + C_v \tag{3.22}$$

After substitution and simplification

$$P_s = \frac{1}{2}\rho\ A\ V_f\ \left[m\ V_f^2 - n_2 V_f V_r + m\ V_r^2\right] \tag{3.23}$$

For maximum power generation, the first derivative should be equal to zero, yielding the optimum velocity as

$$\frac{dP_s}{dV_1} = 0 \tag{3.24}$$

$$V_{opt1} = \frac{2mV_f + V_f\sqrt{4m^2 - 3n^2}}{3m} \tag{3.25}$$

$$V_{opt2} = \frac{\left(2mV_f - V_f\sqrt{4m^2 - 3n^2}\right)}{3m} \tag{3.26}$$

The power on the shaft will be maximized if the drag on the convex side is zero, that is, the drag coefficient by the returning blade is nil, so in that case,

$$C_{v1} + C_{v2} + C_{v3} = C_v = 0 \tag{3.27}$$

$$m = n = c_1 \tag{3.28}$$

Hence the optimum rotational speed is

$$V_{r\,opt} = \frac{1}{3} V_f \tag{3.29}$$

Accordingly, optimum power generated is

$$P_{s\,opt} = \frac{2}{27} \rho \; C_c \times A \times V_f^3 \tag{3.30}$$

The power $P_{s\theta}$ at specific azimuthal position can be found by

$$P_{s\theta} = \frac{2}{27} \rho((C_c)_\theta - (C_v)_\theta) \times A \times V_f^3 \tag{3.31}$$

From the above equation, the power generated by the TST at any instant of azimuthal angle can be calculated by computing the drag coefficient of buckets at that position.

3.5.3 *Computational Optimization of Telescopic Savonius Turbine Buckets*

The numerical study aims to explore the three-dimensional steady flow over buckets of different thickness as understanding the flow pattern is crucial for optimizing the bucket thickness. Since the incidence angle varies from 0°–360° during each rotation, the drag coefficient has to be predicted for the same to emulate a rotating turbine. The computational domain is shown in Figure 3.28. The far field domain is extended by $4d_1$ on both upstream and downstream, where d_1 is the maximum bucket diameter, so that the pressure outlet approaches the atmospheric pressure.

$$\rho\bar{u}_i\frac{\partial\bar{u}_i}{\partial x_j}=\rho\bar{f}_i+\frac{\partial}{\partial x_i}\left[-\bar{p}\delta_{ij}+\mu\left(\frac{\partial\bar{u}_i}{\partial x_j}+\frac{\partial\bar{u}_j}{\partial x_i}\right)-\rho\overline{u'_i u'_j}\right] \tag{3.32}$$

RANS model was used to solve the model:

$$\frac{\delta_y}{y^+}=8.6LRe_L^{\frac{-13}{14}} \tag{3.33}$$

A hybrid mesh is adopted for the current domain with structured quadrilateral elements in the vicinity of the bucket surface to capture the vortex and unstructured triangular elements in the far field. The computational domain is discretized with 418,150 elements and 226 nodes. The bucket surface is clustered with 20 inflation layers with a growth rate of 1.1. Utmost effort has been dedicated to refine the mesh to capture the flow separation at the bucket surface as shown in Figure 3.28.

The cell spacing is obtained through the y^+ value from Eq. (3.33). δ_y is the distance from the wall to the center of the first cell. The value of y^+ obtained is well within the allowable maximum value of 1.0. The boundary conditions are imposed with constant inlet velocity of 6 m/s and outlet to the atmospheric pressure. The ambient air is inherently turbulent and the magnitude of turbulence is exacerbated by the buildings when

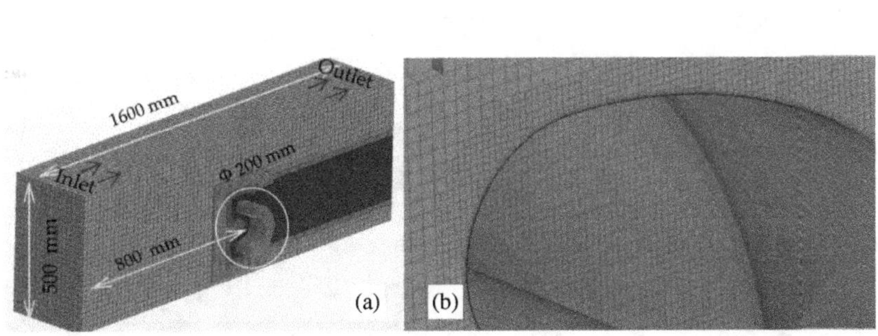

Figure 3.28 CFD domain characteristics and mesh details on Savonius bucket.

installed on rooftops. Hence a suitable turbulence model selection is essential to simulate actual conditions. For the current study, realizable $k-\varepsilon$ model was adopted and represented in Eq. (3.34).

$$\frac{\partial(\rho\epsilon)}{\partial t}+\frac{\partial(\rho\epsilon u_i)}{\partial x_i}=\frac{\partial}{\partial x_j}\left[\frac{\mu_t}{\sigma_k}\frac{\partial k}{\partial x_j}\right]+2\mu_t E_{ij}E_{ij}-\rho\epsilon \tag{3.34}$$

3.5.4 *Discussion on Computational Results of Telescopic Savonius Turbine Buckets*

The simulations are performed through the commercial software code, Star CCM+. The computational studies so far carried out on the CST Savonius rotor rely on three vital assumptions. First is that the cross-section is constant for the entire span of curved or straight blade. The second assumption is the derivative of the first assumption in that the drag coefficient is unaltered for the entire blade span under investigation. The third assumption considers that the blade as a surface of negligible thickness in relation to the diameter of the blade. The current study stands out from the rest by deviating from all the above assumptions and in fact the study is to establish the relation between these assumptions through computational results. Though the simulations compute the drag coefficient for 0°–360° at an increment of 5°, the flow patterns at

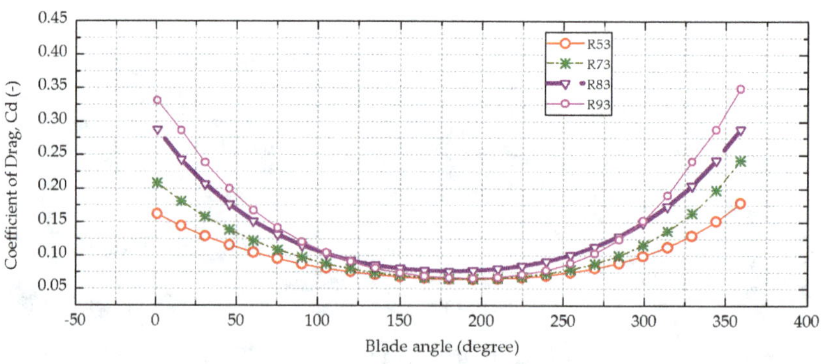

Figure 3.29 Drag coefficient variation with AoA of Telescopic buckets.

critical angles of 0° and 90° are presented in Figure 3.30. The computed values of drag coefficient for the buckets may deviate when simulated as a rotor, due to the interaction of flow from the other buckets. Nevertheless, the current study is appropriate in an effort to determine the drag

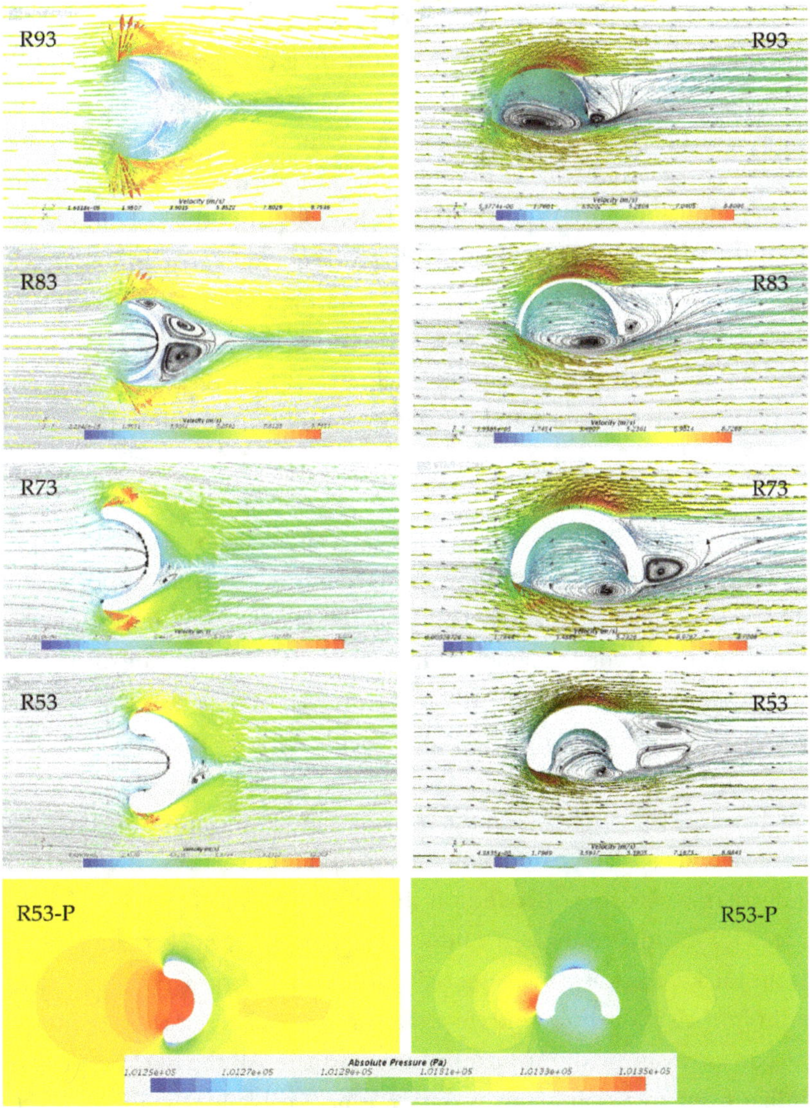

Figure 3.30 Flow pattern on the bucket at 0° (left) and 90° incidence angles (right).

variation in relation to the thickness of the buckets. In order to validate the numerical study, the results are compared against past experimental studies of similar Re range.

The study on hollow hemispherical body by Heinrich and Haak[15] attains a drag coefficient of 0.3 for the matched Re of R93 mm radius bucket. The computational results show a higher drag coefficient of 0.33. The discrepancies may be due to the fact that the flow in wind tunnel is more turbulent than the assumed turbulence value and the test models are influenced by wind tunnel boundary layer growth. Since the current study is to get an insight of the flow behavior rather than obtaining an absolute value of drag coefficient, the current mesh structure and cell quantity are retained. For the 0° incidence angle there is a marked difference in the flow pattern. The formation of large-scale vortices is prominent with the lower bucket thickness (R93, R83). These vortices are trapped and recirculate to form a low-pressure zone behind the convex surface.

The drag difference between the concave and convex surfaces is thus maximized due to the formation of this low-pressure zone and is attributed to the higher drag coefficient of 0.26. The flow will undergo considerable change when the buckets are subjected to rotation due to varying AoA and unsteady velocity. Sufficient time may not be available for the vortex to develop and hence the drag coefficient may differ.

Furthermore the gap between the buckets will significantly alter the flow on the bucket surface, disturbing the formation of vortices. The position and size of the vortices depend on the impingement of gap flow. In the case of 90° incidence angle, the flow pattern for all the buckets are similar. As the bucket thickness increases (R73 and R53) the formation of vortices is diminished as the flow tends to attach on the bucket surface and merges into downstream without forming any vortices. All the buckets project similar front areas, yet a flow variation is noticeable for the R53 bucket.

The bucket tends to project itself as a streamlined body and as a result the net drag force available decreases which reduces the net torque on the shaft. Two vortices are formed for slower thickness buckets (R93, R83, R73), one inside the bucket and another on the downstream surface, whereas for R53, both the vortices are formed outside the bucket surface creating more drag on the flow direction. The computational results for the buckets are portrayed in Figure 3.29.

The least drag coefficient of 0.08 is displayed by the R53 bucket behaving close to a flat plate, with minimal difference for 360° incidence angle as it has higher thickness that accommodates the rest of the buckets, whereas for a thinner bucket (R93) the drag coefficient is 0.33. The reduced drag coefficient of thick buckets tends to generate lower moment on the turbine shaft, leading to power loss. The outcome of this study will compel structural designers to come up with thinner blades for better efficiency.

3.5.5 *Experimental Study on Telescopic Savonius Turbine*

The buckets are 3D printed and coated with wax to improve the surface roughness. Each bucket is guided by two aluminum rods fixed in place by the shaft holder attached to the end plate made of polycarbonate. Top and bottom buckets are fixed to top and bottom end plates, respectively. The end plates in TST have added functionality of transmitting the torque to the rotor shaft other than preventing the airflow from flowing over the rotor. The rotor shaft is supported by a pair of deep groove ball bearings on bottom and self-aligning ball bearing on top to accommodate any shaft misalignments. TST can be retracted or extended manually and the position can be preserved by locking the top shaft holder to the rotor shaft. The mounted test rotors are shown in Figure 3.31 and the experimental setup is shown in Figure 3.32. The TST was subjected to experimental investigation to determine the Cp, dynamic Ct, static Ct and static thrust coefficient.

Figure 3.31 Telescopic Savonius model in wind tunnel setup.

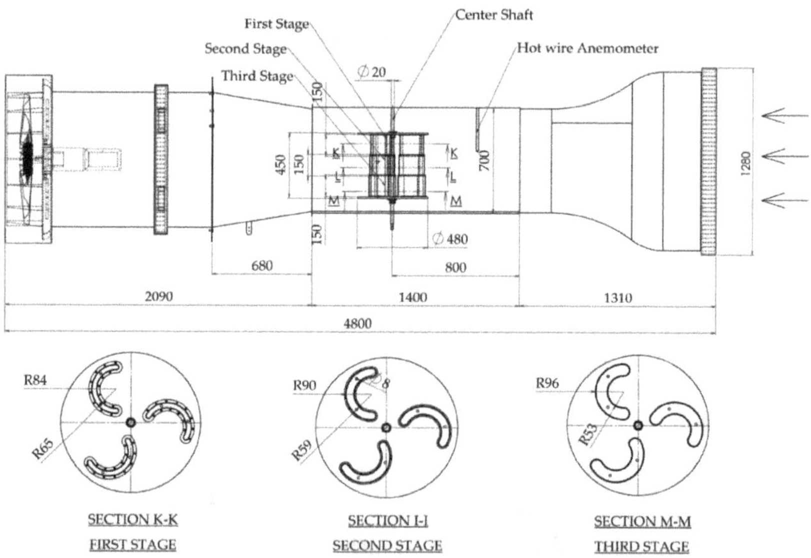

Figure 3.32 Wind tunnel schematic with Telescopic Savonius rotor.

For the current study, investigating these geometric parameters are out of scope. The dynamic performance characteristics were obtained by varying freestream wind speed and constraining the rotor speed for different TSRs by applying braking torque. For the static characteristics such as static Ct and static thrust coefficient, the azimuthal angle of the rotor is oriented from 0° to 360°, both in extended and retracted configurations at varying wind speed.

3.5.6 *Dynamic Performance of Telescopic Savonius Turbine*

The dynamic characteristics of TST includes Ct and Cp variation with TSR. An optimal TSR is critical in determining the maximum Ct. Low TSR rotors are unable to extract the available energy from the fluid unless they are equipped with more blades. A higher TSR than optimal will result in the condition where the blades are operating in a highly turbulent wake of preceding blade, expending energy. The dynamic characteristics are obtained at extended configuration, as retracted configuration is immaterial in power extraction mode. Ct was computed from the measured torque by Eq. (3.4). The Re was varied from 1.2×10^5 to 2.6×10^5 corresponding to the freestream wind speed of 4–9 m/s. The variation of Ct with TSR is shown in Figure 3.34. The Ct at low Re and low TSR cannot be measured as the torque generated by the rotor is not sufficient to overcome the bearing friction and the windage losses, which makes it difficult to stay at prescribed TSR.

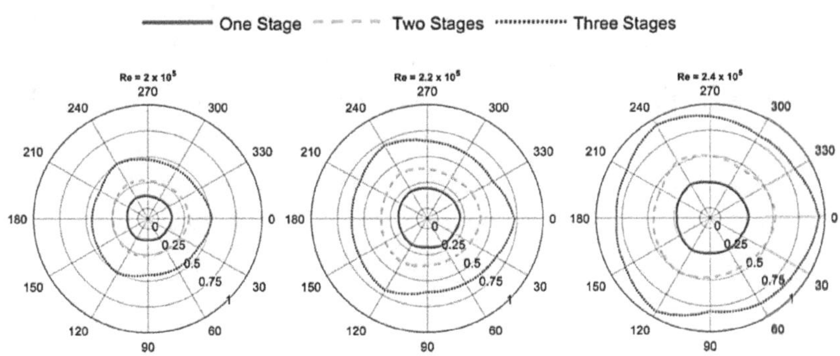

Figure 3.33 Thrust load (kg) vs azimuthal angle for various Re.

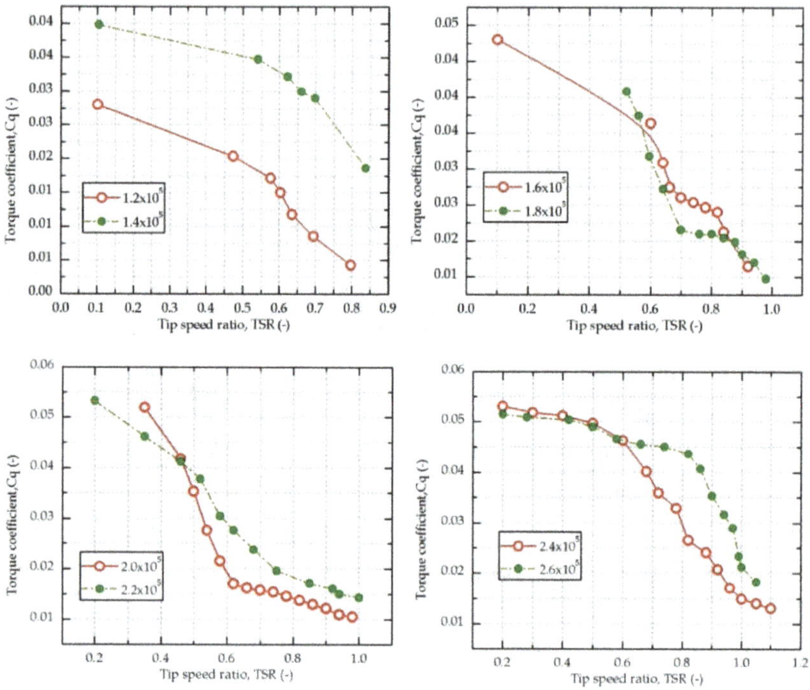

Figure 3.34 Ct vs TSR for various Re of TST.

It is evident from Figure 3.34 that the maximum Ct achieved is 0.05 for Re 2.6×10^5. The peak Cp achieved was 0.14 for Re 2.6×10^5, which is substantially less than the CST C_p of 0.17 as reported in the literature. The reason is evident by poring over the numerical results. The net drag force between the concave and convex sides of the bucket is reduced by 23% for the third stage bucket and 18% for the second stage bucket, compared to first stage bucket in extended configuration. The bucket thickness can be reduced in the order of CST blade thickness (2 to 5 mm), if the turbine is scaled up to higher kW.

The Ct depicts a decreasing trend with increase in TSR, similar to the behaviour of CST but at a lower magnitude. The maximum and minimum torques generated influence the smoothness and stability of the turbine operation. The variation of Cp for different Re is shown in Figure 3.35. Following a causal trend, the Cp value increases to a certain TSR, and after that a steady fall for further increase in TSR. Hence there exists an optimal Cp for every TSR, a characteristic for every turbine. For Re 2.6×10^5, the maximum Cp of 0.13, 0.13 and 0.14 is observed for the corresponding TSR of 0.5, 0.6 and 0.7. Though a direct comparison with CST is not sensible, the Cp obtained is generally low for the prototype model of similar size. Considerable optimization is required to further enhance the efficiency of TST during the development of full-scale modeling, especially on the blade thickness.

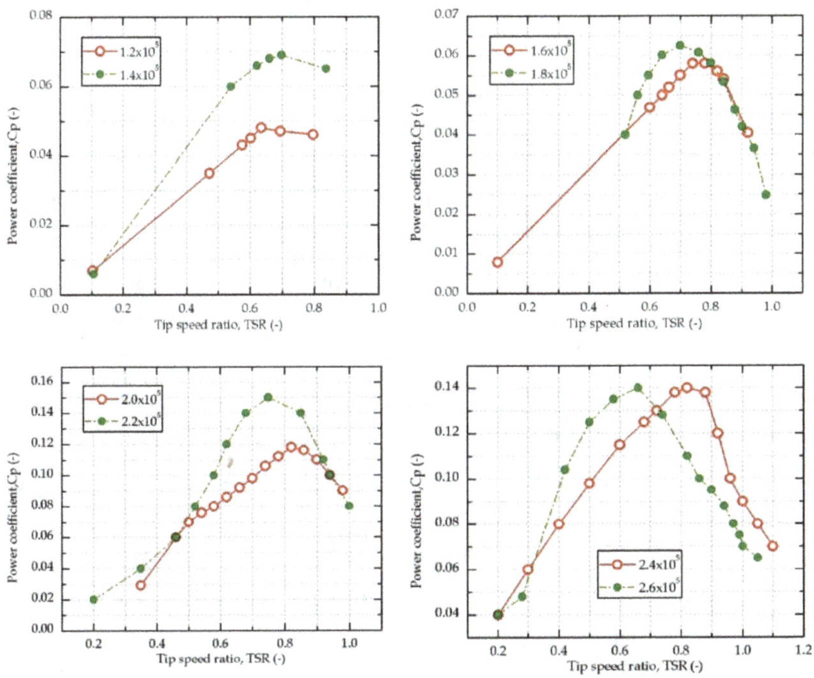

Figure 3.35 Cp vs TSR for various Re of TST.

3.5.7 *Static Performance of Telescopic Savonius Turbine*

The static Ct measurement is imperative for two reasons. Firstly, the starting characteristics of a turbine is solely decided by the static torque generated, and the second reason is to determine the structural requirements for the bucket at survival wind speed with respect to azimuthal angle. The static torque for the TST is measured for the wind speed of 3–5 m/s for every 30° increment from 0°–360° azimuthal angle. Since the TST under investigation is three-bladed, the static torque curve repeats every 120°. The static torque measurement is performed for each stage to ascertain the relative torque generated. Since TST is a high aspect ratio turbine, a lightweight rotor is possible for the same solidity compared to CST. Figure 3.36 illustrates the static Ct variation for 3 m/s wind speed. The maximum static Ct value of 0.16 is achieved when the rotor is at 0° azimuthal position under fully extended configuration.

3.5.8 *Thrust Load*

The torque in a Savonius rotor is essentially generated through drag and invariably it creates thrust. Among the vertical axis turbines, Savonius rotor has high solidity that presents maximum obstacle to the wind even in parked conditions. Additional power-augmenting accessories such as deflectors and partial enclosures to reduce the drag on the returning blade will contribute significantly to additional thrust. As per IEC 61400-2, the ultimate thrust load is calculated for 50-year high wind, but the probability of encountering such high wind is low. The total thrust load can be calculated from Eq. (3.6). The turbine has to be structurally sound to withstand the load generated in their 20-year life period. This demands additional cost and results in high inertia of the rotor, increasing the cut in wind speed of the turbine substantially. The thrust force generated is proportional to the exposed area and one possible way of reducing the thrust load is to keep the exposed area to a minimum. The drag coefficient varies for each stage and for every azimuthal angle. Hence this segment of the study is focused on evaluating the thrust load on TST. Loads such

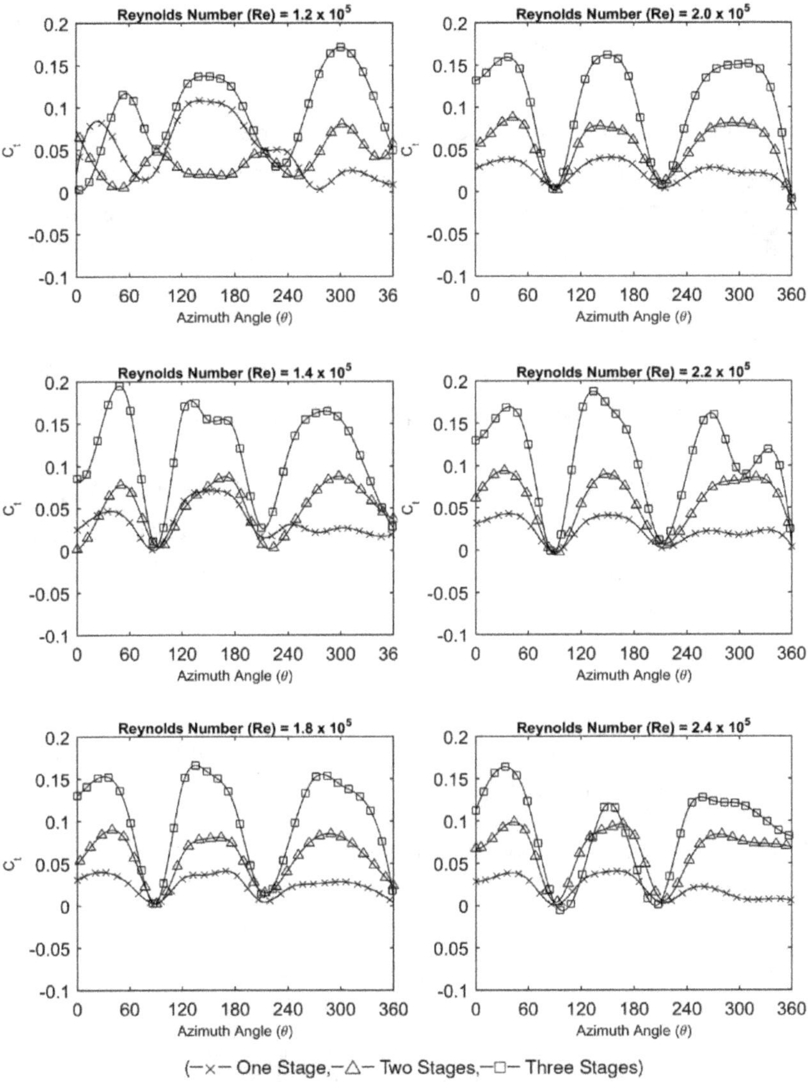

Figure 3.36 C_{ts} vs TSR for various Re of TST.

as fatigue and gust loads are inevitable during operation and the turbine should be designed to withstand these loads during their lifetime. The thrust load on the rotor was measured from the load cells attached to both ends of the shaft. The rotor was positioned at every 30° azimuthal angle and the thrust load was computed by adding the values from the top and bottom load cells.

Thrust loads for the wind speeds 6–8 m/s are measured as the loads are not significant for low wind speeds and rotor will be under braking above the rated wind speed of 8 m/s. From Figure 3.33 it is clear that thrust coefficient varies periodically for every 120°, as the turbine is three-bladed immaterial of the stages. It is also evident that the thrust load will reduce by 72.4% for the first stage and 43.5% for the second stage compared to fully extended configuration. Same pattern in the thrust load is exhibited for the wind speed of 7 m/s, except for the variation in the magnitude. The highest value of thrust coefficients obtained for the first, second and third stages is 0.15, 0.29 and 0.54, respectively, at $Re = 2.6 \times 10^5$. It can be noticed that even when the turbine is completely retracted, there exists a significant parasitic drag due to guide rods and end plates, and the magnitude is relatively high due to low aspect ratio of the test rotor in which the above said components present a considerable frontal area.

3.5.9 *Experimental Study on Hybrid Darrieus Telescopic Savonius Turbine*

A scaled version of the Telescopic Savonius rotor is fabricated to integrate with the Darrieus rotor as shown in Figure 3.37. The Darrieus rotor is 400 mm in diameter and 300 mm in height and consists of three blades of NACA 0018 with a chord length of 100 mm. The HDTS rotor is free of struts and the Darrieus rotor blades are fixed in place by the top and bottom end plates. The Telescopic Savonius rotor is 180 mm in diameter and consists of four blades. Each blade has three buckets of varying diameter. It is printed via fused deposition modeling as a single piece as the buckets will not slide with respect to each other. The Telescopic Savonius rotor has its end plates provided with the holes to fix on the Darrieus rotor end plates facilitating the torque transfer. Since the diametrical difference between the buckets is less for such miniaturized prototype, the Telescopic Savonius rotor is almost similar to the

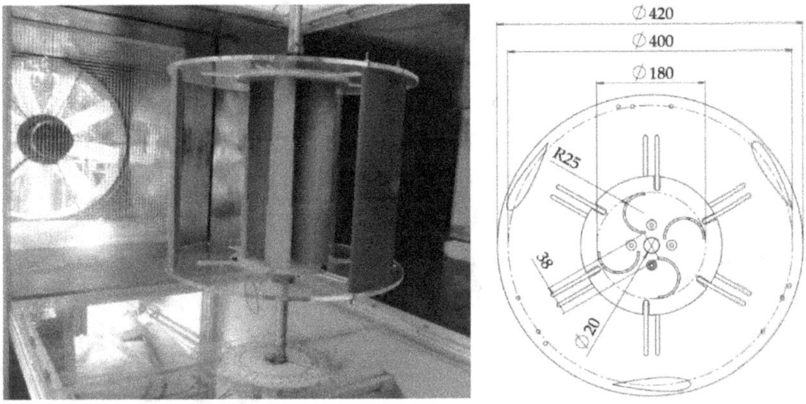

Figure 3.37 Hybrid Darrieus Telescopic Savonius rotor with critical dimensions.

conventional Savonius rotor with projections, and it is anticipated that the investigation in the wind tunnel will yield similar results to the HDTS rotor. Since the Telescopic Savonius rotor has four blades and the Darrieus rotor has three blades, an angular offset may have some influence, which may be less pronounced for the Cp comparison. The performance analysis of the HDTS rotor with two stages and single stage is not performed as it is cumbersome to design sliding buckets for such small-sized Savonius rotors. Moreover, as the current study is concerned with the power performance of the Darrieus rotor at high TSR, the two and single stages are only deployed to curtail the performance at high winds.

3.5.10 *Dynamic Performance of the Hybrid Darrieus Telescopic Savonius Rotor*

The fabricated HDTS rotor is subjected to wind tunnel investigation and compared with the conventional Darrieus rotor in a similar environment. The investigation is carried out at high Re as Savonius rotor dominates in the low wind speed regime. The tested wind speeds are 6, 7, 8 and 9 m/s as the objective is to find the influence of the Telescopic rotor on the conventional Darrieus rotor. The experimental results are in correlation with past studies that the Savonius rotor starts to brake when the TSR is 1~1.5.[1] For all the tested wind speeds the HDTS rotor displays higher Cp than the conventional Darrieus as shown in Figure 3.38. HDTS displays a similar trend of steep drop in the Cp for all the wind speeds, whereas the Darrieus rotor displays a smooth drop due to smooth stall. The peak Cp of the Savonius rotor is reached around 1.2 for the tested wind speed. The peak Cp of the Darrieus rotor increases in a smaller value from 0.11 for 6 m/s to 0.13 for 9 m/s, whereas the HDTS rotor increases from 0.21 for 6 m/s to 0.23 for 9 m/s. It can be concluded from the wind tunnel test for the tested wind speeds that the HDTS rotor is able to achieve 80% higher Cp than the conventional Darrieus rotor, and it is apparent that the torque developed by the Savonius rotor largely contributes. The Darrieus rotor Cp may exceed for higher TSR when tested in the higher wind speeds. The steep drop in HDTS rotor Cp may induce vibrations as the rotor accelerates.

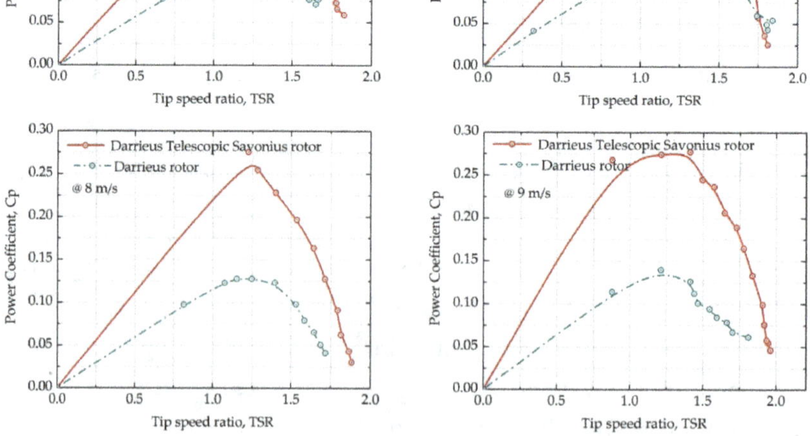

Figure 3.38 Cp comparison of HDTS and conventional Darrieus rotor.

The four potential concepts are subjected to preliminary computational optimization and their power performances are investigated through wind tunnel test by comparing with conventional Darrieus rotors. While all the four concepts show improved performance in low wind speed, the extent to which they influence the performance of the Darrieus rotor at rated wind speed is key. The KF-N-21 stepped airfoil is able to generate reverse drag in the cavity but the vortex formation is unlikely to occur due to rapid change in AoA. Even a formed vortex will be unstable due to change in apparent wind speed leading to reduced lift and torque. Dual rotor displays the lowest performance among the four concepts. The coupled two rotors reach their peak Cp at different TSRs and the wake from the inner rotor severely degrades the performance at most Re. The NACA 0018TC-39 airfoil with trailing edge cavity shows minimal improvement in the starting torque as the cavity is relatively small, generating low drag force. The Telescopic Savonius rotor shows higher performance among the lot and its excellent performance at low wind speed is due to larger drag surface area.

A redefined Hybrid Darrieus-Savonius rotor with minimal structural components is the best solution with the potential to implement commercially at minimal cost.

Table 3.4 Feasibility check on concepts.

Concept	Results of Feasibility Check	Option
Airfoil with step (KF-N-21)	Poor performance at higher TSR	
Dual Darrieus rotor	Poorer performance than conventional Darrieus rotor	
Airfoil with trailing edge cavity (NACA 0018TC-39)	Improvement in starting torque, higher drag at high Re	
Hybrid Darrieus Telescopic Savonius rotor (HDTS)	Higher starting torque for the investigated Re, structurally complex	
Hybrid Darrieus-Savonius rotor	Structurally simple and overall better aerodynamic characteristics	Yes

Chapter Nomenclature

A_s	Swept area of the turbine
B_T	Total blockage
C_p	Power coefficient
CT_s	Static torque coefficient
C_H	Coefficient of thrust
C_t	Dynamic torque coefficient
P_{rotor}	Power generated by the rotor
P_{wind}	Power available in the wind
R	Radius of the turbine
R_e	Reynolds number
T	Dynamic torque
T_S	Static torque
T_H	Total thrust on rotor
V_c	Blockage-corrected wind speed
V_c	Freestream wind speed
ρ	Density of air
ω	Rotational speed of the turbine
ε_s	Solid blockage
ε_w	Wake blockage
q_c	Blockage-corrected dynamic pressure
q_u	Uncorrected dynamic pressure

References

1. Fujisawa, N. On the torque mechanism of Savonius rotors. *Journal of Wind Engineering and Industrial Aerodynamics* **1992**, *40*, 277–292.
2. Pope, A.; Harper, J. J. *Low Speed Wind Tunnel Testing.* John Wiley and Sons **1966**.
3. Kline, R. L.; Fogleman, F. F. Airfoil for aircraft having improved lift generating device. **1977**.

4. Cox, M.; Avakian, V.; Huynh, B. Performance of a stepped airfoil at low reynolds numbers. *Proceedings of the Proceedings of the 19th Australasian Fluid Mechanics Conference*. Melbourne **2014**.
5. Krishnamurthi, J. A Numerical design simulation of a novel notched airfoil. *Proceedings of the The 37th Middle Atlantic Regional Meeting*. New Jersey **2005**.
6. Mishriky, F.; Walsh, P. Effect of step depth and angle in Kline-Fogleman (KFm-2) Airfoil. *Global Journal of Research In Engineering* **2016**, *16*, 1–9.
7. Law, S.; Gregorek, G. Wind tunnel evaluation of a truncated NACA 64–621 airfoil for wind turbine applications. **1987**.
8. Standish, K.; Van Dam, C. Aerodynamic analysis of blunt trailing edge airfoils. *Journal of Solar Energy Engineering* **2003**, *125*, 479–487.
9. Tangier, J.; Somers, D. NREL Airfoil Families for HAWTs. *Proceedings of the American Wind Energy Association Windpower Conference*. Washington, DC **1995**.
10. Baker, J.; Mayda, E.; Van Dam, C. Experimental analysis of thick blunt trailing-edge wind turbine airfoils. *Journal of Solar Energy Engineering* **2006**, *128*, 422–431.
11. Chong, W.; Fazlizan, A.; Poh, S.; Pan, K.; Hew, W.; Hsiao, F. The design, simulation and testing of an urban vertical axis wind turbine with the omni-direction-guide-vane. *Applied Energy* **2013**, *112*, 601–609.
12. Bangalore Gangadharacharya, K. Vortex shedding and aerodynamic drag on truncated trailing edge airfoil. **2015**.
13. Baker, J. P. *Drag Reduction of a Blunt Trailing-Edge Airfoil* (PhD Thesis). University of California, Davis **2009**.
14. http://www.ebay.com/itm/Vertical-Axis-3KW-Wind-Turbine-Generator-EOLO-3000-hybrid-Darrieus-Savonius-VAWT-/111332254933.
15. Heinrich, H. G.; Haak, E. L. The drag of cones, plates, and hemispheres in the wake of a forebody in subsonic flow. NASA Technical Reports **1961**.

Chapter 4

Mathematical Modeling of Adaptive Hybrid Darrieus Turbine*

---

* Parts of this chapter are adapted with permission from Kumar, P. M.; Ajit, K. R.; Srikanth, N.; Lim, T.-C. On the mathematical modelling of adaptive Darrieus wind turbine. *Journal of Power and Energy Engineering* 2017, *5*, 133–158.

4.1 Introduction to Adaptive Hybrid Darrieus Turbine

Of several solutions that are discussed in the previous chapter, hybrid Darrieus and Savonius turbine is a potential candidate that may be redesigned to improve the low wind speed performance and over speed regulation. The conventional hybrid Darrieus turbine integrates a Savonius rotor to a common shaft to effectively utilize the high torque generated by the Savonius for rotor acceleration. Similar to other concepts, the hybrid Darrieus-Savonius also suffers from poor performance when the Darrieus rotor accelerates beyond 1.

The optimum TSR for a two-bladed Darrieus rotor lies between 3 and 5[1] and that for Savonius rotor is 1. The Savonius rotor tends to generate resistive torque and in fact energy must be expended to rotate it for TSR above 1. The mismatch between the optimum TSRs for the two rotors severely degrades the performance at higher TSR. Practically, a conventional hybrid Darrius-Savonius rotor will not accelerate beyond 1.5 resulting in an iota of improvement in annual energy output. Hence a novel design has been put forward to minimize the influence of the Savonius rotor beyond TSR 1.

The strategy is to transform the Savonius rotor into a shape that leaves minimum wake downstream without any resistive torque at higher TSR. Two-bucket Savonius rotor can be transformed into a nominal cylinder if they are able to slide. The wake behind the downstream is axisymmetric with minimum width compared to other shapes. The wake width and the kinetic energy imbibed dictate the performance of the Darrieus rotor. Two-stage two-bucket Savonius rotor offset at 90° can improve the directional starting. The three operating configurations are shown in Figure 4.1(a, b, c). At low wind speed the Darrieus rotor torque ($M_d = +ve$) and the Savonius rotor torque ($M_s = +ve$) are in the same direction when the Savonius buckets **a** and **b** are arranged as shown in Figure 4.1(a).

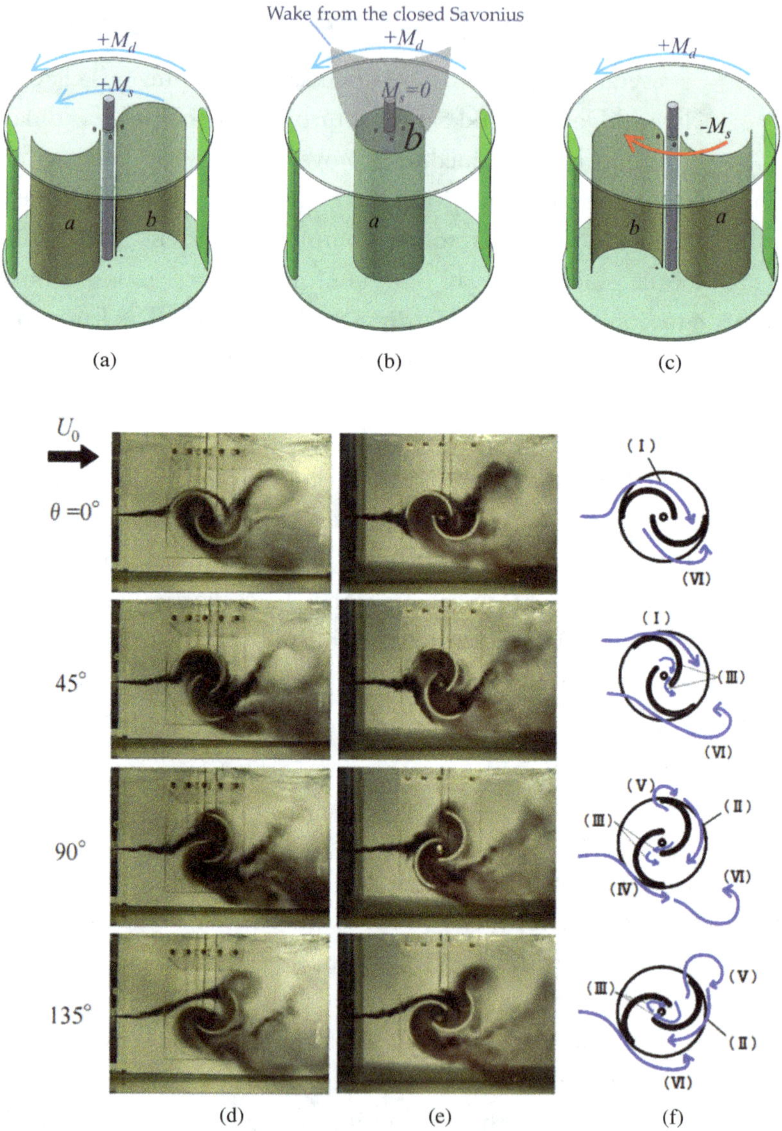

Figure 4.1 (a) ADWT rotor in open configuration. (b) Closed configuration. (c) At high wind braking configuration. (d) Flow over single-stage Savonius.[2] (e) Flow over double-stage Savonius. (f) Flow sequence on single-stage Savonius.

As the TSR reaches 1 the buckets slide towards the axis of rotation to form a cylinder without any torque generation ($M_s = 0$) as shown in Figure 4.1(b). In extreme wind conditions and above the rated rotor rpm, the Savonius buckets slide in opposite directions generating resistive torque ($M_s = -ve$), decelerating the rotor as shown in Figure 4.1(c).

The number of blades on the Darrieus rotor, orientation of the rotor to the oncoming wind, angular offset between the Savonius buckets and the Darrieus rotor, and ratio between the diameters of Darrieus and Savonius rotors are the crucial design parameters that determine the starting characteristics.

The angular offset between Darrieus blades and Savonius buckets will have minimal impact on the performance, as two stages are arranged offset at 90°. Hence from the structural perspective, the Savonius buckets can slide on the Darrieus blade-connecting struts, eliminating the requirement of additional structures. Thus, this adaptive hybrid Darrieus turbine (ADWT) has the capability to start at low wind speed, operate with minimal effect on the Darrieus rotor, and decelerate the rotor when it rotates beyond the rated rpm. The construction and mechanical arrangement are less complex, making this concept commercially implementable.

4.2 Analytical Model of Adaptive Hybrid Darrieus Turbine in Open Configuration (Open Savonius)

The analytical model of the ADWT in open configuration is similar to the conventional hybrid Darrieus-Savonius rotor. In order to simplify the development of the analytical model, the Savonius buckets are arranged in line with the blades of the Darrieus rotor. The simplified configuration for ADWT is with two Savonius buckets without an overlap mounted on the same axis as the Darrieus rotor with two blades. The velocity diagram of the ADWT is shown in Figure 4.2.

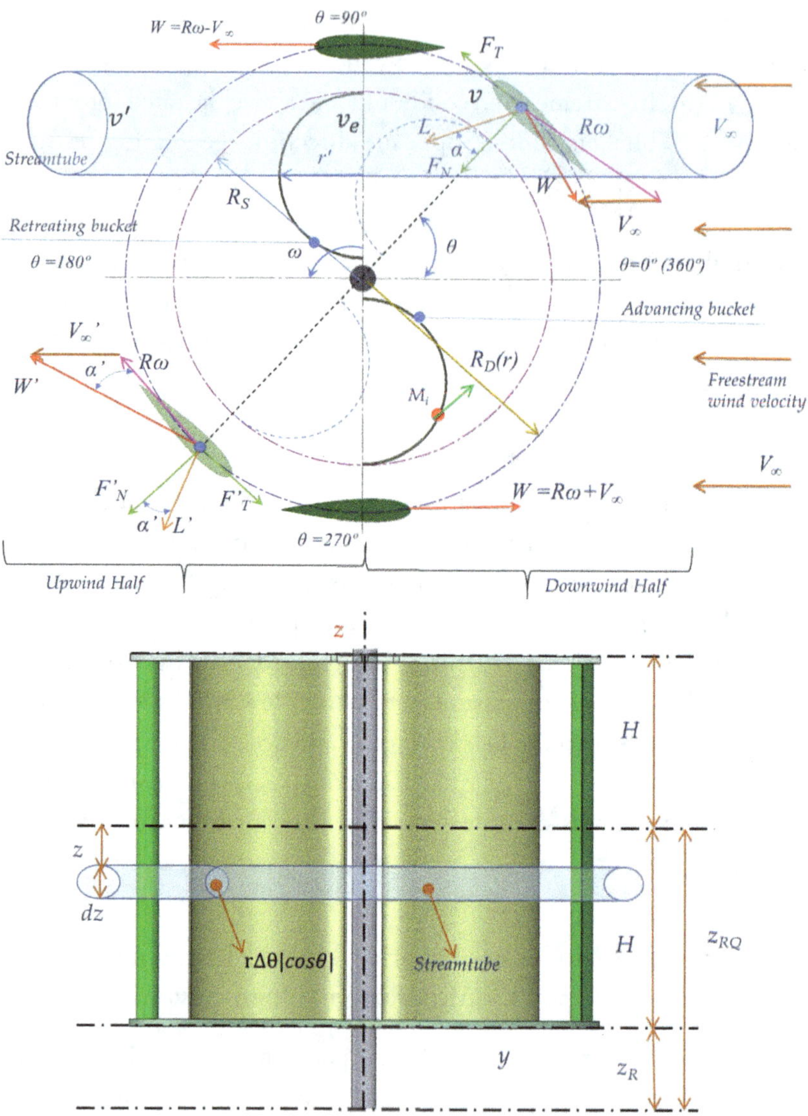

Figure 4.2 Velocity diagram of ADWT rotor.

4.2.1 *Wake of Savonius Rotor in Open Configuration (Conventional Two-Bucket Savonius Rotor)*

Before proceeding with the analytical solution it is indispensable to predict the wake flow pattern of a two-bladed Savonius rotor. Countless studies focused on the flow over the Savonius rotor rather than on the downstream wake. A detailed flow investigation and adequate knowledge are a must to generalize the wake structure on the downstream. It is possible to deduce the wake flow pattern on the downstream within the flight path of the Darrieus blades by examining the flow leaving the Savonius buckets. Typical flow of a two-bucket Savonius rotor is presented in Figure 4.1(d, e, and f). The flow is characterized by a sequence in which flow is first attached to the convex side of the advancing blade. As the rotation advances, the flow transfers from the advancing blade's convex surface to the returning blade's concave side. The flow simultaneously enters in between center shaft space, followed by the upstream flow on the convex side of the returning blade. The vortex shedding from the returning and advancing blade tips contributes to increased wake width. The wake pattern is highly turbulent due to alternating suction and pressure zone occurrence on the concave and convex sides of the blade. The flow on a single-stage Savonius rotor is different from the double-stage rotor.[2] The flow tends to jump from the high-pressure zone of one stage into the low-pressure zone of the adjoining stage. A span-wise flow is induced due to the flow movement from one stage to another, and it widens the wake width when it leaves the rotor.

The unsteady and highly turbulent wake when interacting with the flight path of the Darrieus blades would have dispersed to a large wake width from the smaller diameter Savonius rotor. As the Savonius rotor diameter increases, the flow in fact has high energy and turbulence. Hence a pragmatic way is to assume the Savonius rotor as a bluff body or as an actuator disk absorbing a part of the kinetic energy, leaving behind low-energy

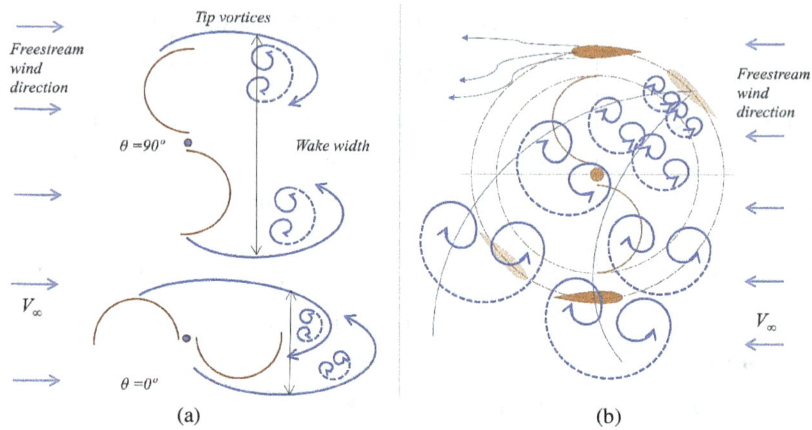

Figure 4.3 Typical wake behind (a) Savonius rotor and (b) Darrieus rotor.

wake. The typical wake pattern of Savonius and Darrieus rotors are shown in Figure 4.3.

The wake width of the Savonius rotor on the Darrieus flight path is assumed to be equal to the Savonius rotor diameter. In reality some portion of the flow may be accelerated due to vortices from the bucket tips. The Darrieus blades may not be able to extract energy from the vortices due to large AoA.

4.2.1.1 *Analytical Model*

The following analytical model attempts to compute the power performance of the AHDT analytically. Both the performance of the Darrieus rotor and Savonius rotors are computed and summed up to arrive at the total power.

$$V_\infty = uV_{\infty i} \tag{4.1}$$

And the equilibrium-induced velocity is

$$V_e = (2u - 1)V_{\infty i} \tag{4.2}$$

With V_e as the input velocity for the downstream half-cycle of the rotor, the induced velocity at the end of the streamtube is

$$V' = u' \, (2u - 1) V_{\infty i} \tag{4.3}$$

The relative velocity for the upstream half-cycle of the rotor, $-\pi/2 < \theta < \pi/2$, is given by the expression

$$W^2 = V^2 \, [(X - \sin \theta)^2 + \cos^2 \theta \cos^2 \delta] \tag{4.4}$$

where $X = \omega r / V$ represents the local TSR. The general expression for the AoA is

$$\alpha = \sin^{-1}\left[\frac{\cos \theta \cos \delta}{\sqrt{(X - \sin \theta)^2 + \cos^2 \theta \, \cos^2 \delta}}\right] \tag{4.5}$$

By equating the blade element theory and the momentum equation for each streamtube

$$f_{up}\left(\frac{V}{V_\infty}\right)^2 = \pi\eta\left(\frac{V}{V_\infty}\right)\left[\left(\frac{V_{\infty i}}{V_\infty}\right) - \left(\frac{V}{V_\infty}\right)\right] \tag{4.6}$$

$$f_{up} u = \pi\eta \, (1 - u) \tag{4.7}$$

$$\eta = \frac{r}{R_D} \tag{4.8}$$

where f_{up} is the function that characterizes the upwind conditions

$$f_{up} = \frac{Nc}{8\pi R_D} \int_{-\pi/2}^{\pi/2} \left(C_N \frac{\cos \theta}{|\cos \theta|} - C_T \frac{\sin \theta}{|\cos \theta| \cos \delta} \right) \left(\frac{W}{V}\right)^2 d\theta \tag{4.9}$$

$$C_N = C_L \cos \alpha + C_D \sin \alpha \tag{4.10}$$

$$C_T = C_L \sin \alpha - C_D \cos \alpha \tag{4.11}$$

Airfoil coefficients C_L and C_D are obtained from the wind tunnel test or from literature and interpolating for local Reynolds number and the local AoA.

Defining the blades' local Reynolds number as Re_b for local conditions given by

$$Re_b = Wc/V_\infty \tag{4.12}$$

the turbine Reynolds number will be

$$Re_b = (Re_t n / X)\sqrt{(X - \sin\theta)^2 + \cos^2\theta\cos^2\delta} \tag{4.13}$$

For each blade in the upstream position, the non-dimensional force coefficients as functions of the azimuthal angle θ are given by

$$F_N(\theta) = \left({}^{cH}\!/_{S}\right)\int_{-1}^{1} C_N \left({}^{W}\!/_{V_\infty}\right)^2 \left({}^{\eta}\!/_{\cos\delta}\right) d\zeta \tag{4.14}$$

$$F_T(\theta) = \left({}^{cH}\!/_{S}\right)\int_{-1}^{1} C_T \left({}^{W}\!/_{V_\infty}\right)^2 \left({}^{\eta}\!/_{\cos\delta}\right) d\zeta \tag{4.15}$$

By integrating for the entire blade

$$T_{up}(\theta) = {}^{1}\!/_{2}\,\rho_\infty c R_D H \int_{-1}^{1} C_T W^2 \left({}^{\eta}\!/_{\cos\delta}\right) d\zeta \tag{4.16}$$

The average half-cycle of the rotor torque produced by $N/2$ of the N blades is given by

$$\overline{T_{up}} = N/2 /_{\pi} \int_{-\pi/2}^{\pi/2} T_{up}(\theta)\, d\theta \tag{4.17}$$

The average torque coefficient will be

$$\overline{C_{Q1}} = \frac{NcH}{2\pi S} \int_{-\pi/2}^{\pi/2} \int_{-1}^{+1} C_T \left(\frac{\eta}{\cos\delta}\right)\left(\frac{W}{W_\infty}\right)^2 d\theta\, d\zeta \tag{4.18}$$

Thus, the power coefficient for the upstream half can be written as

$$C_{P1} = \frac{\omega R_D}{V_\infty} \overline{C_{Q1}} \tag{4.19}$$

It is similar for the downstream half-cycle. The relative velocity for the downstream half-cycle of the rotor, $\pi/2 < \theta < 3\pi/2$, is given by the expression

$$W'^2 = V'^2 (X' - \sin\theta)^2 + \cos^2\theta \cos^2\delta \tag{4.20}$$

where $X' = \omega r/V'$ represents the local TSR. The general expression for the AoA is

$$\alpha' = sin^{-1}\left[\frac{\cos\theta\cos\delta}{\sqrt{(X' - \sin\theta)^2 + \cos^2\theta\cos^2\delta}}\right] \tag{4.21}$$

By equating the blade element theory and the momentum equation for each streamtube

$$f_{dw}\left(\frac{V'}{V_\infty}\right)^2 = \pi\eta\left(\frac{V'}{V_\infty}\right)\left[\left(\frac{V_{\infty i}}{V_\infty}\right) - \left(\frac{V'}{V_\infty}\right)\right] \tag{4.22}$$

$$f_{dw}\, u' = \pi\eta(1 - u') \tag{4.23}$$

$$f_{dw} = \frac{Nc}{8\pi R}\int_{\pi/2}^{3\pi/2}\left(C'_N \frac{\cos\theta}{|\cos\theta|} - C'_T \frac{\sin\theta}{|\cos\theta|\cos\delta}\right)\left(\frac{W'}{V'}\right)^2 d\theta \tag{4.24}$$

$$F'_N(\theta) = \left(\frac{cH}{S}\right)\int_{-1}^{1} C'_N \left(\frac{W'}{V_\infty}\right)^2 \left(\frac{\eta}{\cos\delta}\right) d\zeta \tag{4.25}$$

$$F'_T(\theta) = \left(\frac{cH}{S}\right)\int_{-1}^{1} C'_N \left(\frac{W'}{V_\infty}\right)^2 \left(\frac{\eta}{\cos\delta}\right) d\zeta \tag{4.26}$$

$$T_{dw}(\theta) = \frac{1}{2}\rho_\infty cRH\int_{-1}^{1} C'_T W'^2 \left(\frac{\eta}{\cos\delta}\right) d\zeta \tag{4.27}$$

$$\overline{T_{dw}} = \frac{N}{2} \Big/ \pi \int_{\frac{\pi}{2}}^{\frac{3\pi}{2}} T_{dw}(\theta)\, d\theta \tag{4.28}$$

The downstream torque coefficient is given by

$$C_{Q2} = \overline{T_{dw}} \Big/ \left(1/2\rho_\infty V_\infty^2 SR_D\right) \tag{4.29}$$

$$C_{Q2} = \frac{NcH}{2\pi S}\int_{\pi/2}^{3\pi/2}\int_{-1}^{+1} C'_T \left(\frac{\eta}{\cos\delta}\right)\left(\frac{W}{W_\infty}\right)^2 d\theta\, d\zeta \tag{4.30}$$

$$C_{P2} = \left(\frac{\omega R}{V_\infty}\right)\overline{C_{Q2}} \tag{4.31}$$

$$C_{P-D} = C_{P1} + C_{P2} \tag{4.32}$$

For the Savonius rotor, the power coefficient can deduced as follows. Suppose pressure difference on retreating side is ΔP_r,

$$\delta Q_p = \Delta P_r \sin\theta \,.\, 2r\cos\theta \times \text{Area} \tag{4.33}$$

Assuming advancing side contributes to negative torque through drag,

$$\delta Q = (\Delta P_r \sin\theta \,.\, 2r\cos\theta - \delta Q_p - \Delta P_a \sin\theta \,.\, 2r\cos\theta)4\, rh \tag{4.34}$$

$$Q = \int_0^{\pi/2} \Delta(P_r - \Delta P_a)\sin\theta\cos\theta \times 4r^2 hd\,\theta \tag{4.35}$$

$$Q = 2r_s^2 H_s \int_0^{\pi/2} \Delta(P_r - \Delta P_a)\sin^2\theta \tag{4.36}$$

Average power P is obtained by integrating torque from 0 to π

$$P_s = \omega.Q = \frac{\omega}{\pi}\int_0^{\pi} Q d\alpha \tag{4.37}$$

Normalized power coefficient can be given as

$$C_{p-s} = \frac{P}{0.5\ \rho\ V_e^3 (4r_s^2\ _s H_s)} \tag{4.38}$$

$$\Delta P_r \big/ 0.5\rho v_{e-s}^2 = S_1 \text{ (Assumption)} \tag{4.39}$$

V_{e-s} is equivalent velocity or relative velocity, v_{is} is absolute velocity.

$$V_{e-s} = v_{es} - v_{is} \tag{4.40}$$

Consider retreating side: ΔPp value is known. Solving the integral,

$$Q_p = 2r^2 H_s \frac{1}{2}\rho s_1 (v_{es}^2) \tag{4.41}$$

$$Q_p = r^2\ H_s \rho s_1 (v_\infty - v_{is})^2 \tag{4.42}$$

$$Q_M = r^2 H_s \rho s_1 (v_\infty^2 + 2r^2 w^2 - \frac{\pi}{2} v_\infty r\omega \cos\alpha - 2v_\infty r\omega \cos\alpha - 2v_\infty r\omega \sin \propto) \tag{4.43}$$

Similarly, for the advancing side:

$$\Delta Pa / 0.5\rho v_\infty^2 = S_2 \tag{4.44}$$

Again, resolving the component will yield the equation of the torque for advancing side.

$$Q_D = 2\rho r^2 \alpha_s^2 H_s S_2 \tag{4.45}$$

Combining all the parts together the final equation for *Cp* is given as

$$C_{P-S} = \left[\frac{S_1\varphi}{4} - \frac{S_1\varphi^2}{2\pi} + \frac{S_1 - S_2}{S_1}\left(\frac{\varphi^3}{4} \right) \right] \tag{4.46}$$

The power coefficient of ADWT rotor in open configuration is

$$C_P = C_{P-D} + C_{P-S} \tag{4.47}$$

4.3 Analytical Model of Adaptive Hybrid Darrieus Turbine in Closed Configuration (Cylinder)

The power coefficient under the influence of Savonius rotor in closed condition can be derived by treating it as a nominal cylinder placed in a steady and homogenous wind flow. Even this simplified assumption gives rise to complex flow wake structures. The objective of the model is to predict the wake width and velocity deficit due to the presence of cylinder. The streamtubes that are influenced by the wake width are identified and the input velocity is modified with the velocity deficit calculated during the iteration.

However, there will be axisymmetric flow acceleration on either side of the wake which is not accounted for. The cylinder can be considered as non-rotating as the cylinder's TSR is low, and the flow field displayed by both rotating and non-rotating flow for low rpm of 80~100 is similar. The current approaches and assumptions will lead to the development of an analytical model that can be well integrated into the existing sub-routine coded in Python.

4.3.1 *Wake of Savonius Rotor in Closed Configuration (Cylinder)*

A wake boundary occurs between two fluids carrying different momentum along with their flow. The momentum deficit in the particular region of unidirectional flow is highly unstable and gives rise to the zone of turbulence mixing layer downstream at a point where the two streams meet for the first time. Though at far downstream the static pressure tends to equalize within the flow and wake, the velocity or momentum deficit continues to travel along with the flow.

Detailed studies by previous researchers have precipitated a very important conclusion which relates the cylinder wake's growth to the ratio of rotational to rectilinear speed ratio. The experimental studies by Prandtl[3] and Díaz *et al.*[4] gave the most vital results which suggested that the eddies behind the cylinder in terms of Karman vortex street disappear at higher rotational-to-rectilinear speed ratio. One of the most dominating reasons for this is, for the low values of rotational speed ratio (TSR), the vortices are shed in alternate fashion behind the cylinder. The cylinder Re dictates the wake pattern behind the cylinder and the wake structure can be established by defining the Re. Alternating eddies are formed and move along the direction of rotation, which progressively decrease until the TSR reaches 1. After that it starts dissipating the turbulent kinetic energy to large eddy structures, leaving behind constantly growing wake. In the case of VAWTs in combination with the central shaft a similar pattern is observed. The flow structure behind the cylindrical shaft rotating at TSR greater than 1 creates a uniformly growing wake without breaking into vortex structures. So for the mathematical modeling, it is safe to assume that the wake structure behind the central shaft is in accordance with the two-dimensional wake behind a single body as suggested by Schlichting *et al.*[5] (Figure 4.5).

4.3.1.1 *Analytical Model*

$$\frac{\partial U}{\partial X}+V\frac{\partial U}{\partial Y}=-\frac{1}{\rho}\frac{\partial P}{\partial X}+\gamma\frac{\partial^2 U}{\partial Y^2} \tag{4.48}$$

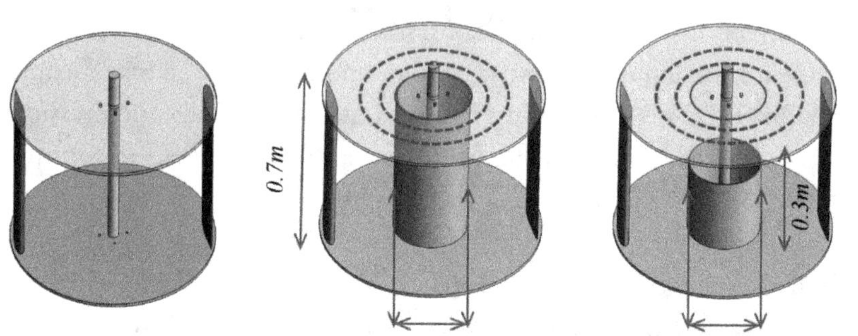

Conventional Darrieus	Full length Savonius	Half length Savonius
Evaluated for different chord length, rotor torque For 2 m/s and 10m/s	Evaluated diameters (0.6m, 0.75m, 1.2m, 2m) For 2 m/s and 10m/s	Evaluated diameters (0.6m, 0.75m, 1.2m, 2m) For 2 m/s and 10m/s

Figure 4.4 Evaluated parameters by analytical model.

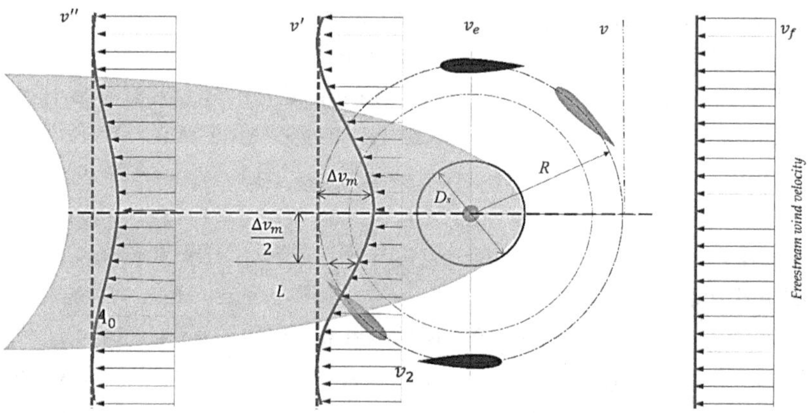

Figure 4.5 Wake deficit of closed Savonius rotor.

The above equation is the governing partial differential equation.

$$-\frac{\partial P}{\partial y} = 0 \tag{4.49}$$

$$\frac{\partial U}{\partial X} + \frac{\partial V}{\partial Y} = 0 \tag{4.50}$$

For cylinder wake which is relatively thick we cannot consider

$$\frac{\partial P}{\partial X} = 0 \tag{4.51}$$

Assuming X is considerably large,

$$\frac{U}{V_\alpha} = 1 \tag{4.52}$$

$$\frac{V}{V_\alpha} = 0 \tag{4.53}$$

The difference between the velocities ($V_\alpha - U$) should be an even function (symmetric wake) and V must be an odd function (asymmetric) in order to get the solution for the governing differential equation.

Drag prediction from Boundary layer assumptions

$$D = \rho_\infty U \int_{-\infty}^{+\infty} (V_\infty - U)dy \tag{4.54}$$

U_d can be neglected from

$$D = \rho_\infty V \int_{-\infty}^{\infty} (V_\infty - U)dy \tag{4.55}$$

$$D = \rho_\infty V \int_{-\infty}^{\infty} U_d dy \tag{4.56}$$

Comparing it with conventional drag equation

$$\frac{1}{2}\rho_{\infty}V_{\infty}.d.C_d = \rho_{\infty}V_{\infty}\int_{-\infty}^{\infty}U_d dy \tag{4.57}$$

$$\int_{-\infty}^{\infty}U_d dy = \frac{1}{2}V_{\infty}.d.C_d \tag{4.58}$$

$$\frac{U_d}{V_{\infty}} \propto \frac{C_D d}{2b} \tag{4.59}$$

From Prandtl's mixing length theory we know that

$$v' = u'.const = const.l.du/dy \tag{4.60}$$

$v' + u'$ are turbulent velocity components. Also rate of increase of width 'b' of mixing zone is proportional to transverse velocity v'.

$$\frac{Db}{Dt} = const.\frac{l}{b}u_d = const.U_{max} \tag{4.61}$$

$$\frac{Db}{Dt} \propto v' = \frac{Db}{Dt} \propto l\frac{\partial u}{\partial Y} \tag{4.62}$$

Average value of $\frac{\partial u}{\partial Y}$, i.e., velocity deficit along Y direction is considered to be proportional to transverse velocity $\frac{U_{max}}{b}$.

$$\frac{Db}{Dt} = const.\frac{l}{b}u_{max} \tag{4.63}$$

Let

$$\frac{l}{b} = \Delta = \frac{Db}{Dt} = const.\Delta.U_{max} \tag{4.64}$$

For two dimensions

$$\frac{Db}{Dt} = const.\frac{l}{b}u_d = const.\Delta.U_d \tag{4.65}$$

Equating above expressions

$$\frac{Db}{Dt} = const\frac{db}{dx} \tag{4.66}$$

$$\frac{db}{dx} \propto \Delta\frac{u}{V_\infty} \tag{4.67}$$

We introduce equation of wake width variation into the drag equation.

We derived previously

$$2b\frac{db}{dx} \propto \Delta C_D d \tag{4.68}$$

Using variable differential form

$$b \propto (\Delta x C_D d)^{1/2} \tag{4.69}$$

We know from 2D incompressible flow equations

$$\frac{\partial U}{\partial t} + U\frac{\partial U}{\partial x} + V\frac{\partial U}{\partial Y} = \frac{1}{\rho}\frac{\partial C}{\partial Y} \tag{4.70}$$

From Prandtl mixing length theory

$$-V_\infty\frac{\partial V_d}{\partial x} = 2l^2\frac{\partial U_d}{\partial Y}\frac{\partial^2 U_d}{\partial Y^2} \tag{4.71}$$

We introduce $K = \frac{Y}{b}$ as independent variable (m = constant).

$$b = m.\sqrt{\Delta}(C_D dx)^{1/2} \tag{4.72}$$

$$U_d = V_\infty\left(\frac{x}{C_D d}\right)^{\frac{1}{2}} f(k) \tag{4.73}$$

Put this function again in the differential equation to get the value of *f(k)*.

$$f(k)=\frac{m\sqrt{\Delta}}{18\Delta^2}(1-k^{3/2)})^2 \tag{4.74}$$

And from momentum equation of experimental data the constant $m\sqrt{\Delta}$ yields the value equivalent to $\sqrt{10\Delta}$.

$$b=\sqrt{10\Delta}(xC_D D)^{1/2} \tag{4.75}$$

$$\frac{U_d}{V_\infty}=\frac{\sqrt{10}}{18\Delta}\left(\frac{x}{C_D d}\right)^{1/2}\left(1-\left(\frac{y}{b}\right)^{3/2}\right)^2 \tag{4.76}$$

According to experiments by Eckelmann,[6]

$$\frac{b'}{2}=\frac{1}{4}(xC_D d)^{1/2} \tag{4.77}$$

Substituting $y = 0$ in the U_d equation, then the velocity deficit is at central region which practically explains the maximum deficit velocity since the distribution is close to Gaussian. From boundary layer theory and momentum equation after neglecting small terms we get

$$\frac{C_{DT}}{4}=\frac{\Delta Vm}{Ve}\cdot\frac{L}{D_T} \tag{4.78}$$

$$(V_e/\Delta V_m)^2=(x+x_0)/(bD_T) \tag{4.79}$$

For small Reynolds number on the order of 10^4, the following result holds true.

$$(L/D_T)^2=c(x+x_1)/D_T \tag{4.80}$$

$$\frac{\Delta Ve}{Ve}=\frac{\Delta Vm.2L}{VeN_T R_T f\theta_0} \tag{4.81}$$

The calculated deficit velocity is used instead of equilibrium velocity in the downstream for the corresponding streamtubes that are influenced by the wake width.

4.4 Discussion on Analytical Predictions

The developed mathematical model is applied to 10 kW ADWT and conventional straight-bladed Darrieus turbine. The dimensional details of the configurations are listed in Table 4.1, and the various configurations are shown schematically in Figure 4.4. The results are evaluated for the starting characteristics, blade and rotor torque, and power coefficient variation. Though the starting characteristics do not reflect the actual starting conditions, it will provide insight into the low Re behavior of cylinder. Another possible configuration that may be of interest for the current study is the height of the Savonius rotor with respect to that of

Table 4.1 Design parameters of 10 kW AHDT rotor.

Description	Value	Unit
Rated power	10	kW
Rated wind speed	10	m/s
Starting wind speed	3	m/s
Rotor height	7	m
Rotor diameter	6	m
Number of blades	2	—
Blade airfoil	NACA 0018	—
Blade chord length	70	mm
Number of struts	6	—
Strut airfoil	NACA 0018	—
Strut chord length	0.1	m
Parametric chord length	20, 50, 70	mm
Parametric closed Savonius rotor diameter	600, 750, 1200, 2000	mm
Parametric height of Savonius rotor	3.5, 7	m

the Darrieus. It is envisaged that the half-length of the Savonius rotor may have minimal influence on the performance of the Darrieus rotor. On the negative aspect, the half-length Savonius may induce uneven loading on the Darrieus blade in the downstream half when it enters the wake zone of the cylinder, yet the advantage that can be reaped at high TSR is attractive to incite an investigation. Blade and rotor torque values are evaluated for the above said architectures and compared with the conventional Darrieus turbine.

4.4.1 *Parametric Study*

A parametric analysis is conducted to evaluate the effect of solidity on the power and torque coefficient on a conventional Darrieus rotor. The solidity is varied by changing the chord length of the blades. The chosen chords for the study are 20 mm, 40 mm and 70 mm. The results are plotted as a function of TSR as shown in Figure 4.6. It is apparent from the result that the peak power coefficient increases with a decrease in solidity but the operating range of TSR decreases. A maximum power coefficient of 0.46 is achieved at 3.8 TSR without accounting for the effect of struts. For 70 mm chord length, the power coefficient decreases by 12.8% to 0.4. Though a low solidity rotor is preferred from the aerodynamic perspective, a thicker blade with increased chord length is favorable from the structural perspective

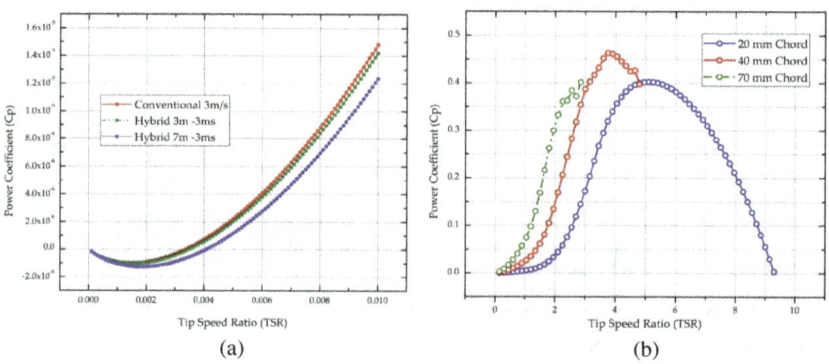

Figure 4.6 (a) Cp vs TSR at startup and (b) Cp vs TSR with different chord length.

and is also desirable for starting characteristics. A longer blade chord with high thickness will enable the blade to withstand alternating fatigue stress and thus extend the operating lifetime with acceptable loss in peak power coefficient. Starting characteristics are compared for the conventional Darrieus and ADWT rotors at a low wind speed of 2–4 m/s. The outcome of the prediction is plotted as Cp vs TSR in Figure 4.6. The conventional Darrieus rotor displays earlier starting than ADWT in closed condition. Low wind speed startup study is to shed additional light on the performance of the conventional Darrieus rotor, since the starting torque will be generated by the Savonius rotor if the ADWT is in open condition.

4.4.2 *Blade Torque and Rotor Torque*

The normal and tangential forces are plotted as a function of azimuthal angles in Figure 4.8. The well-established pattern of double peak is displayed by both conventional and ADWT rotors. The non-dimensional force coefficients F_N and F_T are computed from the predicted normal and tangential forces, respectively. The normal force coefficients are less in the downstream half. The tangential and normal forces are predicted for both full- and half-length Savonius rotor. From the figure it is evident that the presence of cylinder increases the AoA resulting in the decrease of lift and increase of drag. The maximum tangential force is obtained at 0° as 90° N. The tangential force difference between conventional Darrieus and the ADWT is less than the difference in normal force. The dynamic stall has significant influence on the normal force. The computed rotor torque is shown in Figure 4.7.

4.4.3 *Power Coefficient and Torque Coefficient*

The power coefficients are evaluated for various diameters of the conventional Darrieus rotor. The diameters are investigated for Darrieus rotor with full-length and half-length Savonius rotor as shown in Figures 4.9 and 4.10 for torque and power curves, respectively. The power coefficient curve follows the same trend for all the diameters investigated and the difference in their magnitudes is diminutive. A maximum reduction in

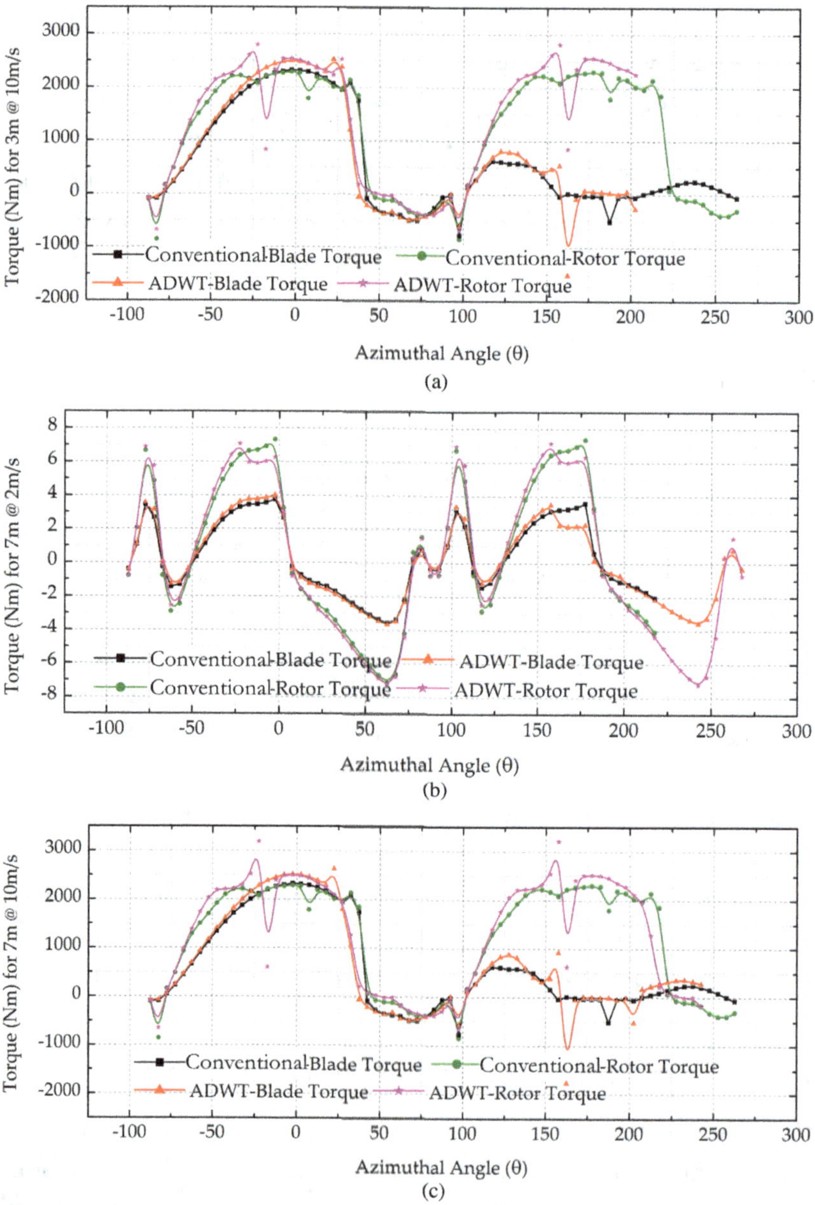

Figure 4.7 Blade and rotor torque comparison of ADWT (AHDT).

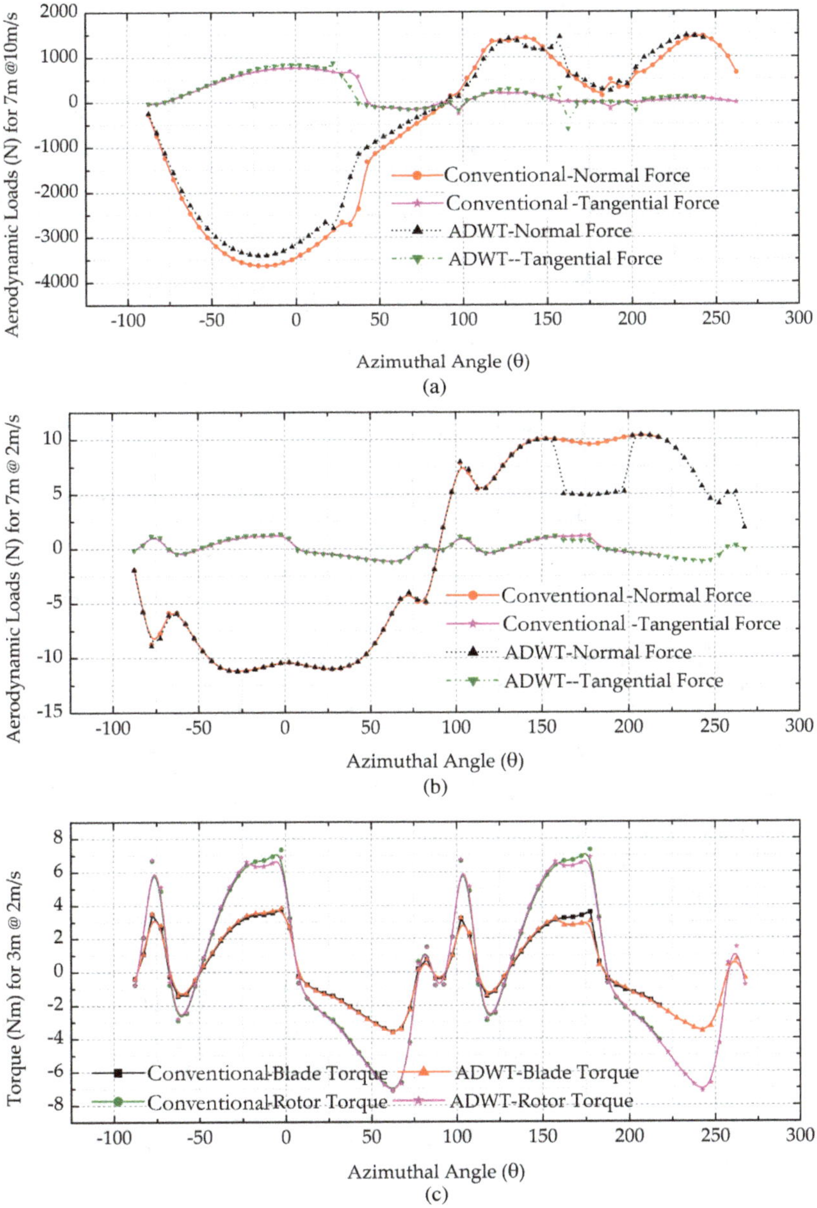

Figure 4.8 (a, b) Aerodynamic loads and (c) rotor torque comparison.

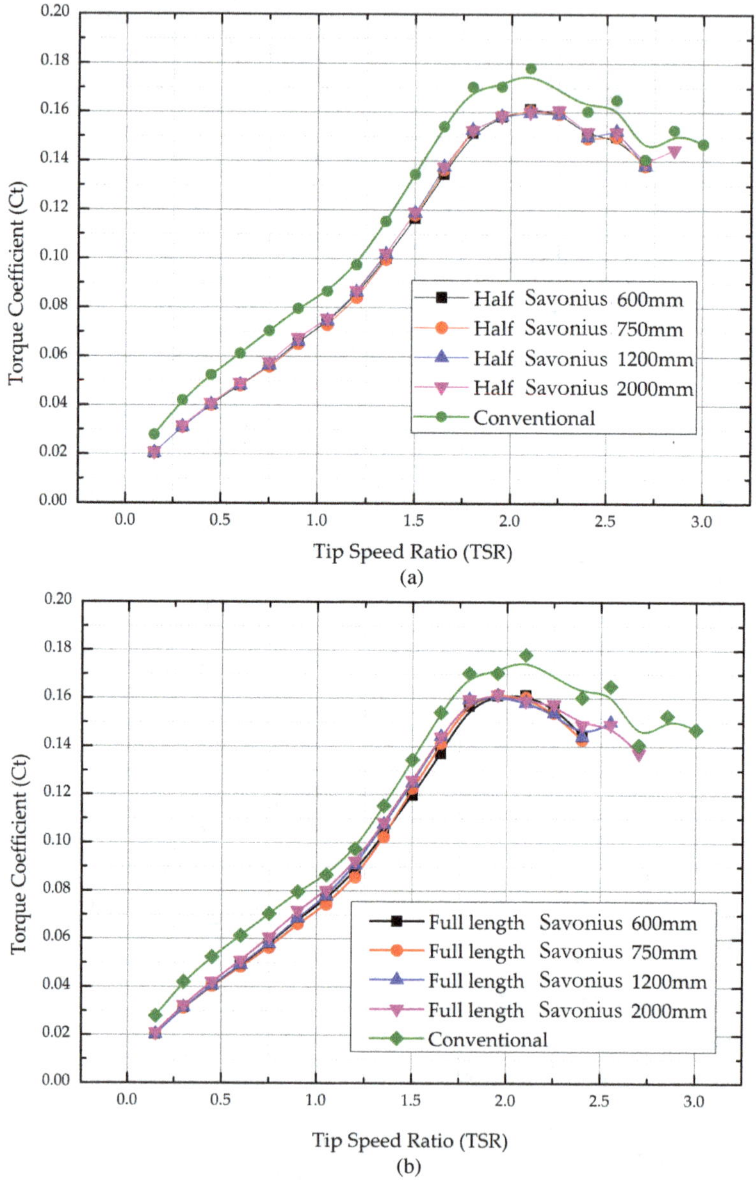

Figure 4.9 Ct vs TSR for (a) half-length and (b) full-length Savonius rotor.

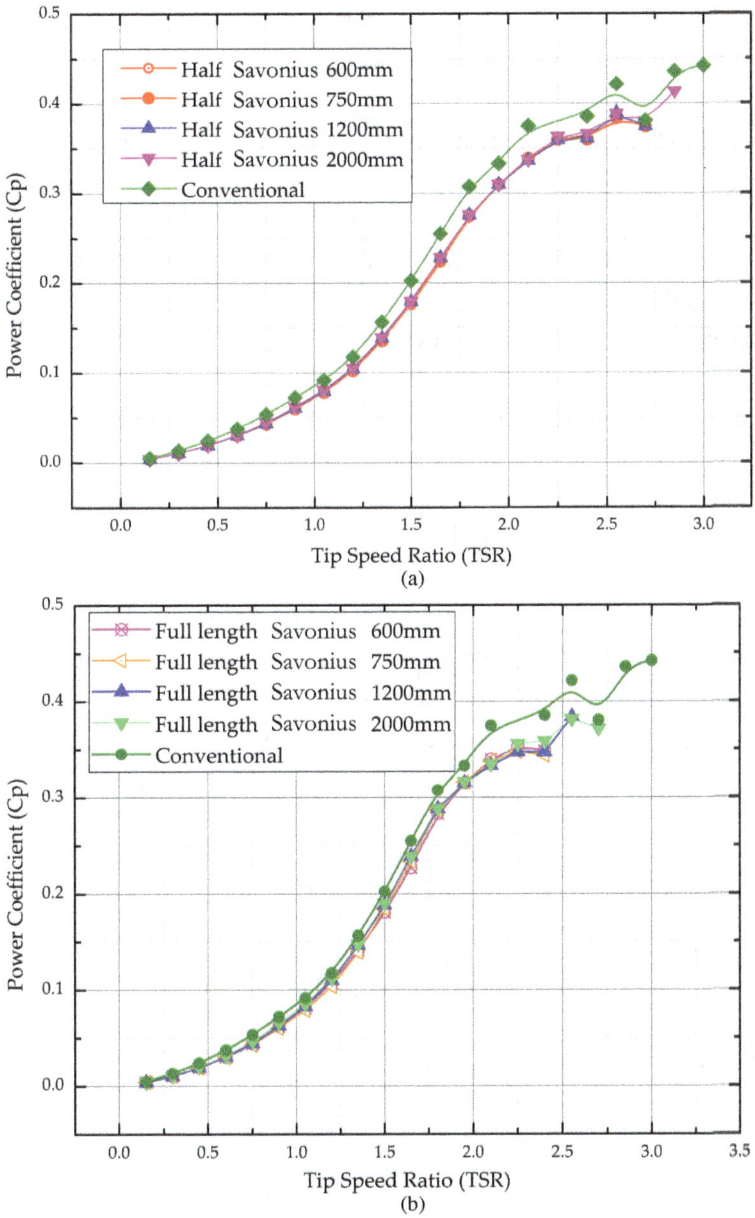

Figure 4.10 Cp vs TSR for (a) half-length and (b) full-length Savonius rotor.

the power coefficient of 5% is reported for the cylinder of 2000 mm compared to the conventional Darrieus rotor.

The peak power coefficient achieved by the conventional Darrieus rotor is 0.42 at TSR 2.5. Also, the difference between the half-length and full-length Savonius rotors is almost negligible. Hence full-length Savonius rotor is preferred aerodynamically from the starting perspective. The power coefficient is anticipated to reduce drastically as explained in past literature, but in reality, the reduction is negligible. A potential reason is strong correlation between the wake width and AoA with respect to azimuthal position. The same trend is followed in the torque coefficient prediction for the half- and full-length Savonius rotors.

The preliminary mathematical modeling sheds light on the possibility of computing the performance of AHDT analytically. The closed configuration model is by far able to predict the rotor torque and forces close to the field test values. The developed model is validated with the available field test values of a 12 kW turbine[7] as shown in Figure 4.11. It should be noted that the analytical model prediction is for a rotor, whereas the

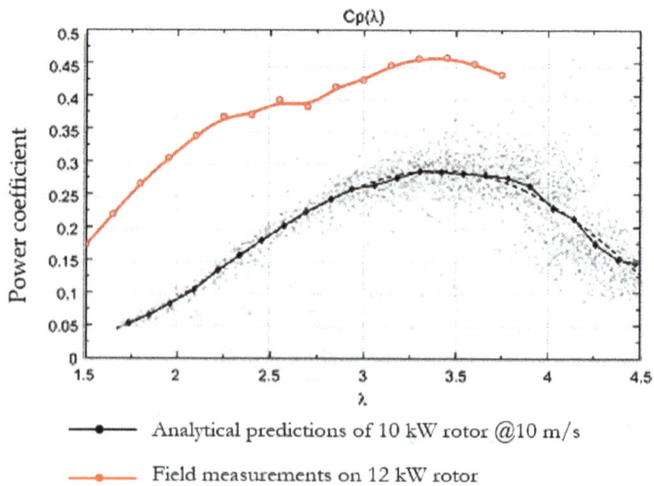

Figure 4.11 AHDT analytical model validation.

field test results are for a turbine. The rotor does not account for any drive train, generator loss or blade tip loss. These losses are significant and by including those losses in the prediction, the value may be closer to the field test results.

Chapter Nomenclature

c	Blade chord, m
CD	Drag coefficient
CFN, *CF′ N*	Upwind and downwind elemental blade normal force coefficients (respectively)
CL	Lift coefficient
CN	Normal force coefficient
$C_{P\text{–}D}$	Darrieus rotor power coefficient
CFT, *CF′ T*	Upwind and downwind blade tangential force coefficients (respectively)
CQ	Rotor torque coefficient
CT	Tangential force coefficient
f	Free vortex frequency
fdw, *fup*	Downwind and upwind functions (respectively)
FN, *F′ N*	Upwind and downwind blade normal force coefficients (respectively)
FT, *F′ T*	Upwind and downwind blade tangential force coefficients (respectively)
H	Half-height of the rotor, m
l	Blade length, m
N	Number of blades
R, R_D	Darrieus rotor radius at the equator, m
Ret	Turbine Reynolds number
Reb	Blade Reynolds number
S	Rotor swept area, m^2
s	Rotor solidity
u, *u′*	Upwind and downwind interference factors (respectively)

V, V'	Upwind and downwind induced velocities (respectively), m/s
Ve	Induced equilibrium velocity, m/s
$V\infty$	Wind velocity at the equator level, m/s
W, W'	Upwind and downwind relative inflow velocity (respectively), m/s
X, X'	Upwind and downwind local tip-speed ratio (respectively)
XEQ	Tip-speed ratio at the equator
z	Local turbine height, m
α,α'	Upwind and downwind local angle of attack, °
β	Rotor maximum diameter/height ratio
δ	Angle between the blade normal and equatorial plane, °
θ	Azimuthal angle, °
$\rho\infty$	Freestream density, kg/m^3
ω	Turbine rotational speed, s^{-1}
η	r/R, non-dimensional Cartesian coordinate
ζ	z/H, non-dimensional Cartesian coordinate
As	Savonius turbine swept area, m^2
C_{P-S}	Savonius power coefficient
r'	Savonius bucket radius, m
H_s	Savonius rotor height, m
N_s	Number of Savonius buckets
Q	Savonius turbine torque, N·m
φ	Savonius turbine tip-speed ratio
P_r	Pressure on the returning Savonius bucket, Pa
P_a	Pressure on the advancing Savonius bucket, Pa
v_{e-s}	Equivalent velocity of Savonius rotor, m/s
v_{is}	Absolute velocity of Savonius rotor, m/s
v_∞	Freestream wind speed, m/s

References

1. Fujisawa, N.; Shibuya, S. Observations of dynamic stall on Darrieus wind turbine blades. *Journal of Wind Engineering and Industrial Aerodynamics* **2001**, *89*, 201–214.
2. Nakajima, S. I.; Ikeda, T. Performance of double-step Savonius rotor for environmentally friendly hydraulic turbine. *Journal of Fluid Science and Technology* **2008**, *3*, 410–419.
3. Prandtl, L. The mechanics of viscous fluids. *Aerodynamic Theory* **1935**, *3*, 208.
4. Díaz, F.; Gavaldà, J.; Kawall, J.; Keffer, J.; Giralt, F. Vortex shedding from a spinning cylinder. *The Physics of Fluids* **1983**, *26*, 3454–3460.
5. Schlichting, H.; Gersten, K.; Krause, E.; Oertel, H. *Boundary-Layer Theory*. Springer **1955**.
6. Eckelmann, H. The structure of the viscous sublayer and the adjacent wall region in a turbulent channel flow. *Journal of Fluid Mechanics* **1974**, *65*, 439–459.
7. Kjellin, J.; Bülow, F.; Eriksson, S.; Deglaire, P.; Leijon, M.; Bernhoff, H. Power coefficient measurement on a 12 kW straight bladed vertical axis wind turbine. *Renewable Energy* **2011**, *36*, 3050–3053.

Chapter 5

Computational Study of Adaptive Hybrid Darrieus Turbine*

* Parts of this chapter are adapted with permission from Mohan Kumar, P.; Surya, M. R.; Sivalingam, K.; Lim, T.-C.; Ramakrishna, S.; Wei, H. Computational optimization of adaptive hybrid Darrieus turbine: part 1. *Fluids* **2019**, *4*, 90.

5.1 Mathematical Formulation of the Computational Fluid Dynamics

5.1.1 *Reynolds-Averaged Navier Stokes Model*

The Reynolds-Averaged Navier Stokes (RANS) method is the fastest CFD approach widely employed to solve flow problems in wind turbines. A new quantity known as Reynolds stress is introduced as the RANS method utilizes time-averaged motion. The RANS models apply eddy-viscosity model or Reynolds stress model to compute this Reynolds stress variable. Equation models such as k-epsilon (k-€), k-omega (k-$\dot{\omega}$) and k-omega SST are most widely used for wind turbines, where k is the turbulence kinetic energy per unit mass, epsilon is the specific dissipation and omega is the specific dissipation rate. The Boussinesq hypothesis is the basis for eddy-viscosity models and concerns the implication of this model. For abrupt flow changes, the Boussineq hypothesis becomes invalid.

However, the implementation of the RANS method is applicable by eradicating the dependency on the Boussineq hypothesis. This can be achieved by using the Reynolds stress method. But still, this approach leads to six more equations which are to be solved, making it more complex and having assumptions similar to the Boussineq hypothesis. As mentioned before, the RANS equations are very complex to solve directly due to their non-linear nature. This is due to the presence of convective acceleration terms. In order to approximate these equations, several numerical methods are available.

$$\rho \bar{u}_i \frac{\partial \bar{u}_i}{\partial x_j} = \rho \bar{f}_i + \frac{\partial}{\partial x_i}\left[-\bar{p}\delta_{ij} + \mu\left(\frac{\partial \bar{u}_i}{\partial x_j} + \frac{\partial \bar{u}_j}{\partial x_i}\right) - \rho \overline{u_I' u_J'}\right] \tag{5.1}$$

The mean velocity fluctuations in the RANS model are calculated for the three Cartesian coordinates as mentioned in Eq. (5.2) for x as instantaneous velocity is the sum of mean velocity and fluctuations.

$$\rho\left(\frac{\partial \overline{u}}{\partial t}+\overline{u}\frac{\partial \overline{u}}{\partial x}+\overline{v}\frac{\partial \overline{u}}{\partial y}+\overline{w}\frac{\partial \overline{u}}{\partial z}\right)=-\frac{\partial \overline{p}}{\partial x}+\mu\left(\frac{\partial^2 \overline{u}}{\partial x^2}+\frac{\partial^2 \overline{u}}{\partial y^2}+\frac{\partial^2 \overline{u}}{\partial z^2}\right)$$
$$-\rho\left(\frac{\partial \overline{u'u'}}{\partial x}+\frac{\partial \overline{u'v'}}{\partial y}+\frac{\partial \overline{u'vw'}}{\partial z}\right)+S_x \tag{5.2}$$

5.1.2 *Turbulence Model*

The k-$\acute{\omega}$ SST model is a blend of k-$\acute{\omega}$ and k-€ models. It is the upgraded form of the baseline (BSL) model which varies linearly between k-$\acute{\omega}$ and k-€ models. This model makes use of k-$\acute{\omega}$ definition in regions where the boundary layer predominates and k-€ definition in regions outside the boundary layer. The transformed k-$\acute{\omega}$ formulation from k-€ model is given by

$$\rho\frac{\partial k}{\partial t}+\rho u_j\frac{\partial k}{\partial x_j}=\tau_{ij}\frac{\partial u_j}{\partial x_j}-\beta_1^{sst}\rho k\omega+\frac{\partial}{\partial x_j}\left[\left(\mu+\sigma_1^{sst}\mu_t\right)\frac{\partial k}{\partial x_j}\right] \tag{5.3}$$

$$\rho\frac{\partial \omega}{\partial t}+\rho u_j\frac{\partial \omega}{\partial x_j}=\alpha^{sst}\frac{\omega}{k}\tau_{ij}\frac{\partial u_j}{\partial x_j}-\beta_2^{sst}\rho\omega^2+\frac{\partial}{\partial x_j}\left[\left(\mu+\sigma_2^{sst}\mu_t\right)\frac{\partial \omega}{\partial x_j}\right]$$
$$+2\rho\left(1-F_1\right)\sigma_2^{sst}\frac{1}{\omega}\frac{\partial k}{\partial x_j}\frac{\partial \omega}{\partial x_j} \tag{5.4}$$

The newly transformed equations due to the combination of the two models are represented by

$$\phi_{sst}=F_1\phi_{\omega}+\left(1-F_1\right)\phi_{\epsilon} \tag{5.5}$$

The blending function F_1 switches between previously defined two methods in desired regions. It is formulated in such a way that the near wake region makes use of the k-$\acute{\omega}$ model and the free shear layer makes use of the k-€ model. Eq. (5.6) mentions the F_1 term and represents the blending function which makes use of both fluid and special terms,

$$F_1 = \tan h\left(\left(\min\left(\max\left(\frac{\sqrt{k}}{0.09\omega y_s};\frac{500v}{y_s^2\omega}\right);\frac{4\rho\sigma_2^{\omega\epsilon}}{CD_{k\epsilon}y_s^2}\right)^4\right)\right) \tag{5.6}$$

$$CD_{k\omega} = \max\left(2\rho\sigma_2^{\omega\epsilon}\frac{1}{x}\frac{\partial k}{\partial x_j}\frac{\partial \omega}{\partial x_j}, 10^{-10}\right) \tag{5.7}$$

The three factors mentioned in the above Eq. (5.7) represent viscous sub-layer, turbulent length scale and small freestream values, respectively. One of the major advantages of the k-ώ SST model in comparison with the BSL model is that the turbulent shear stress is accounted for by limiting eddy viscosity. The turbulent shear stress is assumed to be proportional to specific turbulent kinetic energy in logarithmic and wake regions of the turbulent boundary layer.

$$\mu_T = \frac{\rho a_1 k}{\max(a_1\omega, SF_2)} \tag{5.8}$$

$$F_2 = \tan h\left(\left(\max\left(\frac{\sqrt{k}}{0.09\omega y_s};\frac{500v}{y_s^2\omega}\right)\right)^2\right) \tag{5.9}$$

The major drawback of eddy viscosity and the Boussinesq hypothesis is that these models assume isotropic turbulence, which would end up generating unrealistic results. In such cases, k-ώ SST model is used.

5.2 Computational Domain and Meshing

It is vital to consider a larger domain size that represents the turbine operating in field conditions and to avoid the blockage. The computational domain is of rectangular size 20 D × 25 D, where D corresponds to the maximum rotor diameter (0.4 m in current study) and rotor is placed at 10 D from the inlet boundary condition. Additionally, the minimum domain size in all directions should be at least 20 times the rotor radius. The exit boundary condition is placed at a distance of 15 D downstream with respect to the rotor in order to allow a complete development of the wake structure.

The inlet boundary has been set as velocity inlet, while outlet is set to pressure outlet. The wind speed at the inlet varies from 4 m/s to 9 m/s. The pressure outlet has been assigned to atmospheric pressure value. For the two side boundaries, symmetric boundary conditions have been used. To maintain the flow continuity in the field, the circular edge in the domain is set as an interface. In order to maintain the mesh linkage between the domain and the turbine, the circumferential edge of the turbine is also set as an interface. The rotating zone revolves at the same angular speed of the rotor, while the domain region is stationary. Since the flow is over a rotating solid, the sliding mesh technique has been used. Figure 5.3 shows the main geometric features and the CFD boundary conditions of the computational domain. A 2D unstructured mesh (triangular type) has been generated in the domain as well as in the rotating zone using ANSYS meshing tool as shown in Figure 5.4. Much attention has been paid to the near wall treatment of the blades by generating cells of smaller sizes with an edge size setting of 0.1 mm. From the blade walls, the mesh cells grow evenly from smaller size topology to higher ones and thus adopt appropriate cell size prevailing in the rotating zone. In the vicinity of the blade, structured mesh has been generated with $y^+ < 1$ to capture the flow near the blades accurately.

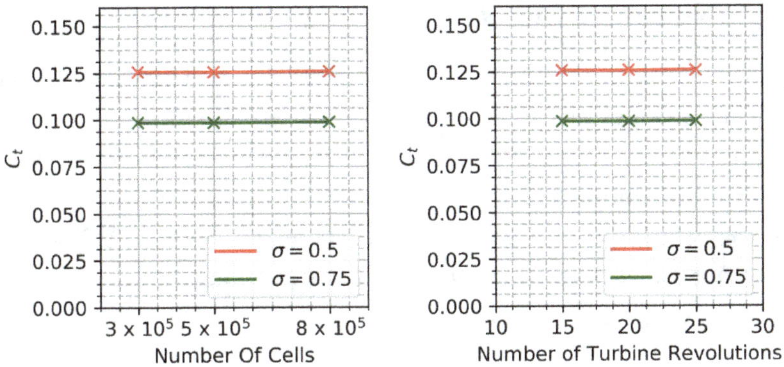

Figure 5.1 Grid independence and time step size.

Considering the flow as incompressible, a global convergence criterion is set when the instantaneous torque coefficient (Ct) values show less than 1% deviation from the previous value. The conventional Darrieus turbine with solidity of 0.5 and 0.75 is subjected to grid independence test as shown in Figure 5.1. The mean value of the instantaneous Ct obtained in the last revolution of the turbine is plotted in Figure 5.1. The mean values are obtained for 0.3, 0.5 and 0.8 million cells. It is found that the deviation in the Ct is less than 1% for the investigated number of cells. Considering the computational cost and time, 0.5 million cells is found to be optimal and it is retained for all further simulations in this study. The variation in the Ct is studied by using different time steps in search of an optimal time step that predicts the Ct within 1% deviation. The turbine is simulated for 20 complete revolutions with a time step size corresponding to 1° increment per time step. CFD solution is obtained by solving the RANS equations using the SIMPLE algorithm. Completed mesh for D_R/D_T ratios of 20, 3.5, 3, 2.5, 2 and 1.5 are shown in Figure 5.4a, b, c, d, e and f, respectively.

5.2.1 *Numerical Model Validation*

In order to validate the set numerical model, the predictions have to be compared with experimental results to have an estimate on the errors. An experimental result of a similar sized rotor obtained from the wind tunnel test is compared with the CFD predictions and found to have a good agreement between them. The plot as shown in Figure 5.2 depicts the comparison between experimental result, previously developed CFD model[1] and the current model.

Figure 5.3(a) shows the boundary conditions and computational domain of the simulation. Figure 5.3(b) and (c) show the mesh structure of two-bladed and three-bladed AHDT, respectively. Inflation layers of the prismatic type were used for the blade surface together with unstructured meshes for the computational domain to get an improved resolution of the flow in boundary layers and to get the wall distance y^+ to be less than one. Figure 5.3(d) and (e) show the unstructured mesh with prismatic layers around airfoil and Savonius blade, respectively.

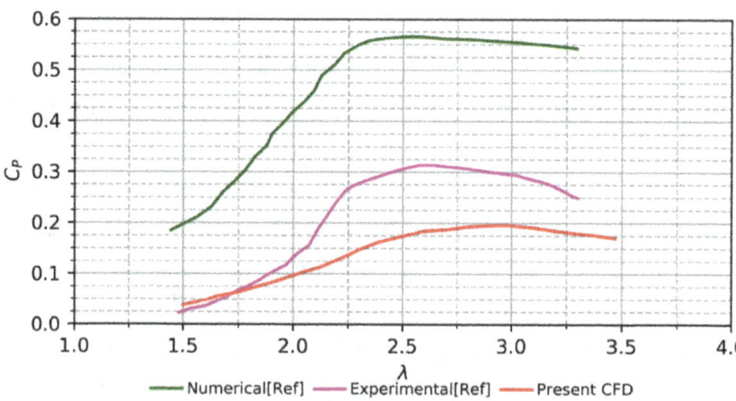

Figure 5.2 AHDT CFD model validation.

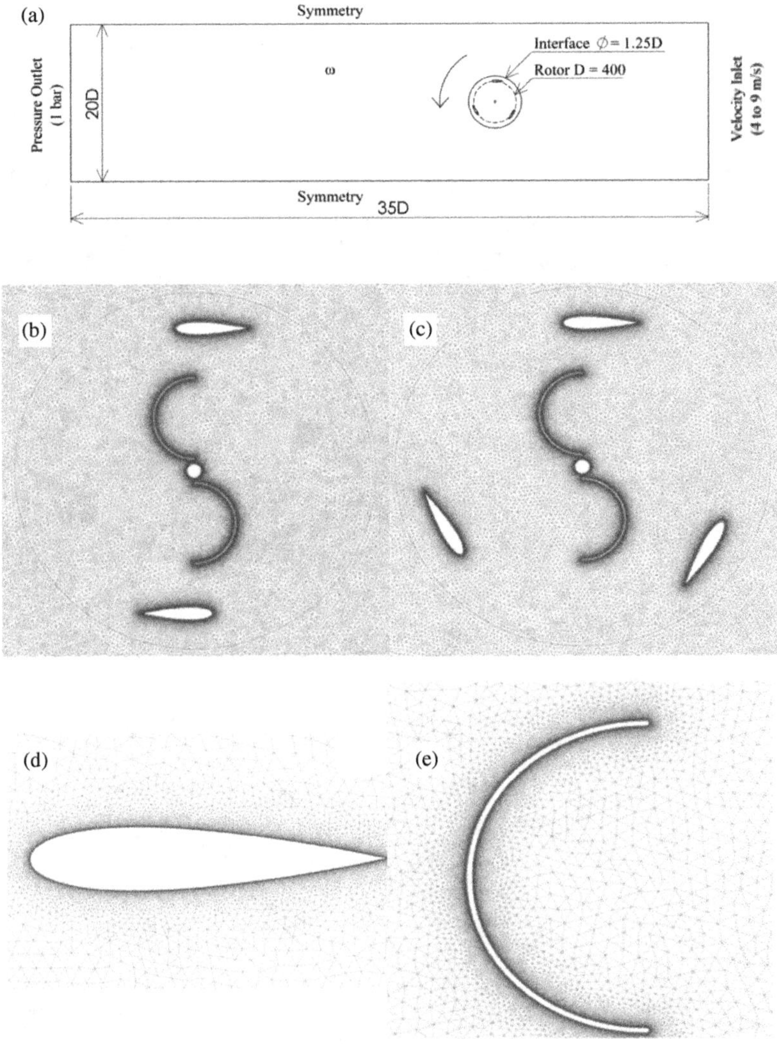

Figure 5.3 (a) Computational domain of AHDT. (b) Mesh structure of two-bladed AHDT. (c) Mesh structure of three-bladed ADHT. (d) Unstructured mesh around airfoil. (e) Unstructured mesh around Savonius blade.

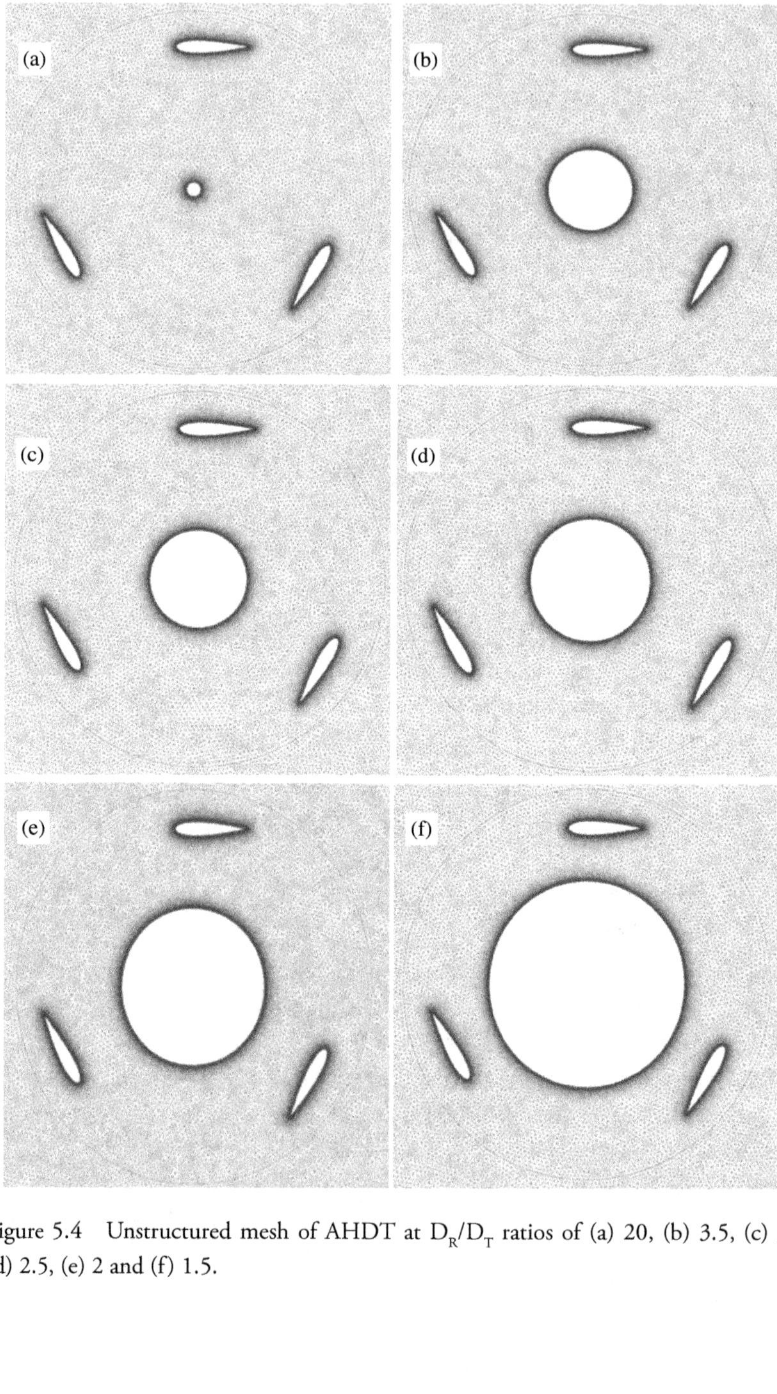

Figure 5.4 Unstructured mesh of AHDT at D_R/D_T ratios of (a) 20, (b) 3.5, (c) 3, (d) 2.5, (e) 2 and (f) 1.5.

5.3 Discussion on Computational Results

5.3.1 *Torque Coefficient Comparison for Different D_R/D_T*

CFD simulations for different Darrieus to closed Savonius rotor diameter ratio (D_R/D_T) are performed to find out the optimum rotor diameters. The AHDT in closed configuration is simulated for five complete revolutions, instantaneously recording the Ct of the blades for every degree of rotation. However, the solution is converged after three revolutions of the turbine. The Ct is computed from the CFD simulations. The coefficient of power (Cp) is calculated from Ct as shown in Eq. (5.10).

$$C_p = C_t \times TSR \tag{5.10}$$

The operating TSR range (λ) considered in the present work is 1 to 4 for two-bladed and 1 to 3 for three-bladed at an increment of 0.2. Computing at a TSR below the mentioned range carries no significance for this optimization study and above the range means operating beyond stall conditions. For high solidities, power and efficiency rapidly decrease at stall conditions.

The freestream wind speed for which the AHDT configurations are simulated is between 4 m/s and 9 m/s and converted to Re for further descriptions in this study. The instantaneous C_t versus azimuth angle is computed for a range of Re = 1.2×10^5 to 2.7×10^5. The results are plotted in Figures 5.5 and 5.6 for two- and three-bladed rotors, respectively. For each Re the various D_R/D_T ratios are evaluated for one complete revolution (0° to 360°). It shows that as D_R/D_T ratio increases, the peak Ct in the cycle increases. Table 5.1 shows the difference in peak Ct between conventional Darrieus and AHDT.

For the solidity of $\sigma = 0.5$ corresponding to the two-bladed rotor, Ct reaches its peak at two different azimuth angles for the investigated Re. This is due to the individual contribution of the blades to the rotor Ct.

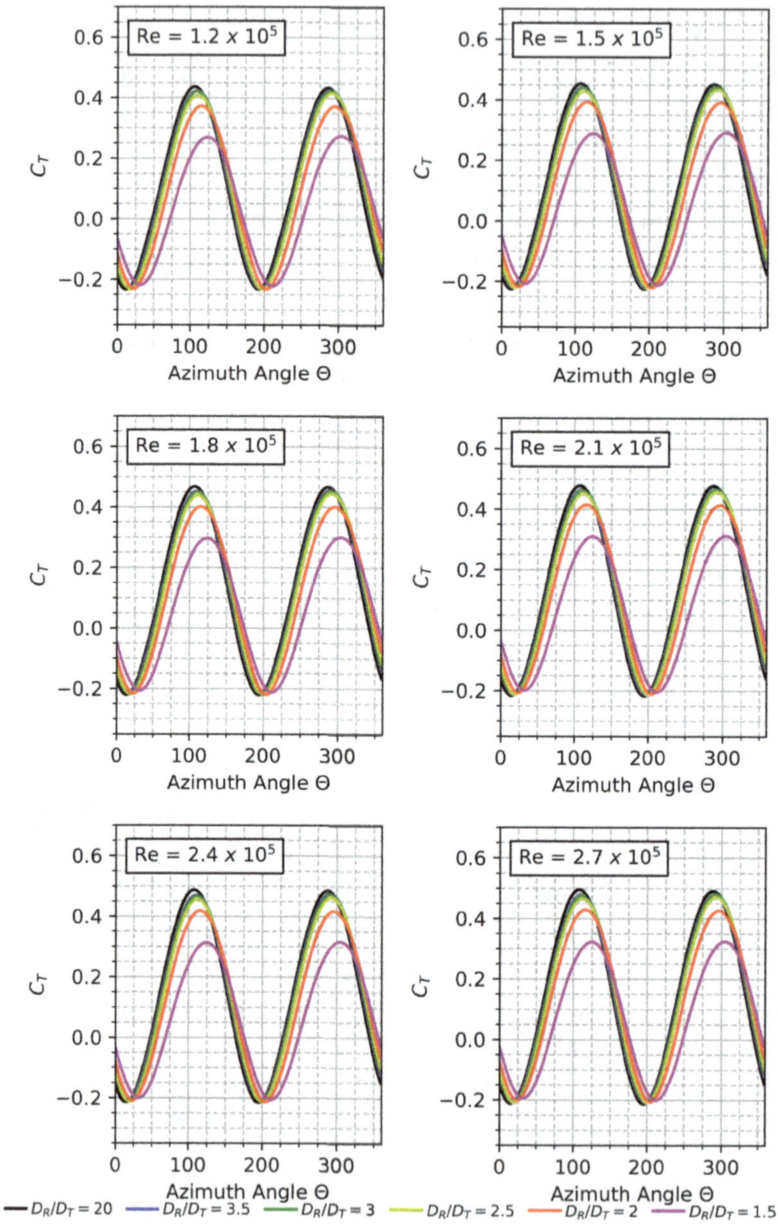

Figure 5.5 Effect of Re on torque coefficient (two-bladed).

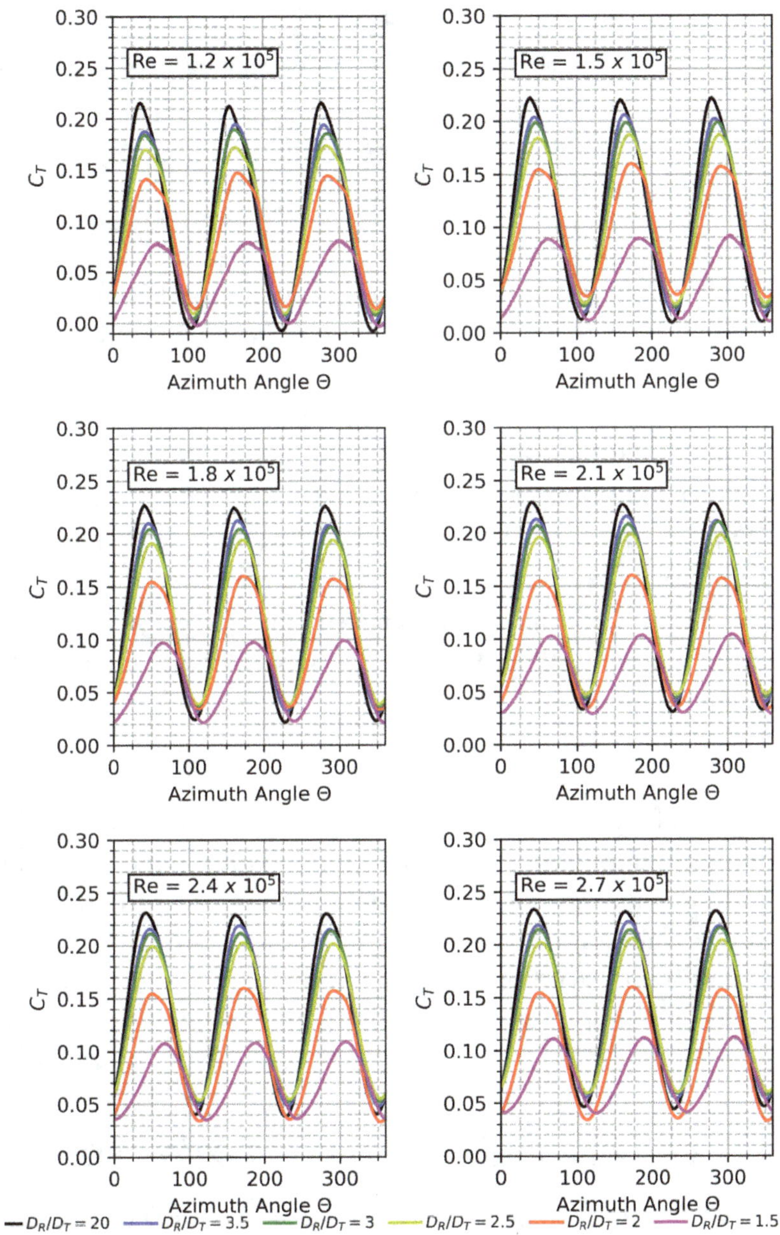

Figure 5.6 Effect of Re on torque coefficient (three-bladed).

Table 5.1 Peak Ct at different D_R/D_T ratios.

D_R/D_T	Peak C_T (two-bladed)	Δ (%)	Peak C_T (three-bladed)	Δ (%)
20	0.436	—	0.238	—
3.5	0.425	2.66	0.206	13.16
3	0.424	2.74	0.198	16.7
2.5	0.415	4.76	0.18	24.27
2	0.373	14.58	0.152	36.18
1.5	0.273	37.33	0.0928	61

$D_R/D_T = 20$ is considered as conventional Darrieus rotor with shaft in this study. The difference in the peak Ct between the Darrieus rotor and the AHDT rotor with D_R/D_T ratios of 2.5, 3 and 3.5 is less than 5%. For $D_R/D_T = 2$ it is 14.58% and for $D_R/D_T = 1.5$ it is 37.33%. It is apparent that for two-bladed AHDT rotor, the peak Ct decreases rapidly for 2 and 2.5 than the rest of the diameters. The percentage variation of the peak Ct between different D_R/D_T ratios is almost the same at all the Re employed in this study.

For the solidity of $\sigma = 0.75$, the peak C_t is attained three times in one revolution of the turbine. Table 5.1 shows the peak C_t comparison for the two- and three-bladed rotors at Re = 1.2×10^5 as a similar trend is followed for other investigated Reynolds numbers. The percentage difference of peak Ct between AHDT with $D_R/D_T = 3.5$ is 13.16% lower than the conventional Darrieus rotor. For $D_R/D_T = 3$, 2.5, 2 and 1.5, the percentage drop in peak C_t is 16.70%, 24.27%, 36.18% and 61%, respectively.

5.3.2 *Power Coefficient Comparison for Different D_R/D_T*

Figures 5.7 and 5.8 show the power curves of the AHDT with rotor solidity $\sigma = 0.5$ and $\sigma = 0.75$. Each subplot shows the power curve (C_p vs λ) for their respective diameter ratio (D_R/D_T). The power curves

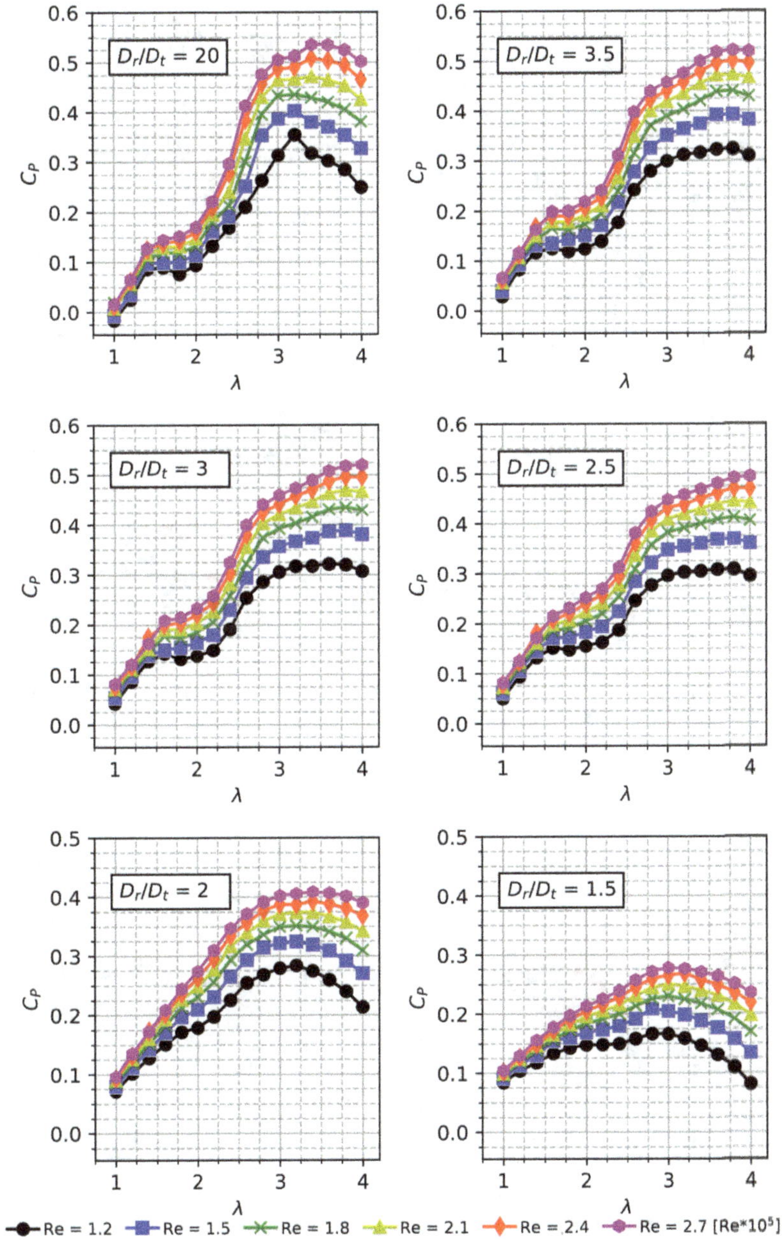

Figure 5.7 Cp vs TSR for two-bladed ($\sigma = 0.5$) for different D_R/D_T.

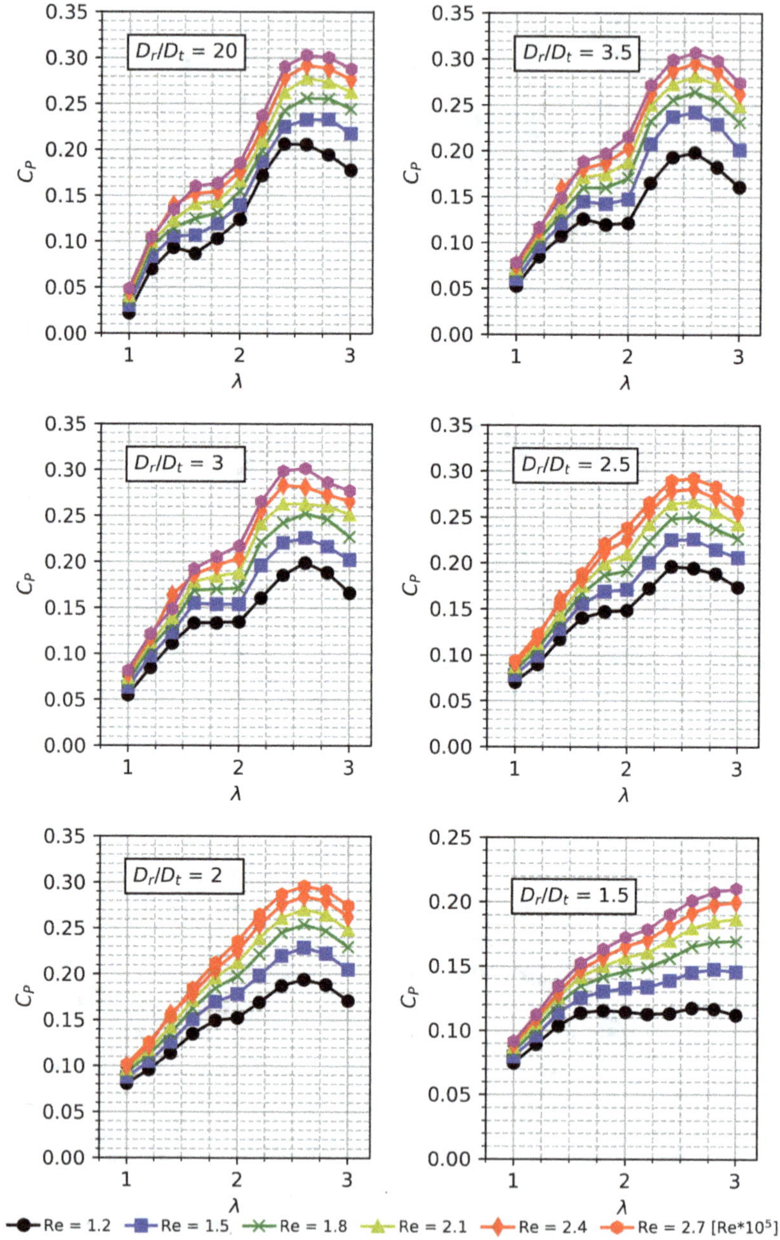

Figure 5.8 Cp vs TSR for three-bladed ($\sigma = 0.75$) for different D_R/D_T.

Table 5.2 Maximum power coefficient at different D_R/D_T ($\sigma = 0.5$).

Re × 10^5	D_R/D_T = 20 C_{pmax}	Δ%	3.5 C_{pmax}	Δ%	3 C_{pmax}	Δ%	2.5 C_{pmax}	Δ%	2 C_{pmax}	Δ%	1.5 C_{pmax}	Δ%
1.2	0.354		0.324		0.321		0.308		0.241		0.109	
1.5	0.401	11.7	0.391	17.1	0.387	17.3	0.368	16.3	0.291	17.2	0.157	30.6
1.8	0.434	18.4	0.438	26.0	0.433	26.1	0.411	25.1	0.328	26.8	0.191	42.9
2.1	0.471	24.6	0.472	31.3	0.468	36.6	0.444	30.6	0.356	32.6	0.216	49.5
2.4	0.506	30.0	0.499	35.1	0.494	35.2	0.470	34.5	0.367	34.6	0.238	54.2

Table 5.3 Maximum power coefficient at different D_R/D_T ($\sigma = 0.75$).

Re × 10^5	D_R/D_T = 20 C_{pmax}	Δ%	3.5 C_{pmax}	Δ%	3 C_{pmax}	Δ%	2.5 C_{pmax}	Δ%	2 C_{pmax}	Δ%	1.5 C_{pmax}	Δ%
1.2	0.205		0.198		0.199		0.194		0.187		0.117	
1.5	0.233	12.1	0.242	18.2	0.226	11.9	0.218	11	0.221	15.4	0.145	19.3
1.8	0.256	19.9	0.264	25	0.252	21	0.251	22.7	0.248	24.5	0.165	29.1
2.1	0.278	26.3	0.282	29.8	0.263	24.3	0.267	27.3	0.254	26.3	0.18	35
2.4	0.292	29.8	0.296	33.1	0.281	29.2	0.278	30.21	0.262	28.6	0.192	39.1

are calculated for a range of Re = 1.2×10^5 to 2.7×10^5. Tables 5.2 and 5.3 show the comparison of maximum power coefficient (C_{pmax}) with investigated Re and D_R/D_T ratio.

5.3.2.1 *Power Coefficient Comparison for Rotor Solidity* ($\sigma = 0.5$)

As mentioned before, D_R/D_T = 20 corresponds to the conventional Darrieus rotor. Figure 5.7(a) shows the Cp curve at different Re ranging from Re = 1.2×10^5 to 2.7×10^5. It can be seen from Figure 5.7(a) that at low Re of 1.2, 1.5 and 1.8 × 10^5, the optimal TSR is 3.2. But for higher Re of 2.1, 2.4 and 2.7 × 10^5, the optimal TSR shifts to 3.4. C_{pmax}

increases by 11.7% at $Re = 1.5 \times 10^5$ when compared to $Re = 1.2 \times 10^5$. At Re = 1.8, 2.1 and 2.4×10^5, the percentage increase in C_{pmax} is 18.4%, 24.6% and 30% higher, respectively, than the C_{pmax} at $Re = 1.2 \times 10^5$. At $Re = 2.7 \times 10^5$, the C_{pmax} value is 0.533, which is 33.6% higher than the C_{pmax} obtained at the lowest Re studied in this work.

5.3.2.2 *Power Coefficient Comparison for Rotor Solidity ($\sigma = 0.75$)*

The effect of D_R/D_T on the Darrieus rotor is studied by increasing the solidity to 0.75 via increasing the number of blades from two to three. Figure 5.8 shows the C_p vs λ curves at different D_R/D_T ratios for the range of $Re = 2.1 \times 10^5$ and 2.7×10^5. C_{pmax} at different D_R/D_T ratios for the investigated Re are shown in Table 5.3. Figure 5.8(a) shows the power curves $D_R/D_T = 20$ which can be considered as a conventional Darrieus rotor with center cylinder of 20 mm diameter and with three blades.

The optimal TSR is 2.6 at which C_{pmax} obtained is 0.3 for $Re = 2.7 \times 10^5$. For $D_R/D_T = 3.5$, at $Re = 1.2 \times 10^5$, C_{pmax} achieved is 0.198 at TSR of 2.6. At 2.7×10^5 the measured C_{pmax} is 35.5% higher compared to $Re = 1.2 \times 10^5$. C_{pmax} attained at the lowest Re used in this study is 0.187. For $D_R/D_T = 2$, as Re increases from 1.2 to 1.5, 1.8 and 2.1×10^5, C_{pmax} increases by 15.4%, 24.5% and 28.6%, respectively. As expected, C_{pmax} is reached at Re of 2.7×10^5 which is 34.2% higher than C_{pmax} obtained at $Re = 1.2 \times 10^5$.

5.3.3 *Comparison of D_R/D_T for Various Re*

The impact of closed Savonius rotor diameters on the performance of the Darrieus rotor can be understood by comparing all D_R/D_T ratios at each wind speed. In this way, the ratio that performs better for the given wind speed can be obtained. The results are plotted in Figures 5.9 and 5.10 for two- and three-bladed rotors.

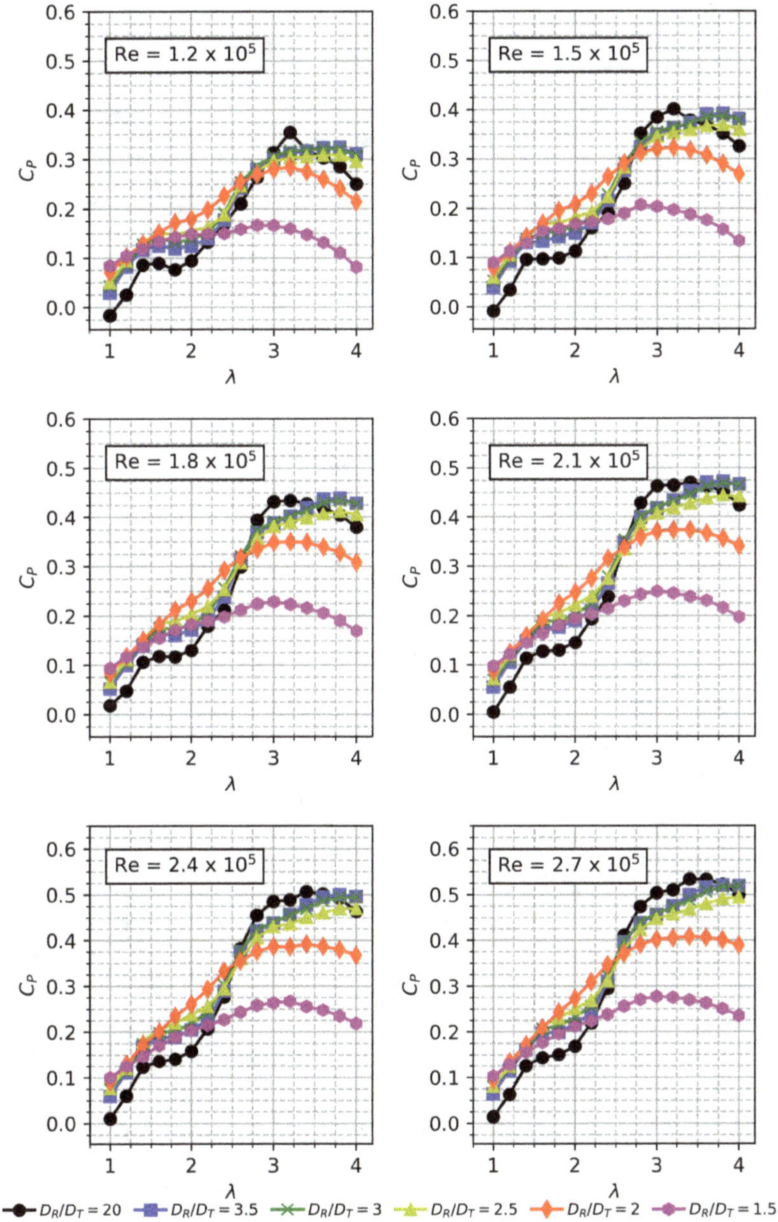

Figure 5.9 Cp vs TSR for two-bladed ($\sigma = 0.5$) for different Re.

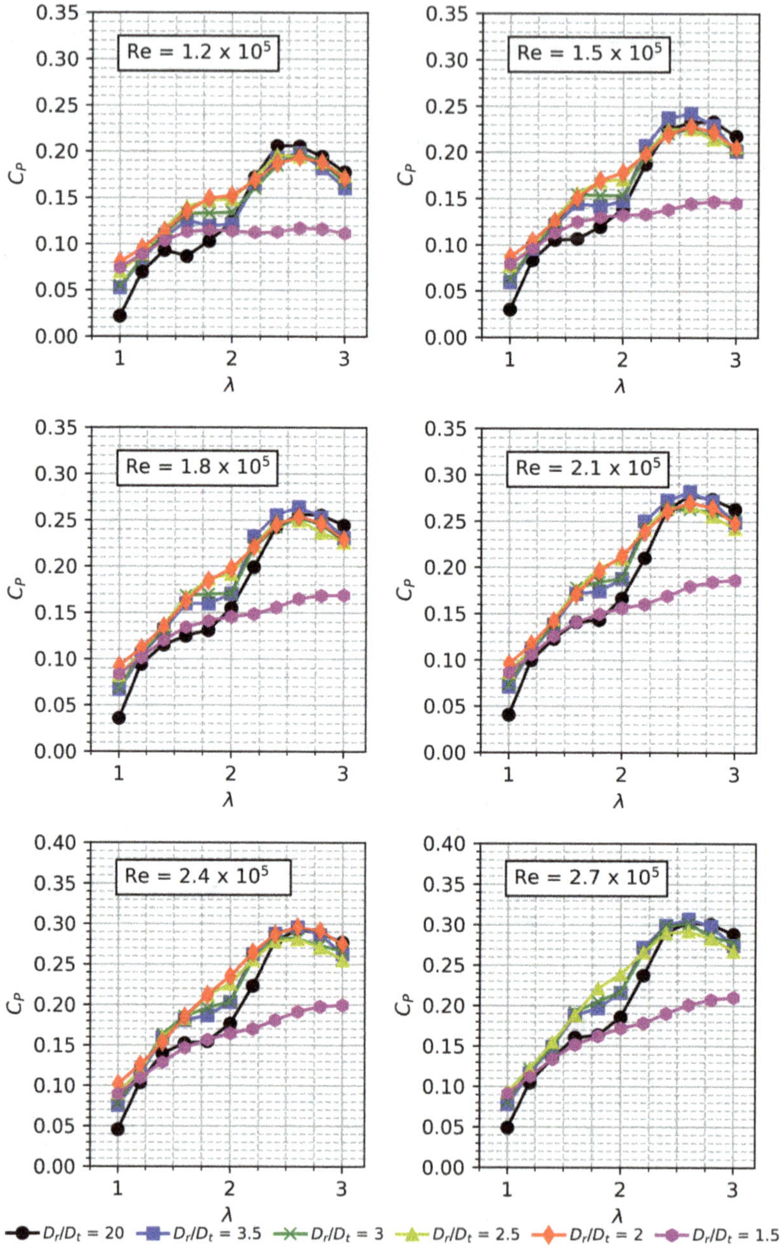

Figure 5.10 Cp vs TSR for two-bladed ($\sigma = 0.75$) for different Re.

5.3.3.1 At Rotor Solidity of $\sigma = 0.5$

For a two-bladed rotor, the obtained peak Cp with respect to the conventional Darrieus rotor are tabulated in Table 5.2 . For Re = 1.2×10^5, the percentage loss of AHDT in C_{pmax} is 8.5%, 9.6% and 13% when D_R/D_T is 3.5, 3 and 2.5, respectively, compared to the Darrieus rotor and the percentage loss increases drastically and reaches 69.2% at the diameter ratio of 1.5. At Re = 1.5×10^5, the conventional Darrieus rotor achieves C_{pmax} of 0.40 and the percentage loss is 2.5%, 3.5% and 8.2% at the diameter ratio of 3.5, 3 and 2, respectively, and can be as high as 60.8% when D_R/D_T is 1.5. For Re = 1.8×10^5, compared to the conventional diameter ratio of 20, the power coefficient loss is 1.4%, 2.8% and 5.3% when the D_R/D_T ratio is 3.5, 3 and 2.5, respectively. Further increase in $D_R/D_T = 1.5$ and the loss increases to 56%. For Re = 2.1×10^5, compared to the conventional diameter ratio of 20 with Cp of 0.476, the loss in diameter ratio of 2 is reported as 25.2%. At Re = 2.4×10^5, when the Savonius diameter is about half of the Darrieus diameter, the power loss percentage is 27.5%. The highest loss percentage in C_{pmax} is about 52.9%. At highest wind speed of 9 m/s, the conventional Darrieus rotor reaches maximum Cp of 0.53, whereas a 53.8% loss is reported for the diameter ratio of 1.5.

5.3.3.2 At Rotor Solidity of $\sigma = 0.75$

Figure 5.10 shows the characteristic power curves of AHDT of solidity $\sigma = 0.75$ with various D_R/D_T ratios. Table 5.3 compares C_{pmax} for the computed Re. At Re = 1.2×10^5, C_{pmax} obtained for conventional Darrieus is 0.205, and as the closed Savonius rotor diameter increases to 3.5, 3 and 2.5, C_{pmax} obtained is 0.198, 0.196 and 0.194, respectively. At Re = 1.5×10^5, the power loss in the AHDT at $D_R/D_T = 1.5$ is 41.5% compared to conventional Darrieus rotor Cp of 0.248. The Cp loss for Re = 1.8×10^5 also exhibits a similar trend with 5% variation between diameter ratios except for 1.5. At Re = 2.1×10^5, the conventional diameter ratio of 20 reaches C_{pmax} of 0.282. For Re = 2.4 × 10^5, the percentage loss in C_{pmax} is 1.4%, 5.1% and 6.1%.

5.3.4 *Discussion on Flow Physics of Adaptive Hybrid Darrieus Turbines*

When the air flows through the AHDT rotor, the flow adheres to the Savonius rotor surface. The flow around the closed Savonius rotor is similar to that around a rotating cylinder. The attached air film is forced towards the direction in which the cylinder rotates. When a rotating cylinder is placed in a fluid, it would create an alternating, vortex-like flow around the cylinder (Von Karman vortices).[2] The induced flow assists the freestream flow over the top of the cylinder surface whereas the induced flow opposes the freestream flow below the cylinder surface. If the cylinder is stationary, the flow will be axisymmetric, whereas wake will be pushed to the direction of rotation if the cylinder rpm is sufficiently higher. As a result, the velocity and pressure fields around the cylinder vary accordingly. This pressure field around the cylinder interacts with the pressure field created by the airfoils when the cylinder diameter is increased. This interaction affects the lift characteristics of Darrieus blades thereby affecting the moment produced by the rotating blades on the upwind side of the rotation. On the downwind side of the rotor, the blades interact with the vortices produced by the rotating cylinder as well as the blades from the upwind side. The vortex strength is proportional to the cylinder diameter and these high-strength vortices set earlier stall in the Darrieus blades and reduce lift due to early separation, leading to significant power loss in the Darrieus rotor.

Figure 5.11 shows the instantaneous C_t of single blade versus azimuthal angle at 5th rotation for various diameter ratios ranging from 5% to 66.25%. It shows the C_t values of blade I in both turbines with a solidity of 0.5 and 0.75. It can be seen that, for both turbines with different solidity, the increasing diameter ratio decreases the Ct of the blade. This

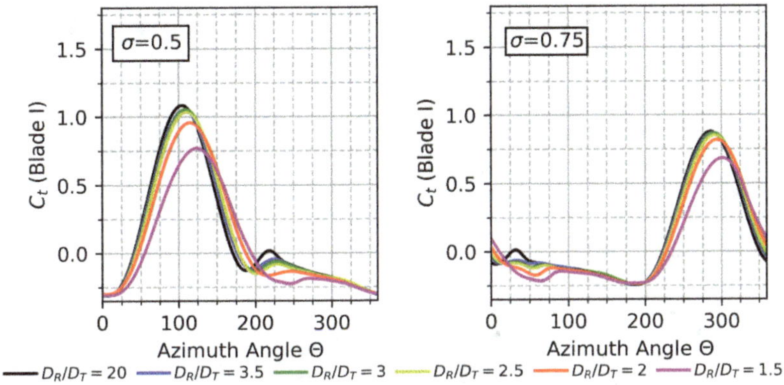

Figure 5.11 Ct variation of a single blade for two- and three-bladed.

effect can be seen on the upwind and downwind sides of the rotation. For a rotor solidity of 0.5, the peak Ct is achieved for azimuth angle in the upwind region between 0° and 180° and for downwind side between 180° and 360°. For the same blade at rotor solidity of 0.75, peak Ct is achieved for azimuth angle between 180° and 360°. This variation in the azimuth angle with rotor solidity is due to position of the blades. For the diameter ratios of 50% ($D_R/D_T = 2$) and 66.25% ($D_R/D_T = 1.5$), the instantaneous C_t decreases drastically in the upwind half. This variation of C_t with the diameter ratios on the upwind half is similar for both solidities of 0.5 and 0.75. The pressure field which is formed around the airfoil is disturbed by the pressure field created by the closed Savonius rotor (cylinder). If the cylinder-to-turbine ratio is below 37.5% ($D_R/D_T = 2.5$), the interaction of the pressure field created by the cylinder with the airfoil pressure field is minimal. As the D_R/D_T increases to 50% and 66.25%, the area between the cylinder surface and the airfoil blades decreases further. At this stage, the pressure field interaction with the airfoil is sufficiently strong that it affects the lift characteristics of the airfoil, thereby reducing the moment and power as shown in Figure 5.12. This pressure field interaction can be seen from the pressure

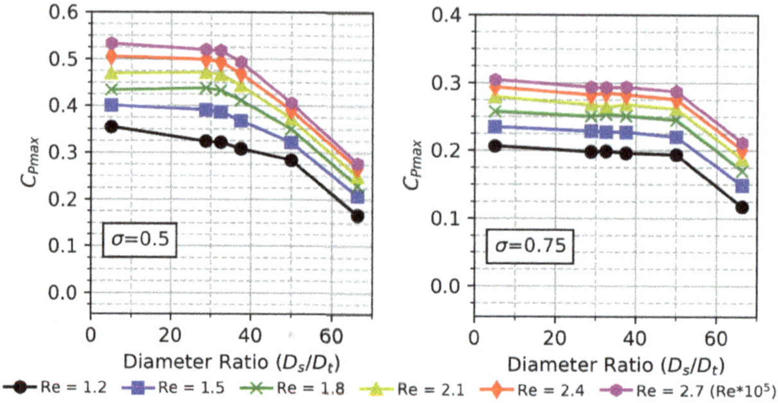

Figure 5.12 Variation of Cpmax for different Re.

contours shown in Figures 5.13, 5.15, 5.17, and 5.19. For the rotor solidities of 0.5 and 0.75, the increasing diameter ratio affects the instantaneous C_t on the downwind half of the revolution. The formation of the vortices for different diameter ratios are shown in Figures 5.14, 5.16, 5.18, and 5.20.

The numerical analysis on AHDT in open and closed configurations is able to quantify the Cp loss of Darrieus rotor for various diameters of Savonius rotors in open and closed conditions. Some of the notable conclusions that can be arrived at based on the computational study are as follows:

1. $D_R/D_T > 5$ is of less significance from the startup perspective, though the Cp loss is minimal.
2. $D_R/D_T < 2$ is of no significance from the performance perspective as it severely degrades the Cp of Darrieus rotor.
3. The optimum diametrical ratio of AHDT lies between $D_R/D_T > 5$ and $D_R/D_T < 2$.

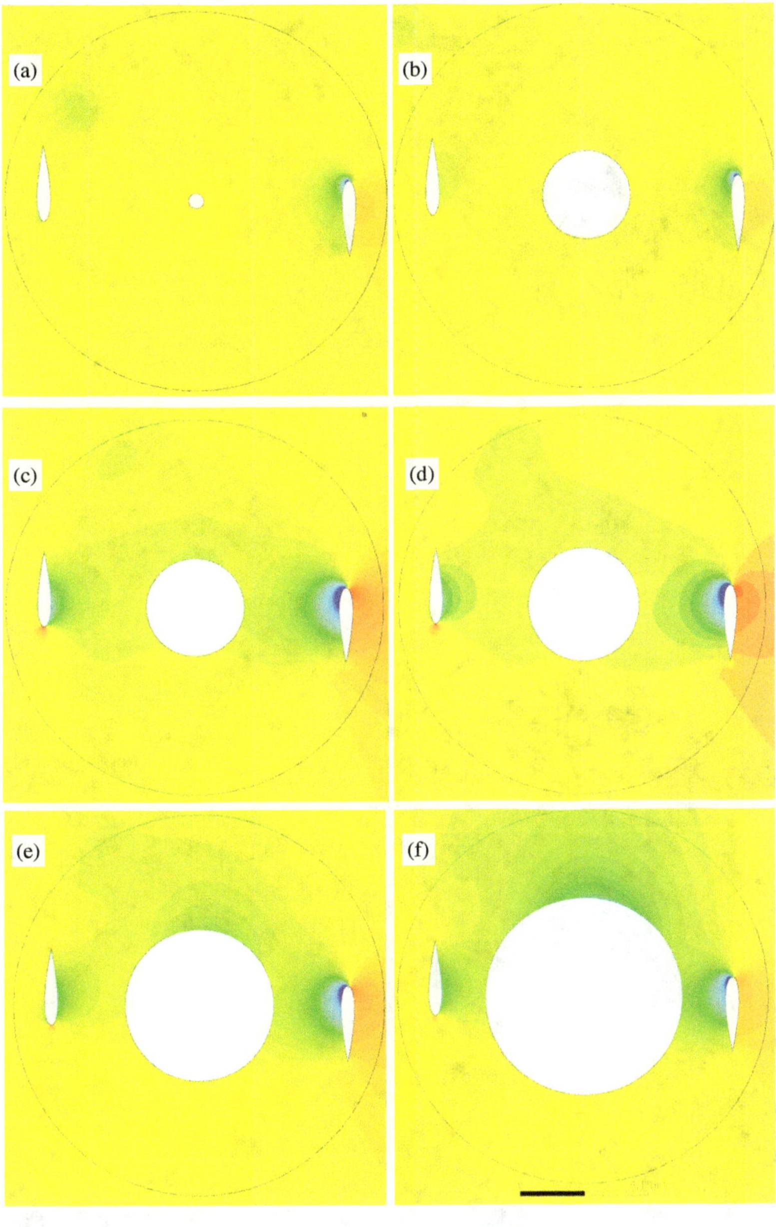

Figure 5.13 Pressure contours of two-bladed AHDT in closed configuration for D_R/D_T ratios of (a) 20, (b) 3.5, (c) 3, (d) 2.5, (e) 2 and (f) 1.5 for the same airfoil position.

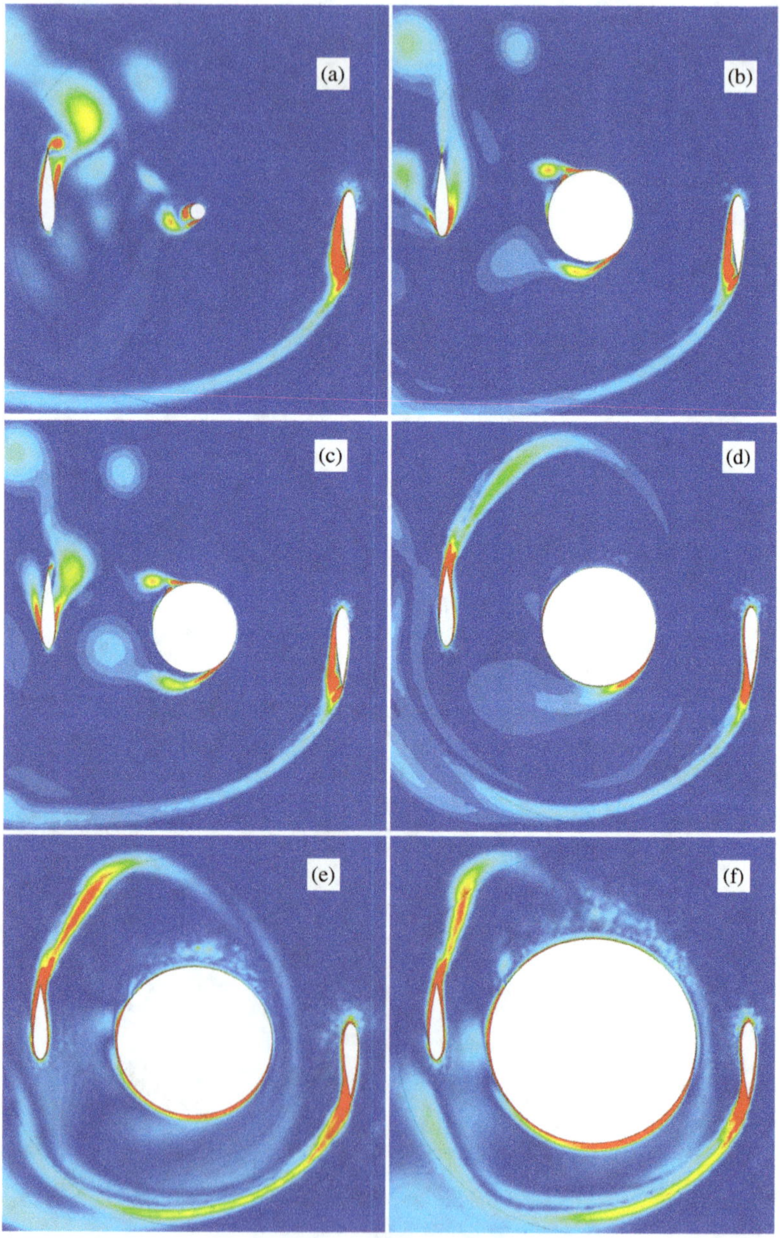

Figure 5.14 Vorticity contours of two-bladed AHDT in closed configuration for D_R/D_T ratios of (a) 20, (b) 3.5, (c) 3, (d) 2.5, (e) 2 and (f) 1.5 for the same airfoil position.

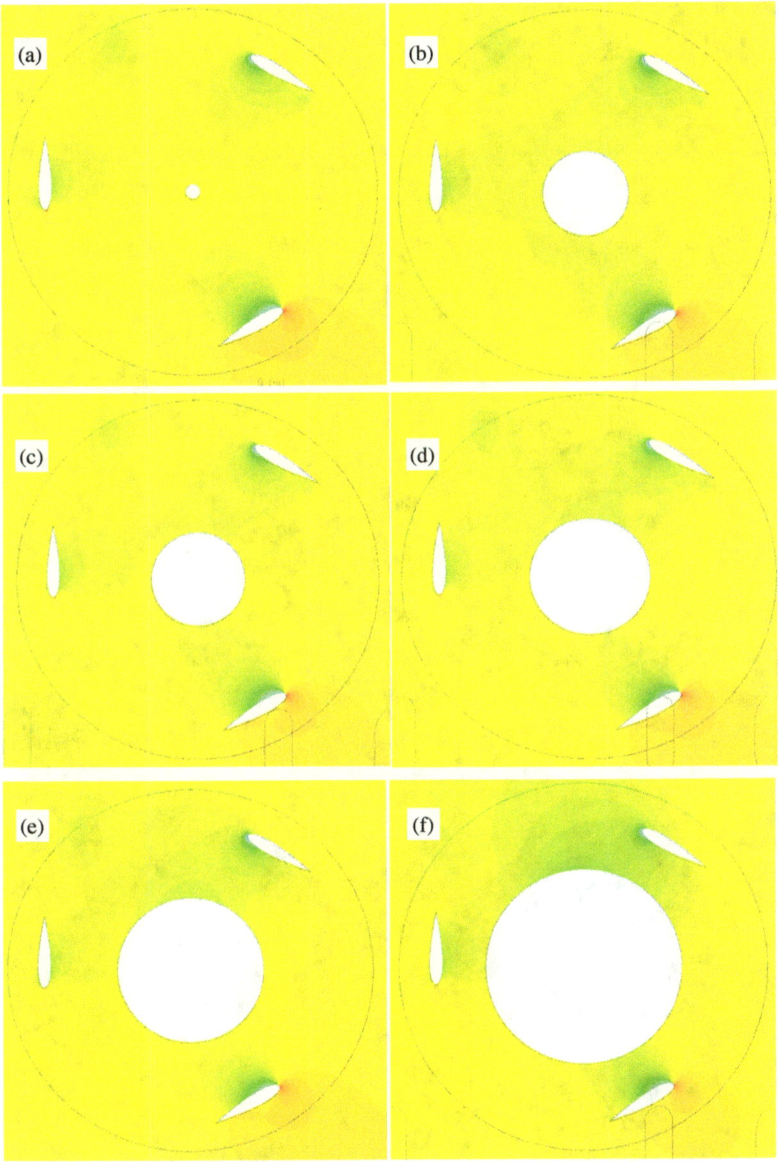

Figure 5.15 Pressure contours of three-bladed AHDT in closed configuration for D_R/D_T ratios of (a) 20, (b) 3.5, (c) 3, (d) 2.5, (e) 2 and (f) 1.5 for the same airfoil position.

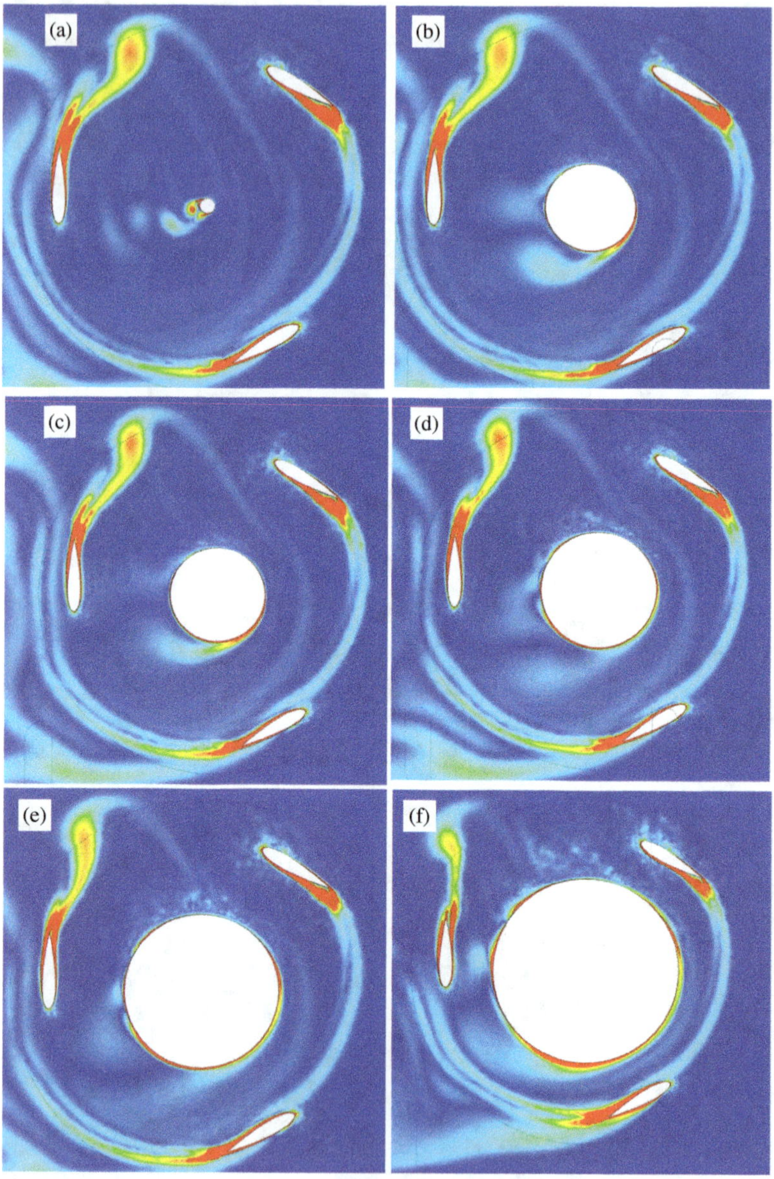

Figure 5.16 Vorticity contours of three-bladed AHDT in closed configuration for D_R/D_T ratios of (a) 20, (b) 3.5, (c) 3, (d) 2.5, (e) 2 and (f) 1.5 for the same airfoil position.

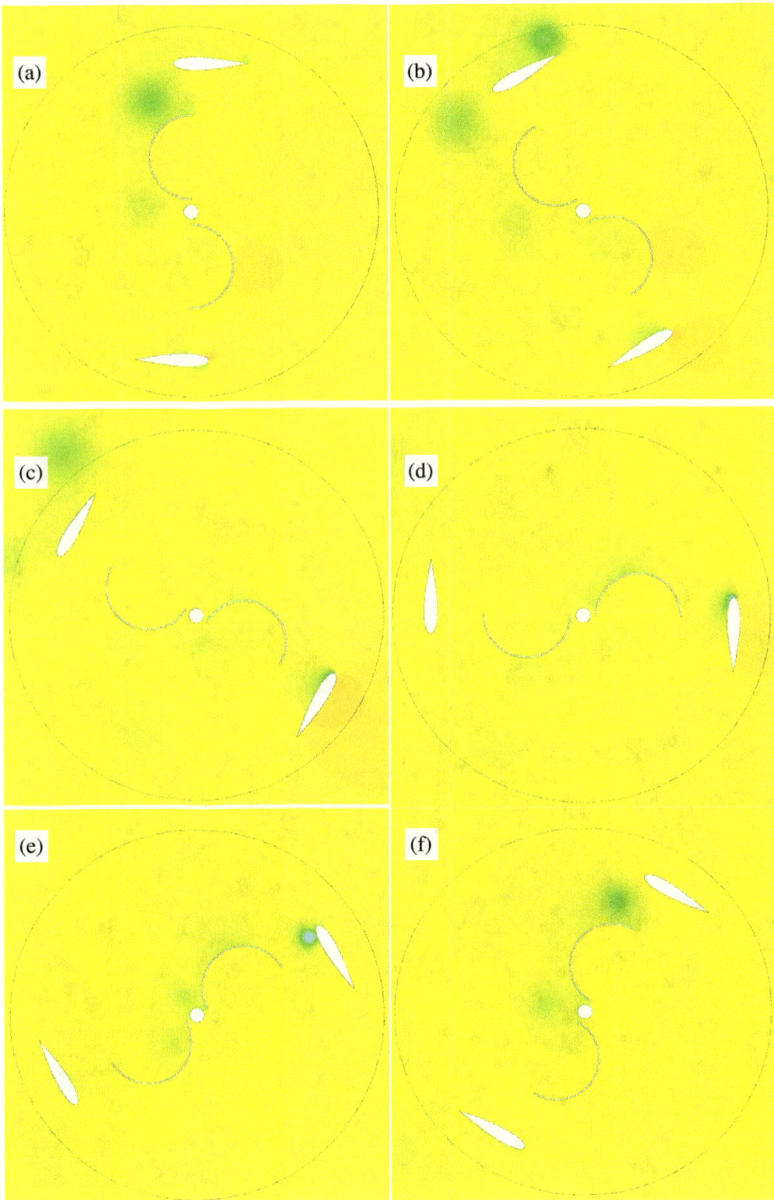

Figure 5.17 Pressure contours of two-bladed AHDT in open configuration for D_R/D_T ratio of 3 at various airfoil positions.

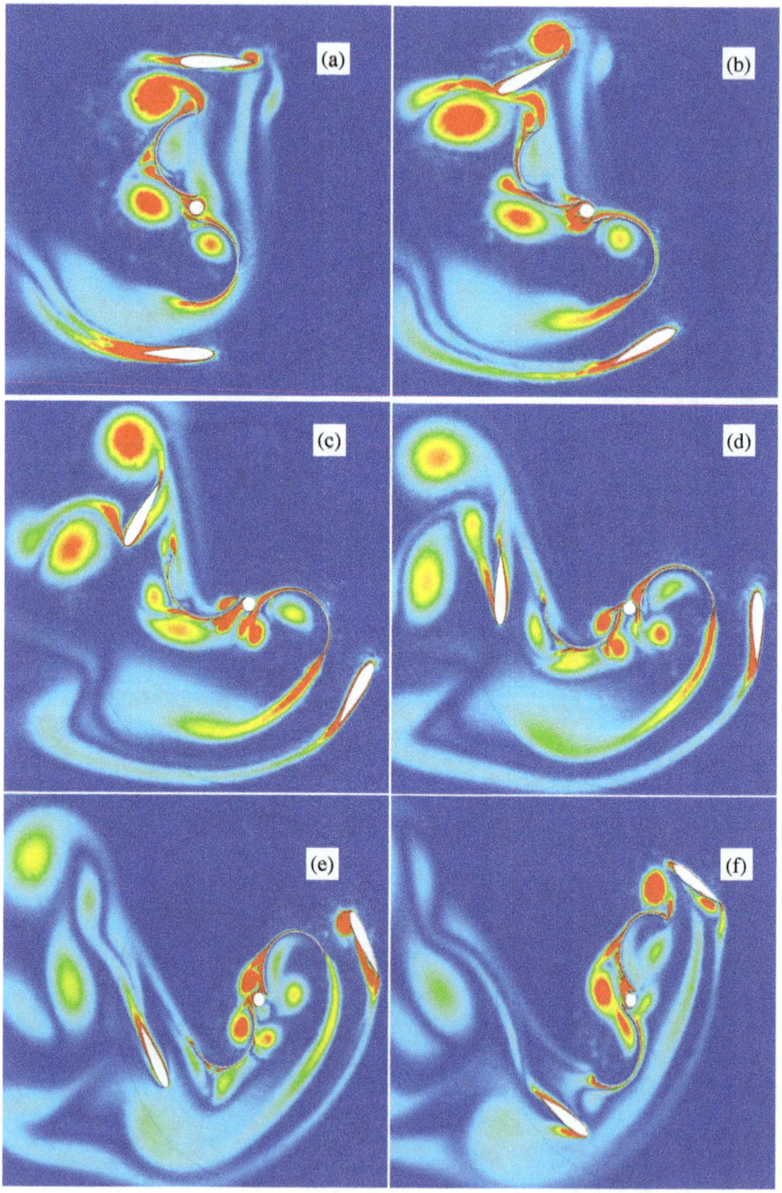

Figure 5.18 Vorticity contours of two-bladed AHDT in open configuration for D_R/D_T ratio of 3 at various airfoil positions.

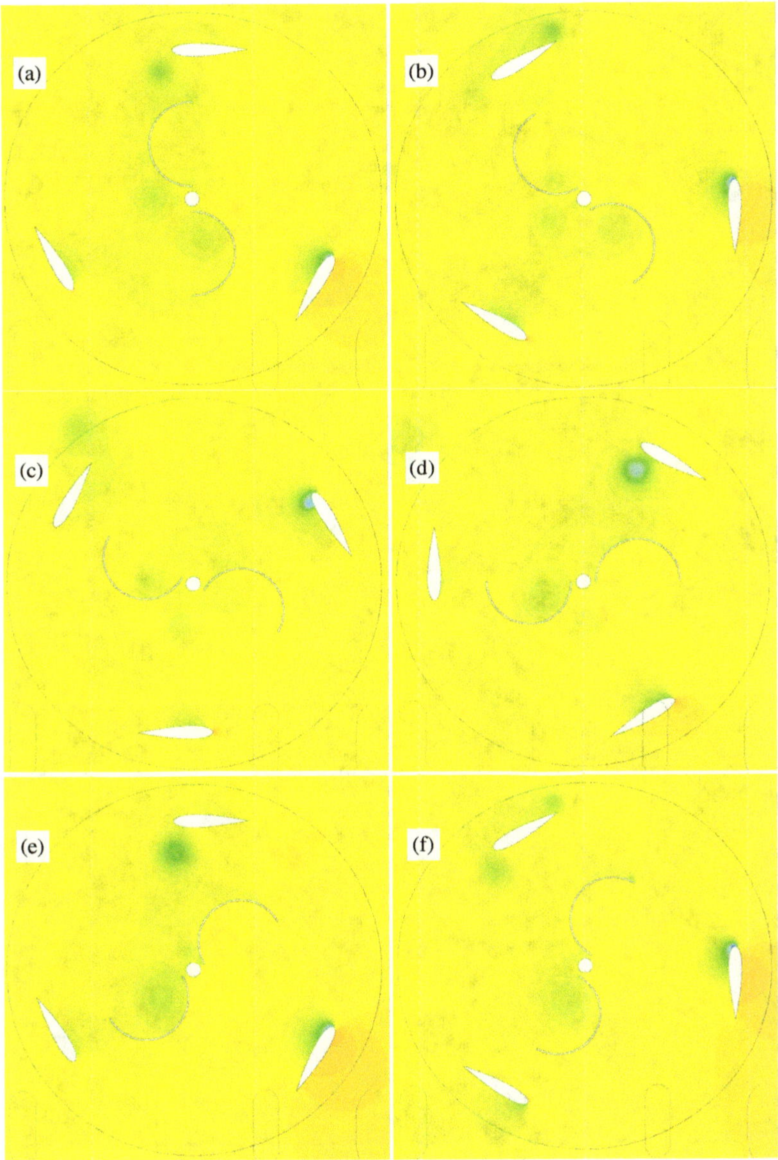

Figure 5.19 Pressure contours of three-bladed AHDT in open configuration for D_R/D_T ratio of 3 at various airfoil positions.

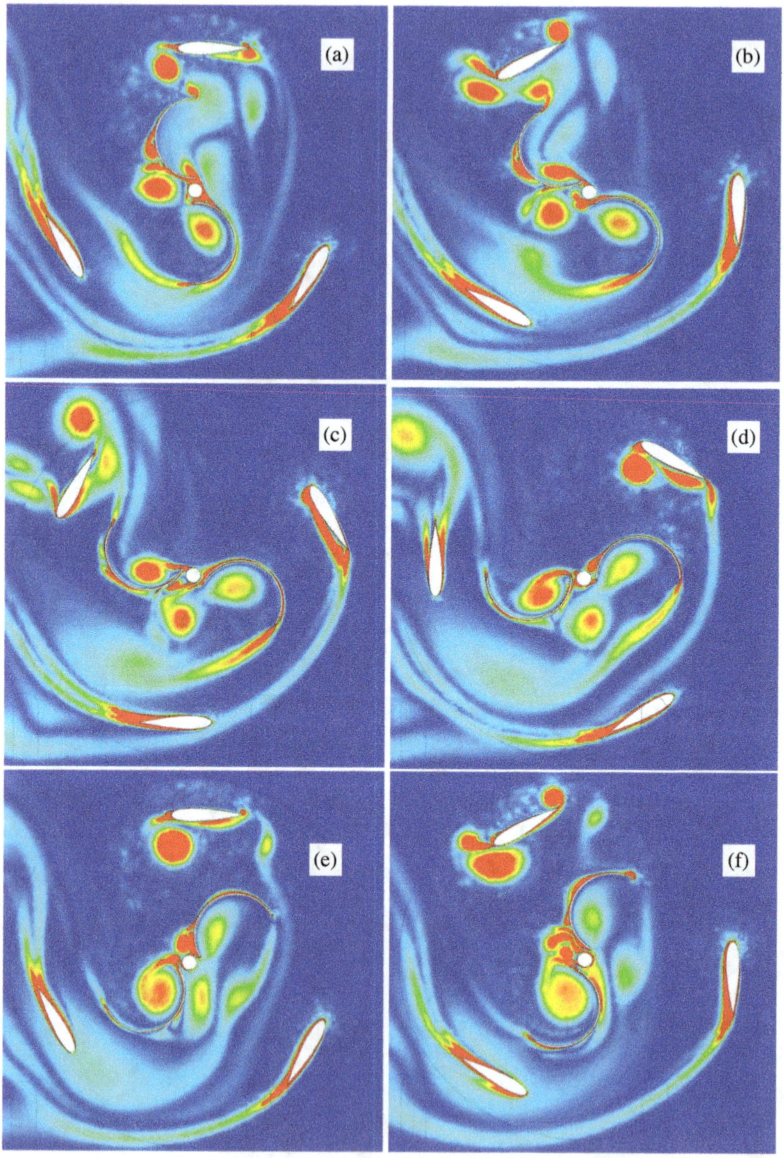

Figure 5.20 Vorticity contours of three-bladed AHDT in open configuration for D_R/D_T ratio of 3 at various airfoil positions.

4. The flow on the closed Savonius configuration (cylinder) is not axisymmetric as assumed in the analytical model.

   From Figure 4.5, the wake deficit on a cylinder was assumed to be symmetrical and the wake width was formulated to reduce the wind velocity for the corresponding azimuthal angles of the H-Darrieus rotor. Figure 5.14 shows that the wake pattern is asymmetrical.
5. The wake of $D_R/D_T = 2$ and lower ratios has significant influence on the upwind half of rotation.

   From Figure 5.11, with reference to the torque generation by the single blade of H-Darrieus, the azimuthal angles that correspond to the upwind half of the rotor generate less torque. The reason is evident from the vorticity contours as shown in Figure 5.16, that H-Darrieus operates in highly turbulent wind formed by the larger cylinder frictional drag, which tends to push the oncoming flow in the rotating direction on the upwind half. It is obvious that large wake due to bigger diameter cylinder on the downwind half will reduce the overall torque generation for one complete cycle.
6. Alternating Van Karman vortices are generated at higher TSR.

   From Figures 5.14 and 5.16, for $2.5 < D_R/D_T < 20$, alternating vortices are formed on the closed AHDT (cylinder). The formation of vortices depends on a number of factors of which critical ones are cylinder Re, cylinder rpm and oncoming wind speed. At TSR > 3, the cylinder Re reaches 40 and above in which van Karman vortices are beginning to appear. This is because of the separation of the fluid boundary layer due to adverse pressure gradient, resulting in reversed flow on the leeward side of a cylinder in a fluid flow.
7. High solidity of H-Darrieus rotor reduces the Cp of AHDT in closed configuration.

   As evident from Figure 5.12, increasing the solidity of H-Darrieus from two to three blades reduces C_{pmax} and TSR of AHDT. This can be attributed to the reduction in Cp of the H-Darrieus rotor. Increase in number of blades reduces the angular velocity and hence the Re. Reduction in lift due to low Re reduces the torque generation. Another notable reason for reduction in Cp is the onset of earlier dynamic stall due to operation of blades in the wake of preceding blades. Since the power generation of AHDT in closed configuration is only through the H-Darrieus rotor, the inherent nature of reduction in Cp with increasing solidity leads to combined rotor Cp reduction.

Chapter Nomenclature

μ_T	Eddy viscosity
k	Turbulence kinetic energy
ω	Specific dissipation rate
S	Invariant measure of strain rate
F_1	First blending function
F_2	Second blending function
y_s	Distance to nearest surface
ρ	Density
γ	Kinematic viscosity
α and β	New closure coefficients
ϕ_{sst}	Variables used in k-$\dot{\omega}$ SST model
ϕ_{ω}	Variables used in k-$\dot{\omega}$ model
ϕ_{ϵ}	Variables used in transformed k-€ model
$\dot{\omega}$	Specific dissipation rate
σ	Closure coefficient
$CD_{k\omega}$	Cross diffusion term

References

1. Mohamed, M. Impacts of solidity and hybrid system in small wind turbines performance. *Energy* **2013**, *57*, 495–504.
2. Karman, T.v. Compressibility effects in aerodynamics. *Journal of the Aeronautical Sciences* **1941**, *8*, 337–356.

Chapter 6

Experimental Optimization of Adaptive Hybrid Darrieus Turbine*

* Parts of this chapter are adapted with permission from Palanisamy Mohan Kumar, S. A.; Narasimalu, S.; Lim, T.-C. Optimization, design, and construction of field test prototypes of adaptive hybrid Darrieus turbine. *Journal of Fundamentals of Renewable Energy and Applications* **2017**, *7*, 245.

6.1 Experimental Setup and Wind Tunnel Models

The wind tunnel campaign is carried out to validate the optimum diameter range resulting from the computational studies. To investigate the influence of the Savonius rotor in closed and open configurations on the Darrieus rotor performance, the wind tunnel investigation is divided into a series of tests. From the computational studies it is evident that the D_R/D_T ratio ranging from 5 to 2 is optimum, hence the wind tunnel tests are restricted to this range. The objective is to search for an optimum diameter that will display better starting characteristics in low wind speed and allow the Darrieus rotor to attain its peak performance at high wind speed.

Since the closed configuration is deployed at moderate to high winds, the wind tunnel tests are carried out for the wind speeds of 6–9 m/s corresponding to the Re of 1.8×10^5, 2.1×10^5, 2.4×10^5, and 2.7×10^5, while the open configuration is tested from low wind speeds of 3–9 m/s. From the computational studies, it is apparent that the Darrieus rotor solidity has global effect on the AHDT performance, and hence an investigation has been carried out by varying the solidity for two- ($\sigma = 0.5$) and three-bladed ($\sigma = 0.75$) of same chord length of 100 mm. Conventional Savonius rotor is also subjected to test to ascertain the individual performance before integrating in AHDT. The configurations evaluated are illustrated in Figure 6.1 and the wind tunnel setup with mounted rotors are shown in Figure 6.2.

The AHDT in closed configuration is evaluated first to eliminate the diameters that severely degrade the Darrieus performance. The promising diameters are then evaluated for performance in the open configuration that corresponds to the low wind speed performance. Subsequently the conventional Savonius and Darrieus performance will provide an insight into their individual performances. The conventional rotor design parameters are assumed to be constant for the present study, though it may

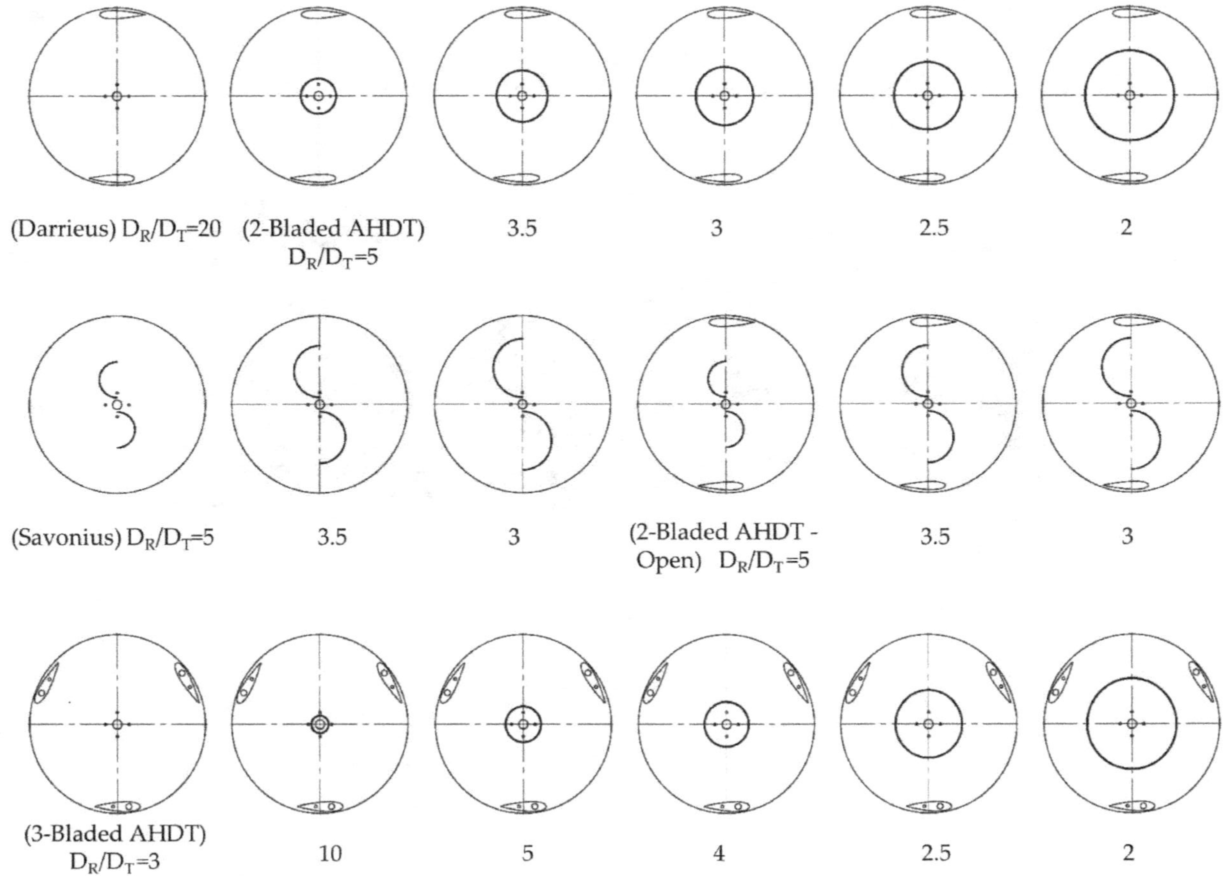

Figure 6.1 AHDT configurations for wind tunnel test.

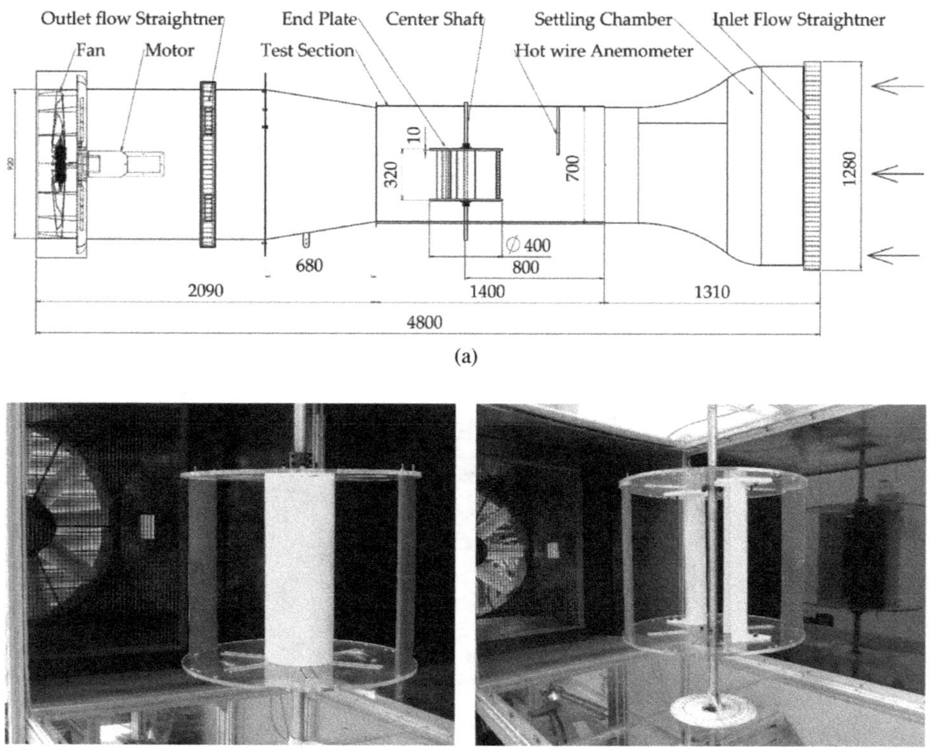

(a)

(b)

(c)

Figure 6.2 (a) Wind tunnel rig. (b) Closed configuration of AHDT. (c) Open configuration of AHDT.

have considerable influence on AHDT performance. Cp is compared against TSR and the measurements from the wind tunnel test are blockage corrected as explained later. The test models are 3D printed and polished. The cylinders are printed in two halves, while the Savonius buckets are printed with holes to mount on the end plates. The end plates are constantly fixed on the shaft, while the Darrieus blades and Savonius buckets can be mounted and dismounted as and when necessary.

6.2 Adaptive Hybrid Darrieus Turbine in Closed Configuration — Two-Bladed ($\sigma = 0.5$)

For the two-bladed rotor, the conventional Darrieus rotor displays higher Cp than AHDT integrated with closed Savonius rotor (cylinder). The investigated cylinder and the two-bladed Darrieus are displayed in Figure 6.3. It is anticipated that the small-diameter cylinder should attain higher Cp than the large-diameter cylinders due to larger wake width, but the plots as shown in Figures 6.4 and 6.5 contradict. At lower Re, the cylinder of $D_R/D_T = 5$ (80 mm) reached lower Cp than the larger-diameter cylinders of $D_R/D_T = 3.5$ and 3, which contradict the CFD

Figure 6.3 Diameters of closed Savonius under test.

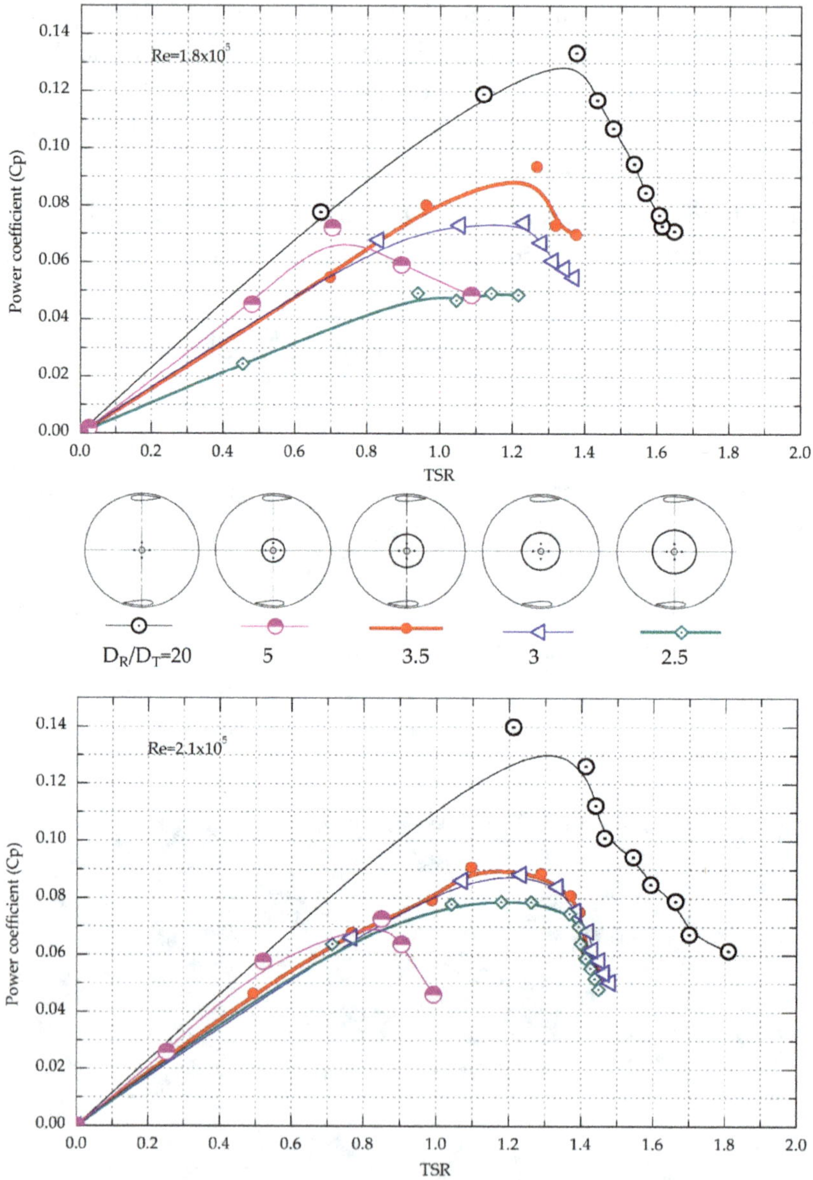

Figure 6.4 Cp comparison of two-bladed AHDT with various D_R/D_T at Re = 1.8 × 10^5 (top) and Re = 2.1 × 10^5 (bottom).

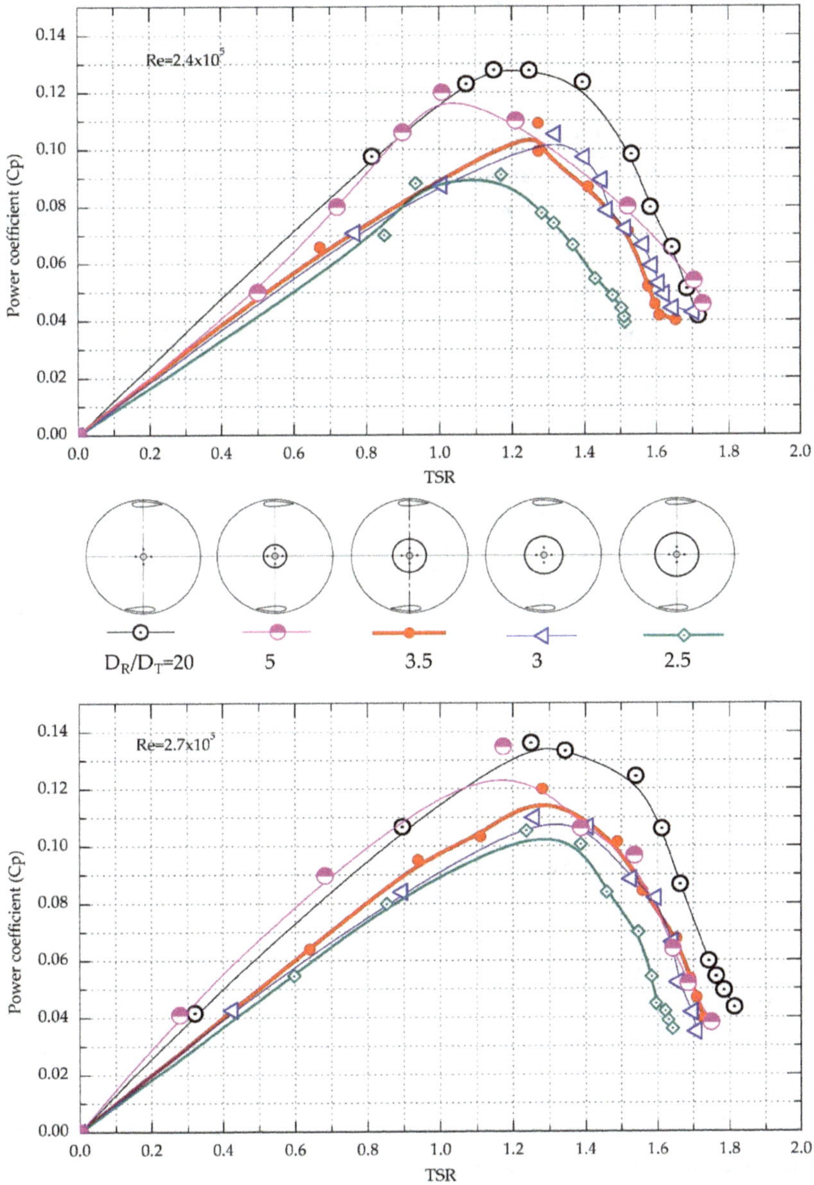

Figure 6.5 Cp comparison of two-bladed AHDT with various D_R/D_T at Re = 2.4 × 10^5 (top) and Re = 2.7 × 10^5 (bottom).

predictions that the loss in the Cp is proportional to the diameter of the cylinder for all the Re. At lower Re of 1.8×10^5, the difference in Cp between the conventional Darrieus and $D_R/D_T = 2.5$ is 62.4%. The nominal TSR at which the conventional Darrieus reached peak Cp is between 1.2 and 1.4, whereas for the AHDT it is between 1 and 1.2. At lower Re the difference in the Cp between Darrieus and AHDT is more pronounced, and as Re increases the difference diminishes. At Re of 2.7×10^5, $D_R/D_T = 5$ (80 mm) reaches almost the same Cp as the Darrieus rotor, whereas the difference in Cp with AHDT of $D_R/D_T = 3.5$ and 3 is 18–20%. It can be concluded that an optimum D_R/D_T should be between 3.5 and 3 for better performance in all Re.

6.3 Adaptive Hybrid Darrieus Turbine in Closed Configuration — Three-Bladed ($\sigma = 0.75$)

A similar trend is also displayed by three-bladed AHDT as in Figures 6.6 and 6.7. As the Re increases the Cp rapidly rises for $D_R/D_T = 5$, whereas for $D_R/D_T = 3.5$, 3 and 2.5 it displays minor increment. $D_R/D_T = 4$ displays better performance in the case of AHDT compared to other diameters.

6.3.1 *Cp Comparison of Savonius Rotor for Various* D_T

Three Savonius rotor diameters of ratio $D_R/D_T = 3.5$, 3.5 and 3 are compared that are found to be optimum from the CFD computations as shown in Figure 6.8. From the plots shown in Figure 6.9, the $D_R/D_T = 5$ displays better performance in both low and high Re. It is imperative to rationalize for such a high Cp, though the geometric parameters are similar except bucket diameter. The vital parameter that can be attributed for the increase in Cp of $D_R/D_T = 5$ compared to $D_R/D_T = 3.5$ and 3 is the end plate. The presence of end plate prevents the leakage of the fluid from the concave side and also the top and bottom of the

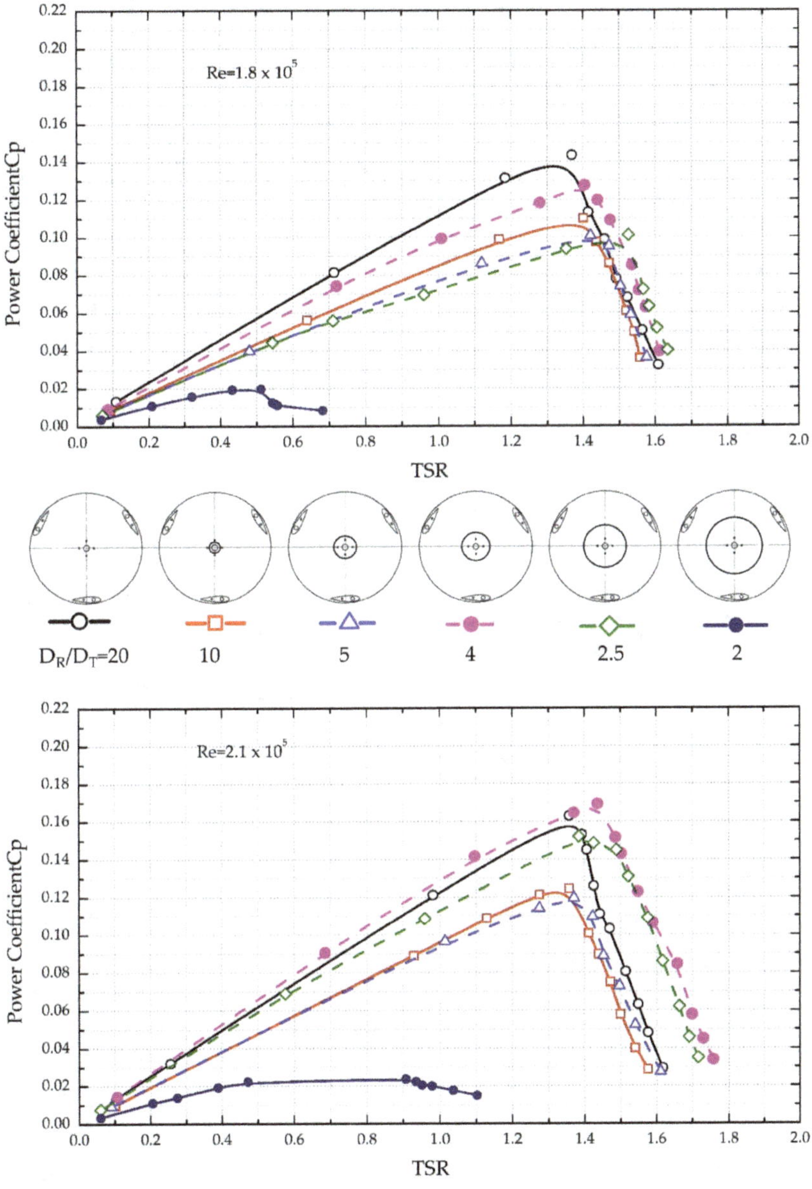

Figure 6.6 Cp comparison of three-bladed AHDT with various D_R/D_T at $Re = 1.8 \times 10^5$ (top) and $Re = 2.1 \times 10^5$ (bottom).

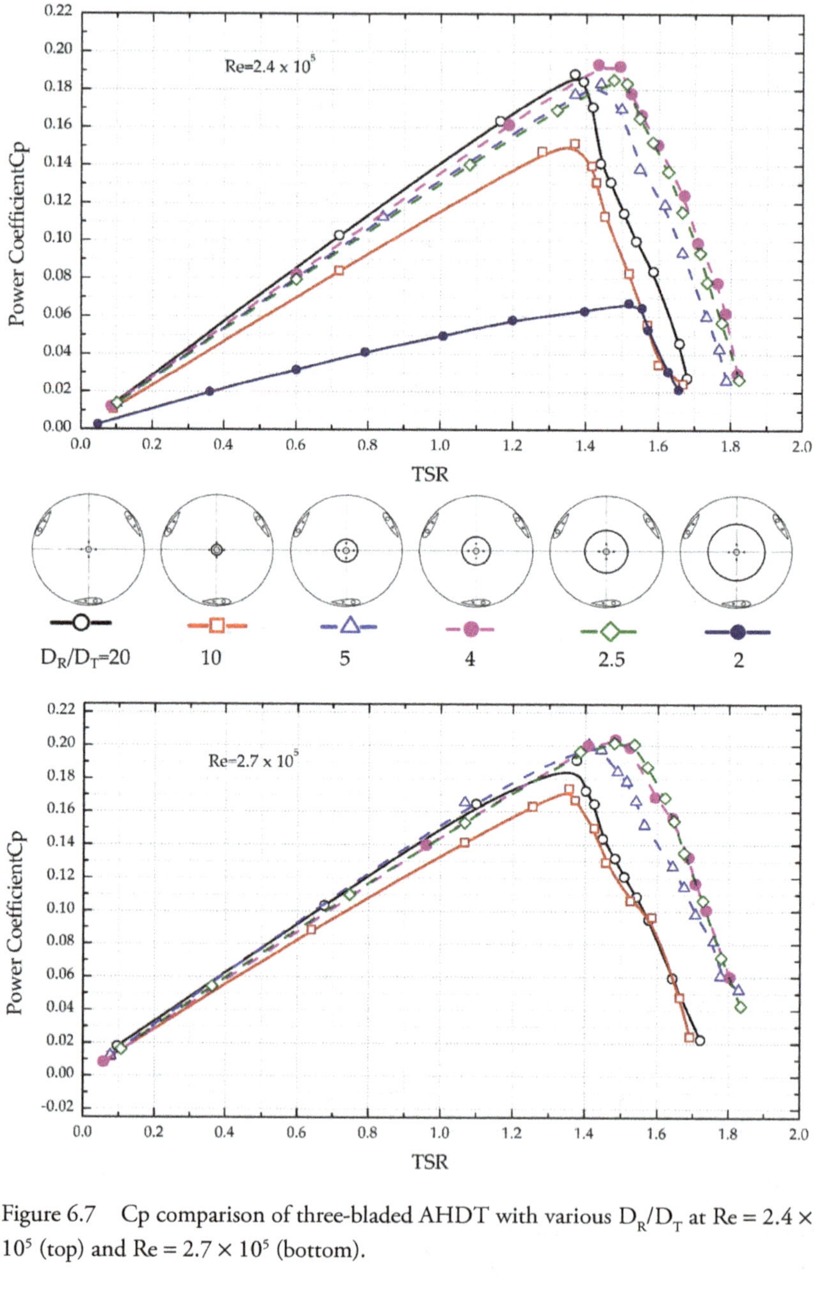

Figure 6.7 Cp comparison of three-bladed AHDT with various D_R/D_T at Re = 2.4×10^5 (top) and Re = 2.7×10^5 (bottom).

Figure 6.8 Diameters of open Savonius rotor under test.

rotor. The previous studies on the influence of end plate found that end plate diameter to rotor diameter ratio of 1.1 is optimum with notable increase in Cp. Hence larger end plate of 2.2 increases the Cp as high as 60% compared to other diameters. A significant decrease in Cp is noted between $D_R/D_T = 3.5$ and 3, with the latter performing better which can be concluded as optimum bucket diameter.

6.3.2 *Cp Comparison of Adaptive Hybrid Darrieus Turbine in Open Configuration for Various D_R/D_T*

The performance of AHDT in open configuration is dominated by the Savonius rotor, following the trend displayed by the conventional Savonius rotor as shown in Figure 6.10. It is evident that at low Re, the $D_R/D_T = 3$ displays Cp of 1.1 at Re 1.5×10^5, higher than other diameter ratios, but as Re increases $D_R/D_T = 5$ performs better. Overall better performance in all the Re is displayed by $D_R/D_T = 3$, whereas lower performance is displayed by $D_R/D_T = 3.5$.

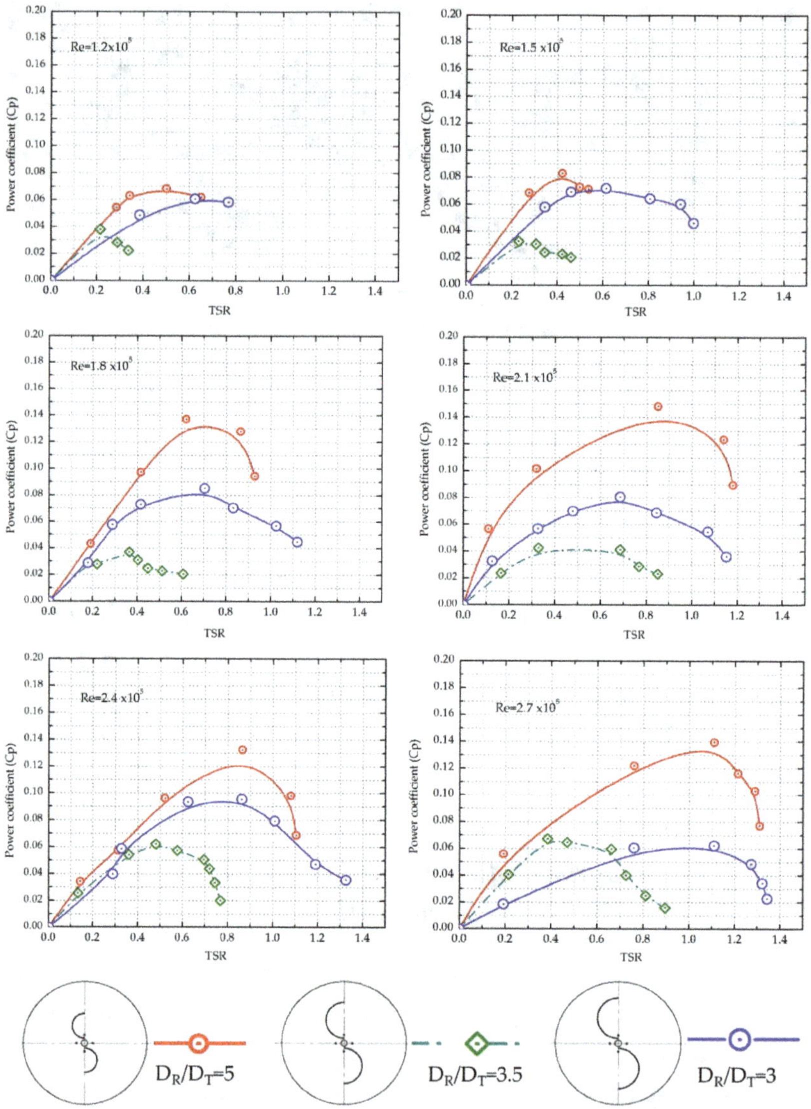

Figure 6.9 Cp comparison of Savonius rotor for various D_R/D_T at various Re.

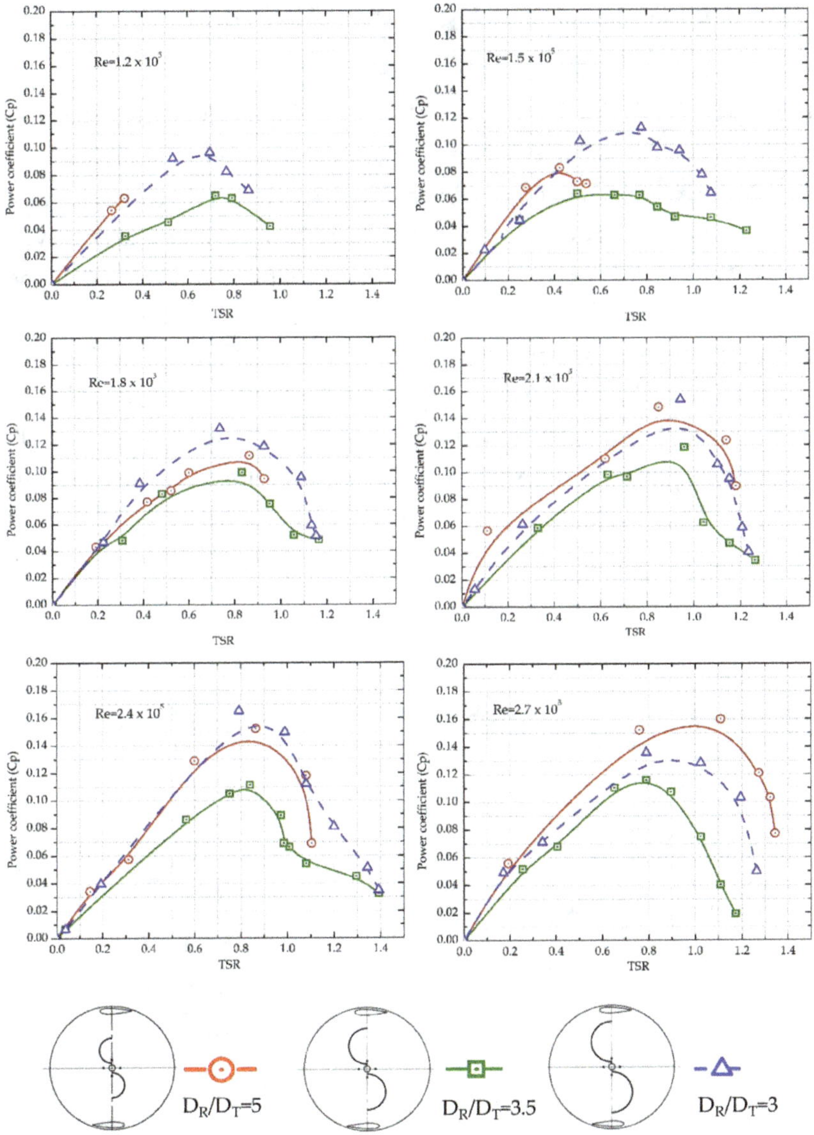

Figure 6.10 Cp comparison of AHDT in open configuration for various D_R/D_T at various Re.

6.3.3 Cp Comparison of $D_R/D_T = 3.5$ for Various Configurations

The $D_R/D_T = 3.5$ corresponds to 115 mm cylinder diameter. The four configurations under investigation are compared as shown in Figure 6.12 with the Cp comparison for the Re of 1.8×10^5 to 2.7×10^5 corresponding to the wind speed of 6–9 m/s as the cylinder configuration is deployed at moderate wind speed. The overview of the plots indicates that the Darrieus rotor displays the better performance and the lowest performance is displayed by the Savonius rotor. The DS-O and DS-C hybrid machines reach higher Cp than the Savonius rotor, revealing that the lift force has significant contribution in boosting the torque. Another notable trend is that the Savonius rotor achieves the peak Cp at a TSR less than

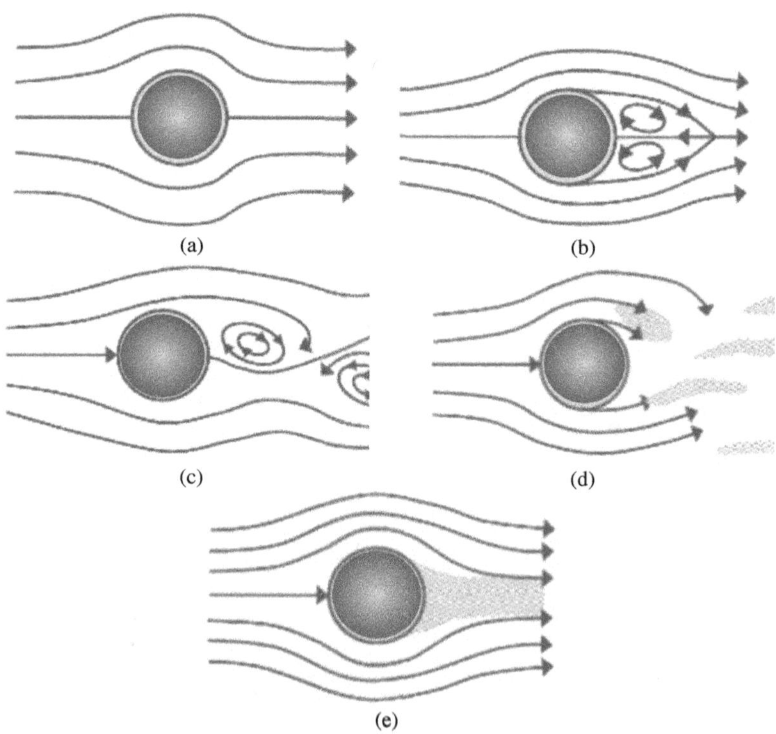

Figure 6.11 Air flow pattern on a cylinder: (a) $Re < 5$, (b) $5 < Re < 40$, (c) $40 < Re < 90$, (d) $300 < Re < 3 \times 10^5$, (e) $3 \times 10^5 < Re < 3.5 \times 10^6$.

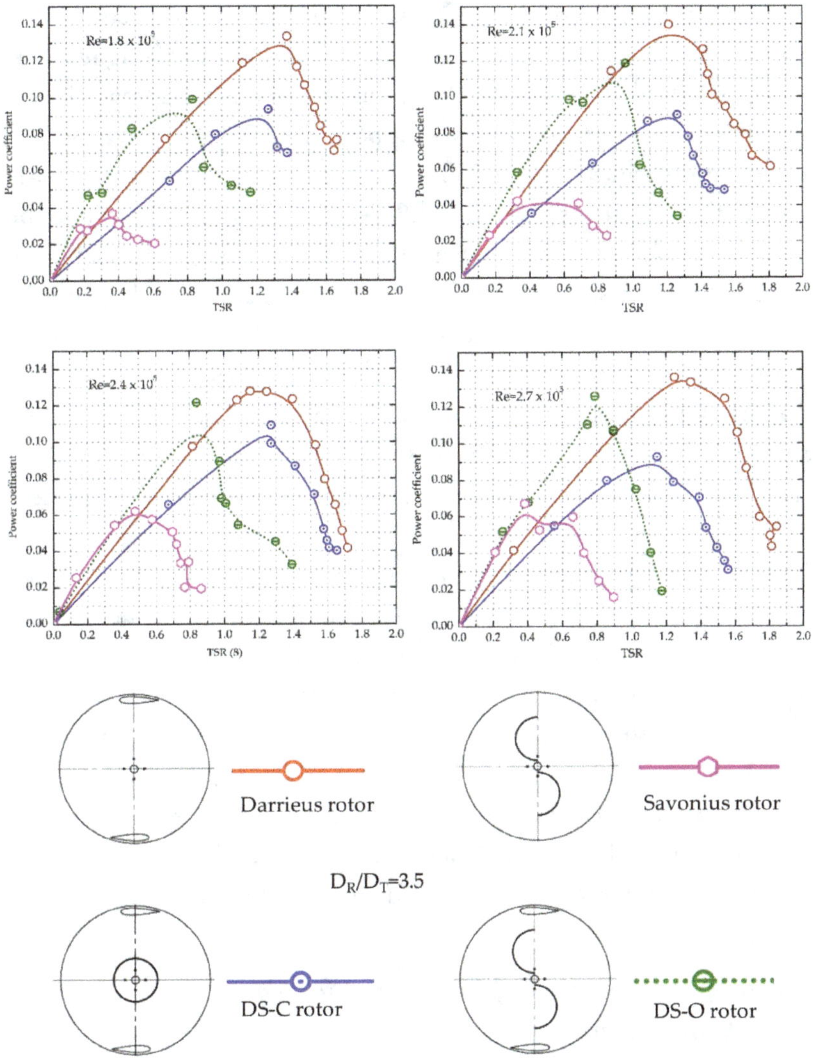

Figure 6.12 Cp comparison of $D_R/D_T = 3.5$ for various configurations.

0.6, whereas for the hybrid machines the peak Cp is achieved in the range of 0.6 to 1.0. For the conventional Darrieus rotor peak Cp is achieved always beyond TSR 1.

The influence of Savonius rotor on the performance of the hybrid rotor can be well understood by comparing the Cp and quantifying the loss thereof. For the Re 1.8×10^5, Cp loss of 24% is reported between the Savonius and Darrieus rotors. For the DS-C and DS-O configurations, the Cp achieved is almost the same, but at a marked difference in the TSR. DS-O configuration achieves the peak Cp at 0.7, whereas the DS-C configuration achieves it at 1.2. Though the magnitude of the Cp is the same for DS-O and DS-C, the difference in TSR demonstrates the dominant force in two rotors. For D_R/D_T of 3.5, and for especially low to moderate wind speed, the power performance is almost the same for both hybrid configurations. But as the wind speed increases the difference in performance deepens, with DS-O gaining an upper hand over DS-C. The peak Cp decreases but the curve becomes flat when operating in a larger TSR range, whereas for the DS-O configuration, the Cp curve becomes sharper and operating range of the TSR is also notably reduced.

From the starting perspective, as the conventional Savonius rotor engages in generating the starting torque, smaller diameter Savonius buckets tend to generate less torque. The maximum Cp of the Savonius rotor for the D_R/D_T of 3.5 is 0.06, which is substantially lower than that for the D_R/D_T of 5. From the plot it can be concluded that $D_R/D_T = 3.5$ is not an optimum diameter due to reduced performance of 72% compared to the conventional Darrieus rotor. The lower performance of the hybrid rotor due to turbulent wake on downstream with $D_R/D_T = 3.5$ indicates higher drag forces on the cylinder and results in high wind loading.

6.3.4 *Cp Comparison of $D_R/D_T = 3$ for Various Configurations*

As the performance of the DS-C configuration significantly drops for ratios other than $D_R/D_T = 3$, a possible explanation is the wake behind the cylinder. The wake flow is fully turbulent due to alternating vortices and their frequencies. The nature of the flow is characterized by the Strouhal number that interacts with the Darrieus blades at various frequencies and in turn with the Darrieus blade wake. The entire flow structure is highly complex to predict due to the interaction of cylinder wake and Darrieus wake for the given Re as shown in Figure 6.11. The difference in Cp between the conventional Darrieus and the AHDT in closed configuration is more than 84% for the low Re of 1.8×10^5 and 2.1×10^5, but as the Re increases to 2.7×10^5 the Cp is almost the same as shown in Figure 6.13.

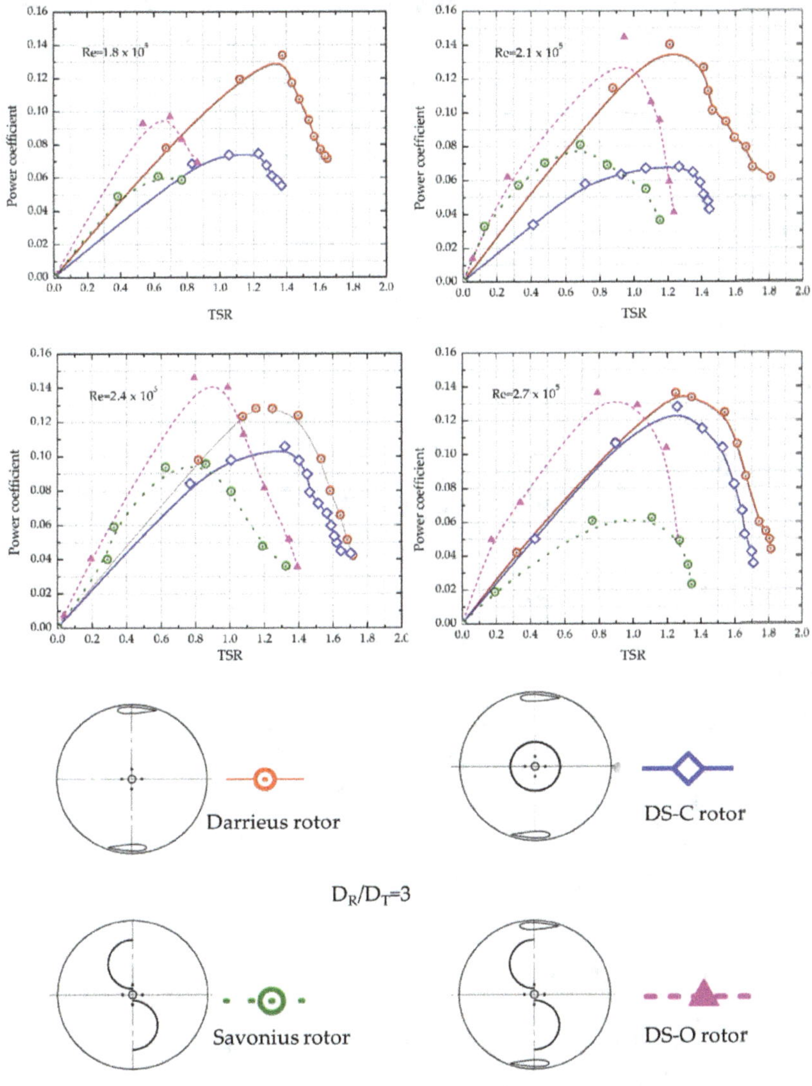

Figure 6.13 Cp comparison of $D_R/D_T = 3$ for various configurations.

6.4 Starting Torque Comparison of Optimum Adaptive Hybrid Darrieus Turbine with H-Rotor

The starting characteristics of the conventional and the optimum AHDT at DR/DT = 3 is shown in Figure 6.14. The rotor with higher static torque coefficient value has the capability to start early with high starting torque. It can be noticed that for displayed graphs the static torque values are cyclic with a periodicity of around 180° for these two-bladed rotors. From the graphs it can also be seen that the AHDT rotor is able to achieve 96.9% higher value than the corresponding conventional Darrieus. Additionally, the positive torque coefficient exists for most of the azimuthal angles. The difference in the static torque coefficient value between

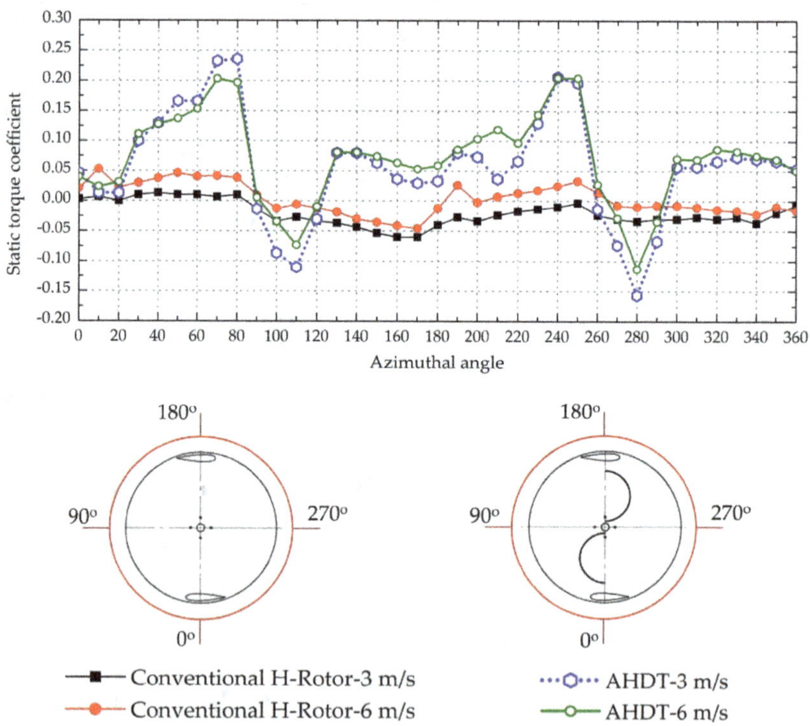

Figure 6.14 Static torque coefficient comparison of $D_R/D_T = 3$.

the measured 3 m/s and 6 m/s is almost the same in the case of AHDT, whereas the difference is about 75.6% in the case of H-rotor. From the plot, undoubtedly the AHDT rotor outperforms the conventional Darrieus rotor in terms of starting early in low wind speed and sustaining the rotation with higher torque.

Chapter Nomenclature

AHDT	Adaptive Hybrid Darrieus Turbine
ADWT	Adaptive Darrieus Wind Turbine
$DS-C$	Darrieus Savonius in closed configuration
$DS-O$	Darrieus Savonius in open configuration
D_R	Darrieus rotor diameter
D_T	Diameter of cylinder (diameter of closed Savonius rotor)

Chapter 7

Overview of Floating Offshore Wind Turbines

7.1 Floating Offshore Wind Turbine

7.1.1 *Overview of Wind Energy*

The principal sources of renewable energy are wind, solar, nuclear, tidal, and biomass. When all of these alternatives are compared, wind energy technology emerges as the most mature and laden with potential. This is due to the fact that it is much less hazardous when compared to nuclear energy, more cost effective than solar energy, more industrial than tidal energy, and more mature than biomass alternatives. These facts mean that wind energy has emerged as the most sustainable source of renewable energy that can be applied on a large scale. As global forces come together to pursue sustainable energy technologies, the majority of the efforts will be focused on wind energy.

From 2001 to 2017 the global capacity for wind power has increased steadily, tracked in Figure 7.1. By the end of 2017, the total installed capacity had reached an estimated 539.1 GW, while the newly installed

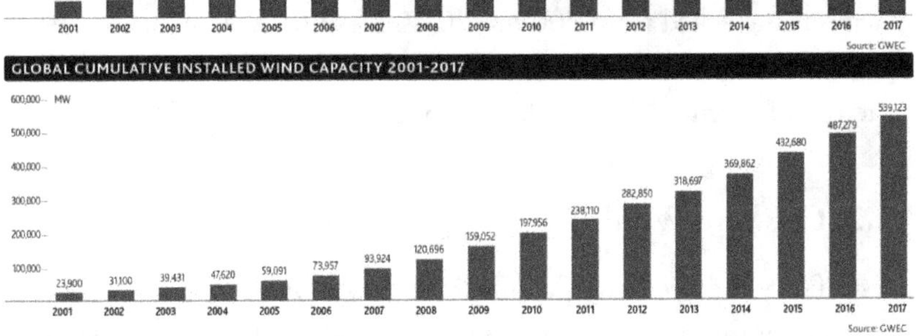

Figure 7.1 Global annual and cumulative installed wind capacity, 2001–2017.[1]

ANNUAL INSTALLED CAPACITY BY REGION 2009-2017

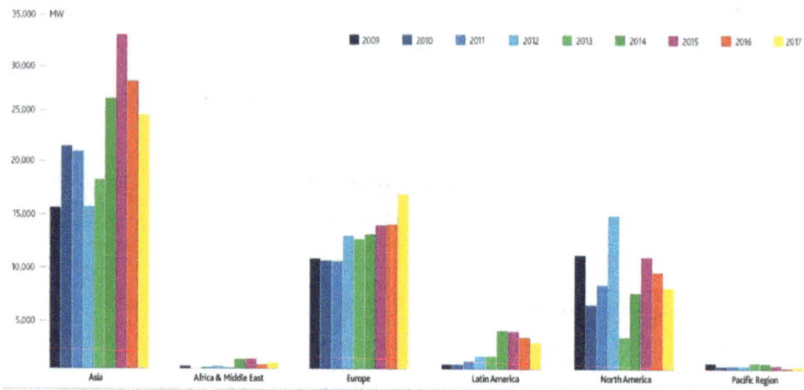

	2009	2010	2011	2012	2013	2014	2015	2016	2017
Asia	15,507	21,481	20,981	15,624	18,252	26,058	33,962	27,721	24,412
Africa & Middle East	251	153	8	131	240	934	953	418	618
Europe	10,660	10,466	10,393	12,862	12,524	12,988	13,831	13,926	16,803
Latin America	471	459	771	1,248	1,240	3,744	3,678	3,078	2,578
North America	11,008	6,208	8,137	14,807	3,112	7,382	10,829	9,359	7,836
Pacific Region	578	294	345	358	655	568	381	140	245

Source: GWEC

Figure 7.2 Total installed wind energy capacity for various regions in the world, 2009–2017.[1]

global capacity was reported at an estimated 52.5 GW. This large growth demonstrates the growing popularity of wind power and is inspiring additional global interest.

From 2009 onwards Asia has cemented itself as the world leader in terms of growth in capacity for wind energy. In Figure 7.2 based on the Global Wind Energy Council report in 2017,[1] the capacity per each of the world's six regions is tracked from 2009 to 2017. Despite the large wind energy capacity of both China and the USA, Europe also developed a large capacity to become the world leader in the wind energy industry.

7.1.2 *Offshore Wind Energy*

In its infancy the wind energy industry focused predominantly on land-based development. As the nascent industry matured, however, the focus shifted from land to sea and offshore wind energy development is now the predominant focus. Compared to onshore wind farming, offshore wind farms offer a range of key advantages.

Firstly, the sea experiences a much higher annual average wind speed than that of land. Studies have demonstrated that the average wind speed 10 km away from shore is 25% higher than that on land. This phenomenon is evidenced in Table 7.1, which outlines the ratio of average wind speeds at sea and on land. It shows that the further from shore, the stronger the wind speed.

Secondly, offshore wind farming offers more predictable and stable wind conditions with lower turbulence and wind shear. An obvious benefit of offshore wind farming is that it does not take up valuable land space and offers a reduced impact on human life.

Finally, wind turbine technology offers a greater unit capacity as well as higher annual utilization hours. Furthermore, offshore wind farms do not require complex transportation conditions and are not restricted by noise regulations.

Europe's offshore wind energy technology is the most mature globally and the nascent technology is moving into the large-scale development phase.

Table 7.1 Ratio of the average wind speed at sea and on land.

Distance from land (km)	Ratio of annual average wind speed	
	On land	At sea
0 (Standard wind speed)	4–6 (m/s)	7–9 (m/s)
25–30	1.4–1.5	1.2
30–50	1.5–1.6	1.4
>50	1.6–1.7	1.5

Predictably, countries with the shallowest water depths (<50 m) and thriving maritime industries — which provide useful oil and gas experience — have dominated the offshore wind market. These advantages, however, are becoming less and less essential: the promise of floating offshore wind structures are helping many other locations to act upon pressure to decarbonize their energy sources and pursue floating offshore wind despite having deep waters (50–200 m). Floating wind farms represent a highly scalable future energy source and the technology is gathering significant attention in markets such as the United States, Japan, and a host of European nations such as the UK, France, Portugal, Norway, and Spain.

7.1.3 *Floating Offshore Wind Sector Forecast*

The spectacular growth in demand for renewable energy sources has outpaced even the most dramatic expectations. Despite this global shift in attitudes, fossil fuels currently account for the majority of the world's energy sources and will continue to do so for decades to come. Nevertheless, the demand for alternative energy sources, such as wind power, will continue to increase rapidly.

A key benefit offered by floating structures is their less intrusive nature. Traditional fixed-bottom structures require agitation to the seabed whereas floating structures have a brighter potential in terms of standardized installation and mass production. The floating wind market is currently composed almost exclusively of the 30 MW park from Hywind Scotland. Technological advances and cost reductions to floating offshore wind farms, however, will see its growth outpace that achieved by onshore and fixed-bottom wind farms at their outset. As floating offshore wind power continues to mature, it will secure thousands of roles which are currently filled to deliver services and goods in the oil and gas industry. The success of Hywind relies upon close collaboration between customers and suppliers in order to reduce costs and secure further deployment.

Statoil's Hywind has key markets on the west coast of the U.S., Europe, and Japan. In fact, the combined technical potential for floating wind power in these regions is 6959 GW. By 2030, it is estimated that high potential markets will grow: Japan would lead with 3.5 GW, France would come second with 2.9 GW, while the U.S. comes in third with 2 GW, and Ireland/UK last with 1.9 GW. See also Figure 7.3.

The North Sea represents the ideal depths to get the most from floating wind, with estimates signaling that over half of it would be suitable for the effective deployment of floating wind power. The European Wind Association has estimated that the energy produced from deepwater turbines would fulfil the EU's electricity consumption by more than four times.

Outside of Europe, however, the largest potential markets for wind power such as the U.S. and Japan do not have appropriate water depths. With just a few sites appropriate for fixed-bottom wind turbines due to deep waters, floating installations would represent a key step toward meeting

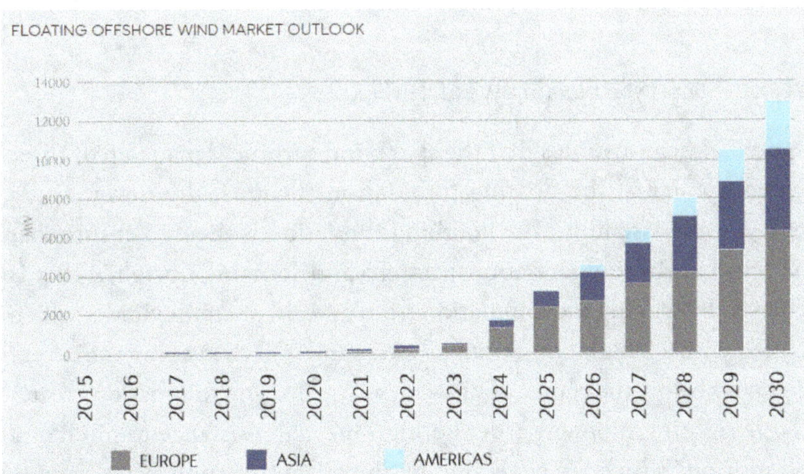

Figure 7.3 FOWT forecast (Google search).

renewable energy goals. California, for example, aims to achieve 50% renewable energy while Japan's shift from nuclear power requires a diverse range of new and sustainable energy sources.

7.1.4 *Floating Platform Configuration*

Offshore wind turbines can come in one of two configurations: fixed-bottom or floating. Fixed-bottom wind turbines are appropriate in shallow or moderately deep waters (less than 50 m), representing a more economical choice. Where seabeds are softer or waters are deeper, floating wind turbines represent a more attractive choice given that they can be used in water as deep as 700 m. Drawing upon classifications from floating offshore platforms used in the offshore oil and gas industry, these platforms have a cheaper anchor installation and come in four main categories:

- Spar buoy type
- Tension-leg platform (TLP) type
- Semi-submersible type
- Pontoon-type (Barge-type)

7.1.4.1 *Spar-Type Floating Wind Turbine*

Figure 7.4 demonstrates that the spar wind turbine is made up of three key components: the floating foundation (floater), the tower, and a rotor-nacelle assembly. The floating foundation is there to ensure that the wind turbine floats appropriately and remains upright. This is achieved by making the foundation from a steel/concrete cylinder which is filled with a ballast composed of water and gravel, keeping the center of gravity below the center of buoyancy. The floating foundation creates a large righting moment arm and high inertial resistance to pitch and roll motions which are generated by the waters, causing a free-yawing effect.

Figure 7.4 FOWT: Spar buoy type.[26]

The first full-scale spar model was deployed off the southwest coast of Karmoy island in Norway. The project was managed by Statoil in the Hywind demonstration project. The turbine is moored by three taut mooring lines going into the deep water, weighs 5300 tons, has a floating foundation draft of 100 m, a hub height of 65 m, a rotor diameter of 82.4 m and a 2.3 MW rating.

7.1.4.2 *Tension-Leg Platform Type*

This turbine is made possible by a floating platform structure which supports the turbine in the water, as shown in Figure 7.5(a). Used extensively in the oil and gas industry, conventional TLP platforms are made up of square pontoons which have columns upon which the topside deck rests. The mini-TLP is a smaller version of this full form and it has been adopted to create the TLP-type floating wind turbine as illustrated in

(a) (b)

Figure 7.5 (a) FOWT TLP-type and (b) oil and gas platform. Image (a) is reproduced with permission from https://ore.catapult.org.uk/wp-content/uploads/2018/01/Floating-wind-technology-assessment-June-2015.pdf. Image (b) is reproduced with permission from https://lubepoint.wordpress.com/2014/12/02/need-to-knowtension-leg-platform-tlphow-does-it-work/.

Figure 7.5(b). A key advantage of this configuration is that it may be assembled and commissioned onshore unlike the spar-type, which must be assembled in the water itself. This means that the TLP-type is far simpler to deploy but has a relatively poorer dynamic response to waves when compared to the other three types. It is also susceptible to a phenomenon known as "pull down" — an increase in draft caused by the platform being offset from its equilibrium position.

7.1.4.3 *Semi-Submersible Type*

This type is made up of three large column tubes which are connected to one another via a series of tubular members. A single wind turbine can be positioned on the column tubes, or multiple turbines could sit on all of the columns. Another configuration could see the wind turbine placed at the geometric center of the column tubes in order to be supported by lateral bracing members. These column tubes are partially filled with water in order to provide a stabilizing ballast. While Principle Power Inc. is promoting this design, there is no semi-submersible floating wind turbine currently in operation. Figure 7.6(a) shows a full-scale prototype of a submersible type which consists of three columns tubes complemented by patented horizontal water entrapment heave plates at the bases. Figure 7.6(b) outlines a concept presented by Ishihara *et al.*,[2] which has three wind turbines upon the tubular columns.

The large pontoon structure supported by this type is capable of supporting a group of wind turbines.[3] Stability is achieved by distributed buoyancy and leveraging the weighted water plane area for the righting moment. While this type can be moored by conventional catenary anchor chains, it is particularly susceptible to roll and pitch motions induced by waves. These motions are also experienced by other ship-shaped vessels which go on the sea. This means that this type can only be placed in calm seas such as in a harbor, lagoon, or sheltered cove. The bulk of research for this type has been conducted by the National Maritime Research Institute (NMRI) in Tokyo and its design concepts — alongside those of Professor Ohta from Kyushu University — can be seen in Figure 7.7.

(a) (b)

Figure 7.6 FOWT: (a) Semi-submersible type — real (Principle Power, 2017). (b) Semi-submersible type — conceptual.[2]

(a) (b)

Figure 7.7 Pontoon-type wind turbine.[3] (a) NMRI concept. (b) Ohta's concept with rotor shroud nozzle.

7.2 Prediction of Aerodynamic Performance of Floating Offshore Wind Turbines

The design and simulation of wind turbine systems is predicated upon predictions of aerodynamic loads. Due to this fact, a range of codes have been developed in order to effectively model wind turbine behavior. The majority of these codes are leveraged in the wind industry in order to validate wind turbine designs, given that there are not a great number of FOWT experiments. There are currently three major approaches which are used to assess the aerodynamic performance of a wind turbine: BEM Theory, Vortex Lattice Method, and Computational Fluid Dynamics (CFD). The larger the size of the wind turbine blade, the more important flexibility and addresses to structural dynamics become. Aero-servo-elasticity is a multidisciplinary technology which is used to assess the interaction between an aircraft's flexible structure, flight control systems, and both the steady and unsteady aerodynamic forces which result from the motion of the aircraft.[4]

7.2.1 *Aero-Servo-Elastic Method*

Aerodynamic models which are capable of determining wind loads and structural models must be used as part of the aeroelastic code to predict the dynamic performance of wind turbines with a high level of accuracy. These codes have been expanded over the years in order to incorporate the dynamics which are specific to offshore installations such as incident waves, hydrodynamics, sea currents, and the foundational dynamics which apply to the support structure. It is important that the accuracy and correctness of these codes are verified given their sophistication and the limited dataset which is available to use for validation purposes. In order to meet this need, the Offshore Code Comparison Collaboration (OC3) was founded, which operates as part of the Subtask 2 of the International Energy Agency (IEA) Wind Task 23. The OC3 tests newly developed codes and has four key activities:[5]

1. Discuss modeling strategies in detail.
2. Cultivate a comprehensive suite of benchmark models and simulations.
3. Run simulations and process the subsequent results.
4. Compare and discuss those results.

These activities are performed through a technical exchange fostered by international participants from universities, research institutions, and industry professionals hailing from across the United States of America, Germany, Denmark, the United Kingdom, Spain, the Netherlands, Norway, Sweden, and South Korea. OC3 is also responsible for testing the majority of the aero-servo-elastic codes which were developed in order to model the dynamic response of offshore wind turbines. In a similar fashion, the author launched a research project at Lloyd's Register (a classification and certification society known as LR). The initial research was conducted in order to assess the unsteady aerodynamic behavior exhibited by FOWTs, and also the implementation of modifications made to the aerodynamic module of the aero-servo-elastic code, FAST.

Thousands of simulations are performed throughout the design review, compliance and certification process through which wind turbines must pass. Performance behaviors of the turbines are assessed using analytic codes like Bladed (commercial and NOT open source) or FAST (non-commercial and open source). The project will investigate the coupled dynamics of a wind turbine which is mounted upon a spar buoy floating platform. FAST — an open-source code developed at the National Renewable Energy Laboratory (NREL)[5] — will be used to analyze wind aerodynamics, wave hydrodynamics, and the response of the structure. A subset of Design Load Case (DLC) simulation results acquired in OC3 Task 23, Phase IV will be verified against the CFD simulation results. Any discrepancies will be met with a thorough investigation to discover their origins and plans will also be made in a big to improve and implement correction factors to the BEM code (AeroDyn) of LRFAST (the LR version of FAST).

7.2.1.1 *BEM Method*

The majority of wind turbine design codes (exclusively open-source code such as FAST) are based predominantly upon the BEM (in the AeroDyn module of FAST). BEM theory brings together blade element theory with momentum theory in order to calculate the local forces which are enacted upon a propeller or wind turbine blade. The theory builds upon Actuator Disk Theory which was first proposed in the late 19th century by Rankine and Froude. BEM theory, on the other hand, is commonly attributed to Betz[6] and Glauert[7], and marks the combination of two distinct theories: blade element theory and momentum theory.

Angular momentum is accounted for in BEM theory which means that the air after interaction with the rotor — referred to as the wake — has angular momentum. That is to say that the air begins a rotation around the z-axis from the moment that it brings to interact with the rotor. This phenomenon is illustrated in Figure 7.8.

Given that the rotor is the device which extracts energy from the wind through its rotation caused by the wind itself, it is important that angular momentum is taken into account. There are two inter-related and simultaneously working components which make up wind turbine dynamics: the rotor aerodynamic forces and the flow field momentum balance. Momentum theory dictates that the rotor is treated as a permeable actuator disk and that there is a discontinuous drop in pressure across it, as demonstrated in Figure 7.8.

To arrive at its conclusions, BEM theory breaks the blades of a wind turbine into a string of elements along the span. Annular regions are traced out by the elements as they rotate in the rotor plan, as outlined in Figure 7.8. It is at these annular regions where the induced velocities created by the wake transform the local flow velocity that is recorded at the rotor plane. BEM provides the tool to analyze the streamtubes through the rotor disk, which have the potential to be smaller than annular regions while providing more computation fidelity. AeroDyn, however, only allows analysis using annular regions as currently written. Figure 7.8

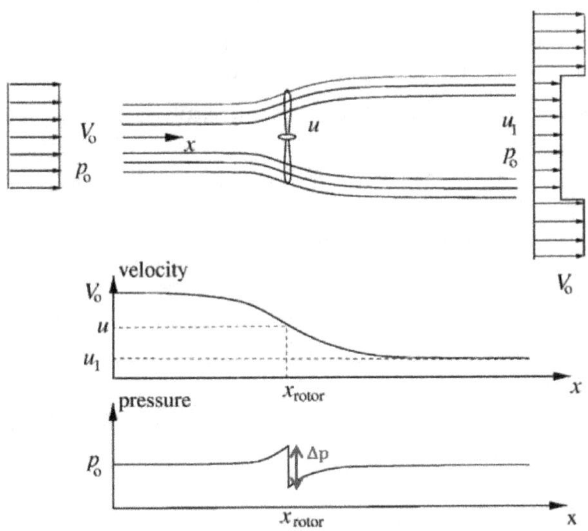

Figure 7.8 Illustration of the streamlines past the rotor, and axial velocity and pressure distributions upstream and downstream of the rotor.[8]

demonstrates that the area of the perpendicular rotor disc multiplied by the pressure drop across the rotor is equal to the thrust on the rotor.

$$T = \Delta p A \tag{7.1}$$

The pressure drop can be found (freestream velocity can be read as V_0 or V_∞) by applying the Bernoulli equation upstream and downstream of the rotor (this avoids the pressure discontinuity).

$$p_0 + \frac{1}{2}\rho V_0^2 = p + \frac{1}{2}\rho u^2 \tag{7.2}$$

$$p - \Delta p + \frac{1}{2}\rho u^2 = p_0 + \frac{1}{2}\rho u_1^2 \tag{7.3}$$

Subtracting Eq. (7.2) from Eq. (7.3), the pressure drop, the thrust, is thus found.

$$\Delta p = \frac{1}{2}\rho(V_0^2 - u_1^2) \tag{7.4}$$

$$T = \Delta p A = \frac{1}{2}\rho A(V_0^2 - u_1^2) \tag{7.5}$$

The law of conservation of mass will yield the mass flow rate if the streamlines in Figure 7.8 form a closed streamtube.

$$\dot{m} = \rho A_0 V_0 = \rho u A = \rho u_1 A_1 \tag{7.6}$$

The law of conservation of momentum is used in order to find the thrust on the rotor.

$$T = \dot{m}\left(V_0 - u_1\right) = \rho u A(V_0 - u_1) \tag{7.7}$$

A relationship between the three velocities is found by equating Eq. (7.5) and Eq. (7.7).

$$u = \frac{1}{2}\left(V_0 + u_1\right) \tag{7.8}$$

Defining the axial-induced velocity as $W_n = V_0 - u$ and substituting into Eq. (7.8) yields:

$$2W_n = V_0 - u_1 \tag{7.9}$$

The calculation of the induced velocity is permitted by substituting Eq. (7.9) into Eq. (7.7).

$$T = 2\rho u A W_n \tag{7.10}$$

$$W_n = \frac{T}{2\rho u A} \tag{7.11}$$

The axial-induced velocity can be non-dimensionalized with respect to the wind speed, V_0. This yields the axial induction factor, a.

$$a = \frac{W_n}{V_0} \tag{7.12}$$

Using the axial induction factor, the velocity at the rotor plane can thus be defined.

$$u = V_0(1 - a) \tag{7.13}$$

Substituting Eqs. (7.12) and (7.13) into Eq. (7.10), a coefficient of thrust, C_T, can be defined for the rotor.

$$C_T = \frac{T}{0.5\rho V_0^2 A} = 4a(1 - a) \tag{7.14}$$

To perform aerodynamic analysis upon a propeller or rotor blade, BEM theory dictates that it can be divided into segments. To determine the thrust and torque that are produced, the aerodynamic forces can be integrated over the whole blade and over one revolution once they are determined. The blade cross-section is observed at the blade element level, and its orientation is related to the flow. The angles involved are the flow angle, φ, the angle of attack, α, and the twist, β, shown in Figure 7.9.

The angle between the relative wind velocity vector and the rotor plane is defined as the flow angle, φ. This is a summation of the other two

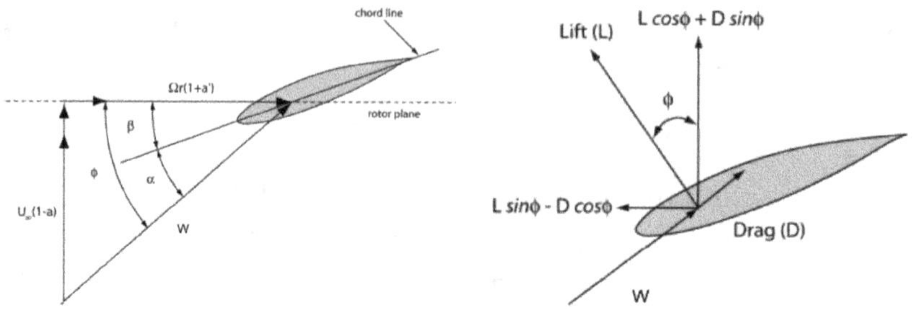

Figure 7.9 Various angles related to a blade element.[11]

angles, the angle of attack, α, which is the angle between the relative wind velocity vector and the airfoil chord, and the twist, β, which is the angle between the airfoil chord and the rotor plane.

The force coefficient in the thrust and rotational directions can be obtained through the use of these angle definitions alongside the coefficients of lift and drag. BEM theory does, however, have its limitations given the simplicity of the theory. The theory assumes, for example, that calculations are static such as the airflow field around the airfoil which is always in equilibrium and the passing airflow which accelerates instantaneously in order to adjust to the changes brought about by the vorticity in the wake. In practice, however, it has been demonstrated that there is an adjustment period necessary in terms of the airfoil response to a changing wake as a consequence of new airflow or turbine operation conditions.[9] AeroDyn recommended the generalized dynamic wake model in order to model this time lag effect more accurately. BEM theory also has another key limitation in that it breaks down at the moment when the blades experience large deflections out of the rotor plane. Due to the fact that the theory assumed momentum is balanced in a plane parallel to the rotor, aerodynamic modeling errors are encountered with any deflections of the rotor. Blade element theory is also a source of further limitations of BEM theory. Due to the fact that the theory assumed forces act upon the blade element in a two-dimensional manner, span-wise flow is neglected.

$$\tan\varnothing = \frac{U_\infty(1-a)}{\Omega r(1+a')} = \frac{1-a}{(1+a')\lambda_r} \tag{7.15}$$

If the blade motion is significant, however, we must include the local velocities in the calculation of the inflow angle, as follows:

$$\tan\varnothing = \frac{U_\infty(1-a)+v_{e-op}}{\Omega r(1+a')+v_{e-ip}} \tag{7.16}$$

Despite the fact that the inflow angle changes with element location, this equation holds for all elements of the blade along the span. The induced velocity components in Eqs. (7.15) and (7.16) are a function of the forces on the blades and we use BEM theory to calculate them. Given that a thorough derivation of these equations can be found in a wide range of existing wind turbine design handbooks,[10] it will only be summarized here. From blade element theory, the thrust distributed around an annulus of width dr is equivalent to

$$dT = B\frac{1}{2}\rho V_{total}^2 (C_l \cos\varphi + C_d \sin\varphi) c dr \tag{7.17}$$

and the torque produced by the blade elements in the annulus is equivalent to

$$dQ = B\frac{1}{2}\rho V_{total}^2 (C_l \sin\varphi - C_d \cos\varphi) c r dr \tag{7.18}$$

Now, to relate the induced velocities in the rotor plane to the elemental forces of Eqs. (7.17) and (7.18) we must incorporate the momentum part of the theory, which states that the thrust extracted by each rotor annulus is equivalent to

$$dT = 4\pi r^3 \rho U_\infty^2 (1-a) a dr \tag{7.19}$$

and the torque extracted from each annular section is equivalent to

$$dQ = 4\pi r^3 \rho U_\infty \Omega (1-a) a' \, dr \tag{7.20}$$

Thus, when we include two-dimensional airfoil tables of lift and drag coefficients as a function of the angle of attack, α, we have a set of equations that can be iteratively solved for the induced velocities and the forces on each blade element. Before the system of equations is solved, however, it is important to make a range of corrections to BEM theory. These corrections include tip- and hub-loss models to account for vortices shed at these locations, the Glauert correction to account for large induced velocities ($a > 0.4$), and the skewed wake correction to model the effects of incoming flow that is not perpendicular to the rotor plane. These three corrections will be described in further depth below.[11]

7.2.1.2 *Tip-Loss Model*

When it comes to major limitations for the original BEM theory, a key one is that there is no influence of vortices shed from the blade tips into the wake on the induced velocity field. Multiple helical structures are created by these tip vortices in the wake, and they go on to play a role in the distribution of induced velocity at the rotor. The impacts on induced velocity are the strongest near the tip of the blades which also happens to be the area that bears the greatest influence on the power which is produced by the turbine. As mentioned by Glauert,[7] AeroDyn makes an effort to compensate for this deficiency by using a theory developed by Prandtl. By modeling the helical vortex wake pattern as vortex sheets that are convected by the mean flow and have no direct effect on the wake itself, Prandtl simplified the wake of the turbine. This theory can be summarized by a correction factor applied to the induced velocity field, F, and is expressed simply in the following way:

$$F = (2/\pi)\cos^{-1}e^{-f} \tag{7.21}$$

where

$$f = \frac{B(R-r)}{2r\sin\varnothing} \tag{7.22}$$

This correction factor is used to modify the momentum part of the BEM equations, replacing Eqs. (7.19) and (7.20) with the following:

$$dT = 4\pi r^3 \rho U_\infty^2 (1-a) aFdr \tag{7.23}$$

$$dQ = 4\pi r^3 \rho U_\infty \Omega (1-a) a' Fdr \tag{7.24}$$

The Prandtl model is frequently used in engineering codes such as AeroDyn due to the reasonable accuracy that it offers when it comes to most operating conditions and easy formulaic implementation. Like other engineering models the Prandtl model does, however, have a set of limitations which affect its accuracy. The first limitation is that it limits its validity to lightly loaded rotors given that the model assumes the wake does not expand. Furthermore, Glauert[7] demonstrates that the accuracy of the model decreases with lower numbers of blades (fewer than three) and higher tip-speed ratios relative to the more accurate and computationally expensive Goldstein solution. In Figure 7.10 an example is given which illustrates that the radial distribution of the tip-loss correction a blade experiences at the inflow angle, φ, is constant along the span

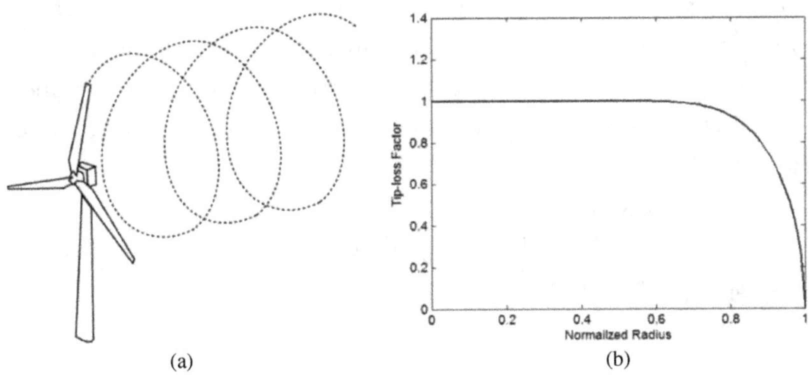

Figure 7.10 (a) On one blade-helical wake pattern with single tip vortex. (b) Tip-loss factor for blade with constant 10° inflow angle along span (optimal twist).[11]

at 10°. With the tip-loss model in use, the tip-loss factor decreases dramatically as the radial position along the blade goes further to the blade tip. This coincides with a significant increase in the induction factor toward the tip. The higher the induction factor, the more dramatically the relative wind speed for a given blade segment decreases along with the angle of attack. This means that the loading (the lift and drag forces) falls near the tip. AeroDyn also uses — in addition to the Prandtl model — an empirical relationship for the tip loss (GTECH) which is based upon the Navier-Stokes solutions of Xu and Sankar,[12] as described in the following equations:

$$F_{new} = \frac{F_{Prandtl}^{0.85} + 0.5}{2} \quad \text{for} \quad 0.7 \le \frac{r}{R} \le 1, \tag{7.25}$$

or

$$F_{new} = 1 - \left(\frac{r}{R}\right)\frac{1 - F_{Prandtl\left(\frac{r}{R}=0.7\right)}}{0.7} \quad \text{for} \quad \frac{r}{R} < 0.7 \tag{7.26}$$

These relationships must be used in conjunction with Eqs. (7.21) and (7.22) given that they are a correction for the Prandtl model. It is important to note, however, that this correction may not be applicable to all turbine configurations given that it was based upon a specific turbine design at one wind speed.[13] It is also physically unrealistic in that the tip blade station experiences a tip-loss factor greater than zero at the tip. The intention of the GTECH (Georgia Tech) model is to provide a better model to measure the effects of the relatively large inflow velocities (compared to the hovering rotors for which the Prandtl model was developed) endured by wind turbine rotors spacing the tip vortex rings farther apart in the wake.

7.2.1.3 *Glauert Correction*

The BEM theory also has another limitation in that the basic theory becomes invalid when the induction factor is great than 0.4. This can occur with those turbines which operate at high tip-speed ratios (constant speed turning at low wind speeds, for example) when the rotor enters a turbulent wake state ($a > 0.5$). Momentum theory dictates that this state is created when a portion of the flow in the far wake begins to spread upstream which is a violation of the basic assumptions BEM theory is built upon. This flow cannot occur physically but rather, more flow enters from outside the wake and the turbulence increases. This means that the flow behind the rotor slows down yet the thrust applied on the rotor disk increases. Glauert[14] developed a correction to the rotor thrust coefficient in order to compensate for this effect. The correction is based upon experimental measurements which were taken of helicopter rotors with large induced velocities. The model was originally developed in order to correct the thrust coefficient of an entire rotor but it is also used to correct the local coefficient of the individual blade elements when used alongside BEM theory. This means that it is essential to grasp the relationship between Glauert's correction and the tip-loss model. The induced velocities are large when the losses near the tip are also high, meaning that there is an increase to the potential for a turbulent wake near the tip. This means that for each element the total induced velocity calculation must be calculated using a blend of the tip-loss and Glauret corrections. Buhl[15] derived a modification to the Glauert empirical relation that incorporated the tip-loss correction as follows:

$$C_T = \frac{8}{9} + \left(4F - \frac{40}{9}\right)a + \left(\frac{50}{9} - 4F\right)a^2 \tag{7.27}$$

or, solving for the induction factor,

$$a = \frac{18F - 20 - 3\sqrt{C_T(50 - 36F) + 12F(3F - 4)}}{36F - 50} \tag{7.28}$$

While the empirical relationship is different from those models of other authors,[10,16] this relationship is necessary in order to ensure numerical

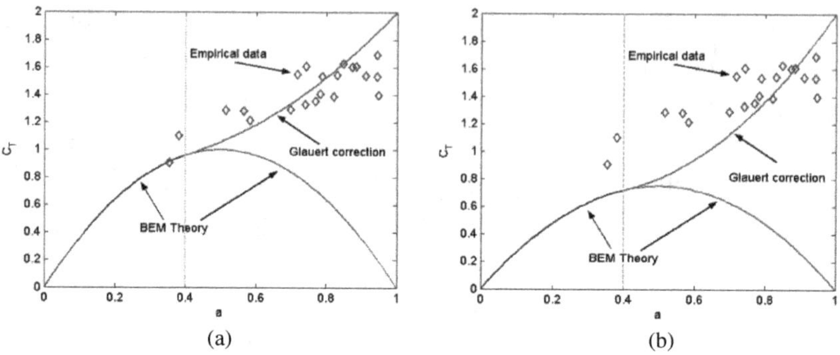

Figure 7.11 (a) Glauert correction for tip-loss factor, F = 1. (b) Glauert correction for tip-loss factor, F = 0.75.[11]

instability is eliminated when using the Glauert to calculate the elemental thrust alongside the tip-loss correction model.

An example of the Glauert correction is shown in Figure 7.11(a) when the tip-loss factor is equal to one. When the induction factor, a, is 0.4 the BEM theory and Glauert correction produce the same value for thrust coefficient of 0.96. At this induction factor the slopes are also equivalent. The BEM theory predicts a much lower thrust coefficient for most induction factors when the tip-loss factor is less than one, e.g., 0.75 as in Figure 7.11(b). This means that the Glauert correction must also adjust so that the value and slopes once again match at the induction factor of 0.4 in order to prevent numerical instability in AeroDyn. This figure demonstrates the sensitivity of the induction factor to the tip-loss factor seen in Eq. (7.28).

It is important to note that the Glauert correction was not originally intended to be applied to a rotor annulus, but rather to an entire rotor disk. An alternative model with BEM theory does not currently exist, however, given that there is a limited amount of experimental data.

Given that NREL FAST (with AeroDyn as a module for aerodynamic prediction) is an open-source tool for the research community and its methods are proven in industrial applications, it has been chosen as the basis for this research. Furthermore, the FAST tool is itself certified for the application.

The NREL 5MW Wind Turbine is chosen herein as an example. There are a wealth of FOWT floating platform concepts available. The report by Jonkman and Musial[5] from IEA OC3 Phase IV showcased a spar buoy concept called "Hywind" which was developed by Statoil of Norway. The data was supplied by Statoil and it has been modified for public dissemination. This concept was chosen given that it has a simple design, it is suitable for modeling, there are comprehensive amounts of data available, and a full-scale prototype exists. Some aspects of the original Hywind platform design were slightly adapted in order to support NREL 5MW baseline turbine which has a few slightly different properties to the Statoil turbine. The support structure (tower and substructure) was altered alongside the control system, while the rotor-nacelle assembly of the NREL 5MW wind turbine, as well as the aerodynamic and structural properties, were the same as in the Hywind project. The new system was referred to as "OC3-Hywind" and has a deeply drafted, slender spar buoy with catenary mooring lines. This can be seen in Figure 7.12.

7.2.2 *Computational Fluid Dynamics*

Aerodynamics is essential to the design and operation of wind turbines. Given that wind turbines need to maximize the extraction of energy that is generated from winds, an accurate understanding of a turbine's aerodynamic performance is essential. The majority of the existing studies on wind turbine aerodynamics have been executed using inexpensive and simple methods based on BEM theory. While this method provides a basic level of insight into the design of turbines and BEM-based methods produce some success, there are significant shortcomings that current BEM models suffer from when it comes to accurately predicting aerodynamic performance. Span-wise and other 3D effects are totally

Figure 7.12 Illustration of the NREL 5MW wind turbine on the OC3-Hywind spar.[11]

neglected using BEM methods, for example, which is straightforward but lacks the accurate physics necessary to resolve the complex flow field around a rotor. BEM theory is also unable to model the resulting development of a turbulent region behind the rotor where it leads to toroidal recirculation normal to the rotor blade, which is expected to be most significant at the blade tips in below-rated wind conditions. CFD techniques, on the other hand, are considered to be capable of mitigating a range of the inaccurate simplifying assumptions which are used in common wind turbine analysis methods.

This chapter places its focus upon a 3D model of the 5MW NREL rotor with rigid body assumption. The rigid body model will initially be considered to test the method of using six DOF of turbine platform motions as per the DLCs simulated in OC3 IEA Task 23, Phase IV, Table 17 on page 57 of Jonkman and Musial.[5] The aero-elastic modeling of the complete turbine would demand unrealistic time and resources in terms of computational capacity, so the rigid body assumption is used in the CFD model with the commercial CFD tool called Ansys CFX.

Given that construction details are unavailable for the blades and hub, they will not be modeled exactly as NREL 5MW, but rather their 3D models will be based upon details which are available in the public domain. Restrictions in terms of computational time mean that the tower is not modeled in CFD. To generate the results for desired values of DLCs, FAST is used. The experimental results of OC3 Phase IV are not available and thus NREL FAST results will only be considered for comparison and used as a reference throughout the primary section of this research study.

The NREL 5MW offshore baseline wind turbine, as described in Jonkman *et al.*,[17] provides the basis for the geometry of the rotor blade considered in this example. The geometry data which was taken from the reference are summarized in Table 7.2. A total rotor radius of 63 m is achieved through a 61 m blade which is attached to a hub with a radius of 2 m. Several airfoil types — provided in the rightmost column of the table — are used to compose the blade and the corresponding profiles are outlined in Figure 7.13. A perfect cylinder composes the first portion of the blade and as the cylinder extends away from the root it is smoothly blended into a series of DU (Delft University) airfoils. At 44.55 m away from the root, the NACA64 profile is used to define the blade all the way to the tip (Figure 7.14). The parameters which remain from Table 7.2 are elaborated upon in Figure 7.14:

- "RNodes" refers to the distance going from the rotor center to the airfoil cross-section in the blade axis direction.
- "AeroTwst" refers to the angle of the twist for a given cross-section, given that the blades are twisted in order to enhance overall aerodynamic performance.

Table 7.2 Wind turbine rotor geometry definition.[17]

RNodes (m)	AeroTwst (deg.)	Chord (m)	AeroCent (–)	AeroOrig (–)	Airfoil
2.0000	0.000	3.542	0.2500	0.50	Cylinder
2.8667	0.000	3.542	0.2500	0.50	Cylinder
5.6000	0.000	3.854	0.2218	0.44	Cylinder
8.3333	0.000	4.167	0.1883	0.38	Cylinder
11.7500	13.308	4.557	0.1465	0.30	DU40
15.8500	11.480	4.652	0.1250	0.25	DU35
19.9500	10.162	4.458	0.1250	0.25	DU35
24.0500	9.011	4.249	0.1250	0.25	DU30
28.1500	7.795	4.007	0.1250	0.25	DU25
32.2500	6.544	3.748	0.1250	0.25	DU25
36.3500	5.361	3.502	0.1250	0.25	DU21
40.4500	4.188	3.256	0.1250	0.25	DU21
44.5500	3.125	3.010	0.1250	0.25	NACA64
48.6500	2.310	2.764	0.1250	0.25	NACA64
52.7500	1.526	2.518	0.1250	0.25	NACA64
56.1667	0.863	2.313	0.1250	0.25	NACA64
58.9000	0.370	2.086	0.1250	0.25	NACA64
61.6333	0.106	1.419	0.1250	0.25	NACA64
62.9000	0.000	0.700	0.1250	0.25	NACA64

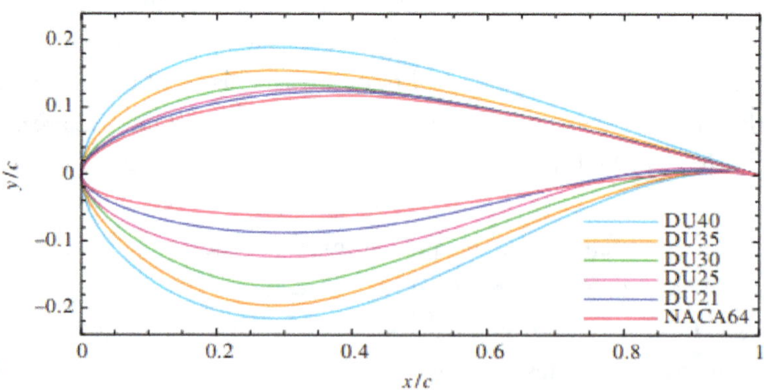

Figure 7.13 Airfoil cross-sections used in the design of the wind turbine rotor blades.

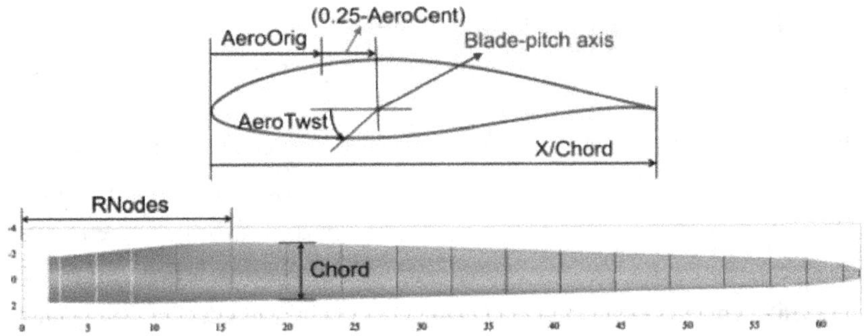

Figure 7.14 Illustration of quantities from Table 7.2.

- "AeroOrig" refers to the location of the aerodynamic center. For the majority of the blade airfoil cross-section, the aerodynamic center is found at 35% of the chord length measured from the leading edge. The aerodynamic center is gradually moved to 50% of the chord length in order to accommodate for the cylindrical shape at the root.

This is not reported in Jonkman *et al.*[17] but mentioned in Kooijman *et al.*[18]

The parameters offered in Table 7.2 have some redundancy. FAST — the aerodynamics modeling software that is typically used for wind turbine rotor computations — receives the variable "AeroCent" as an input. FAST is based around look-up tables and can provide blade cross-section steady-stage life and drag forces based upon the airfoil type, relative wind speed, and angle of attack. Empirical relationships are used to model the effects of the hub, trailing edge turbulence, and blade tip. FAST makes the assumption that the blade-pitch axis passes through each airfoil section at 25% of the chord length, and defines AeroCent – 0.25 to be the fractional distance between the aerodynamic center and the blade-pitch axis along the chordline, positive toward the trailing edge. AeroOrig + (0.25 – AeroCent), therefore, gives the location where the blade-pitch axis passes through each airfoil cross-section. This added complexity is unnecessary in this instance but the naming system is being used for the purposes of backwards compatibility with reference reports.

Quadratic non-uniform rational basis spline (NURBS) is used to represent the 2D airfoil form for every blade cross-section. The weights of the NURBS features are set to unity. In order to symbolize the round cross-sections of the blade in perfect detail, the weights are adjusted near the basis. The move-sections are lofted in the blade axis direction, also the use of quadratic NURBS and harmony weights. The power of isogeometric illustration is seen here, as a clean rotor blade surface is produced using an especially small range of input parameters. Figure 7.15(a) shows the final form of the blade in conjunction with the airfoil cross-sections. Figure 7.15(b) clarifies the twisting of the go-sections with a top view.[19]

There are three steps involved in CFD analysis procedures: pre-processing, simulation, and post-processing. The tool can overcome disadvantages of the potential theory and it can be used, for example, to solve complex fluid flow problems. Given that CFD requires a great deal of computation time to execute, it is only used for the analysis of very specific problems which throw doubt on theories.

The fundamental governing equations of fluid dynamics — the continuity, momentum, and energy questions — act as the cornerstone of CFD. These three equations are the three fundamental physical principles upon which the entirety of fluid dynamics is based: (1) conservation of mass, (2) Newton's second law, and (3) conservation of energy. Each of the three fundamental principles can be expressed as mathematical equations, in their general form most usually as partial differential equations.[20]

(a)

(b)

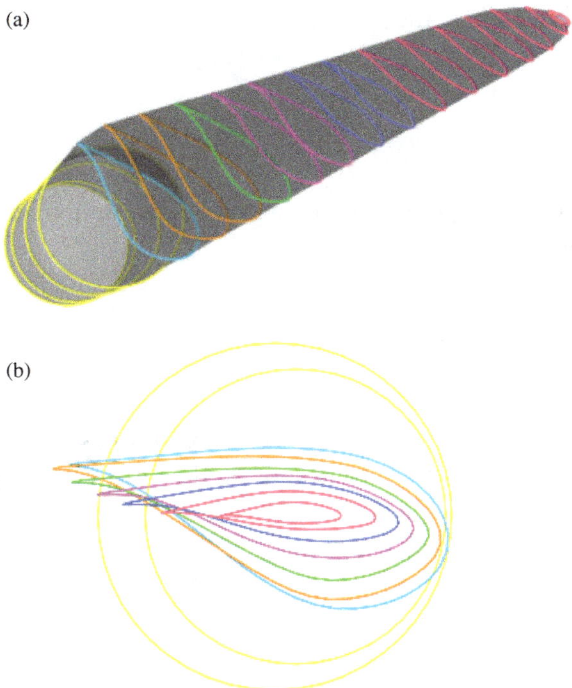

Figure 7.15 (a) Airfoil cross-sections superposed on the wind turbine blade. (b) Top view of a subset of the airfoil cross-sections illustrating blade twisting. Hexahedral meshing scheme is selected for meshing the geometry model.[19]

Mass conservation equation is known as the continuity equation:

$$\frac{\partial \rho}{\partial t} + \frac{\partial(\rho u_i)}{\partial x_i} = S_m \tag{7.29}$$

This equation is a normal form of mass conservation equations which is suitable for compressible and incompressible flow. ρ means the flow density. S_m is a continuous quality which is added from two dispersed items.

Continuity equation in the Cartesian coordinate system can be described as:

$$\frac{\partial \rho}{\partial t}+\frac{\partial (\rho u)}{\partial x}+\frac{\partial (\rho v)}{\partial y}+\frac{\partial (\rho w)}{\partial z}=S_m \quad (7.30)$$

In the i direction of inertial coordinate system, the momentum conservation equation is:

$$\frac{\partial (\rho u_i)}{\partial t}+\frac{\partial (\rho u_i u_j)}{\partial x_j}=-\frac{\partial p}{\partial x_i}+\frac{\partial (\tau_{ij})}{\partial x_j}+\rho g_i+F_i \quad (7.31)$$

where p refers to the static pressure, while g_i and F_i define the gravity volume force and external volume force in the i direction, respectively.

Stress tensor can be defined as:

$$\tau_{ij}=\left[\mu\left(\frac{\partial u_i}{\partial x_j}+\frac{\partial u_j}{\partial x_i}\right)\right]-\frac{2}{3}\mu\frac{\partial u_l}{\partial x_l}\delta_{ij} \quad (7.32)$$

7.2.2.1 *Turbulence Model for Reynolds-Averaged Navier-Stokes Equations*

Reynolds average is a method which divides the instantaneous variables of the Navier-Stokes equation into two parts: time average value and turbulent fluctuation value. The velocity can be defined as:

$$u_i=\overline{u}_i+\overline{u}_i' \quad (7.33)$$

where $\overline{u}_i$ is the time-averaged velocity and $\overline{u}_i'$ is the turbulent fluctuation velocity.

Similarly,

$$\varnothing=\overline{\varnothing}+\varnothing' \quad (7.34)$$

where $\varnothing$ denotes pressure, energy and so on.

Substituting the equations into the instantaneous continuity equation and momentum equation gives the time-averaged results. Therefore, we can obtain the continuity equation and momentum equation under Cartesian coordinates.

$$\frac{\partial \rho}{\partial t}+\frac{\partial(\rho \bar{u}_i)}{\partial x_i}=0 \tag{7.35}$$

$$\rho\frac{D\rho}{Dt}=-\frac{\partial p}{\partial x_i}+\frac{\partial}{\partial x_j}\left[\mu\left(\frac{\partial \bar{u}_i}{\partial x_j}+\frac{\partial \bar{u}_J}{\partial x_i}-\frac{2}{3}\delta_{ij}\frac{\partial \bar{u}_i}{\partial x_l}\right)\right]+\frac{\partial\left(-\rho\overline{u_l' u_J'}\right)}{\partial x_j} \tag{7.36}$$

The two equations above are both Reynolds-Averaged Navier-Stokes (RANS) equations. They have the same form as the instantaneous Navier-Stokes equation. These two methods differ in that the velocity and the remaining solving variables are changed into the time-averaged variable. The extra expression $-\rho\overline{u_l' u_j'}$ indicates the Reynolds stress which describes the effect of turbulence. If we want to solve this equation, the expression $-\rho\overline{u_l' u_j'}$ must be simulated to enclose this equation. Boussinesq assumption is a normal method to simulate $-\rho\overline{u_l' u_j'}$ which supposes that Reynolds stress and average velocity gradient are directly proportional:

$$-\rho\overline{u_l' u_J'}=\mu_t\left(\frac{\partial u_i}{\partial x_j}+\frac{\partial u_j}{\partial x_i}\right)-\frac{2}{3}\delta_{ij}\left(\rho k-\mu_t\frac{\partial u_i}{\partial x_j}\right) \tag{7.37}$$

Turbulence models are required to modify the RANS equations to simulate the turbulent flow. There are three categories of turbulence models.

First is the turbulent transport coefficient model which was proposed by Boussinesq in 1877 for solving the 2D flow problem. The second-order correlation amount of speed pulsation is defined as the product of the average velocity gradient and turbulent viscosity coefficient.

$$-\rho\overline{u_1'u_2'} = \mu_t\left(\frac{\partial u_1}{\partial x_2}\right) \tag{7.38}$$

Then he extended it to three dimensions and described this using Cartesian tensors.

$$-\rho\overline{u_i'u_j'} = \mu_t\left(\frac{\partial u_i}{\partial x_j}+\frac{\partial u_j}{\partial x_i}\right)-\frac{2}{3}\rho k\delta_{ij} \tag{7.39}$$

The objective of the model is calculating the turbulent viscosity coefficient μ_t. According to the number of differential equations which were needed in the modeling establishment, it can be divided into zero-equation, one-equation and two-equation models.

Boussinesq assumption is applied in the Spalart-Allmaras one-equation model and k-ε two-equation model. Boussinesq offers a distinct advantage in that it is capable of reducing the computation time which is required to calculate the turbulent viscosity coefficient. Otherwise, it considers that the turbulent viscosity coefficient μt is an isotropic scalar. This is not strictly true especially for some complex flow. So this assumption is a limitation.

Another method is to calculate every component of turbulent stress in the transport equation. It needs to solve an extra scalar equation such as the turbulent dissipation rate equation, which means that four extra and seven extra transport equations should be solved for 2D and 3D turbulent flow problems, respectively. Obviously, it requires more time and more internal memory of the computer.

Third is large-eddy simulation. The first two methods mentioned above were statistical average of all vortices which were on the foundation of statistical structure of turbulence. However, large-eddy simulation divides the turbulence into large-scale and small-scale turbulence. By solving the

3D modified Navier-Stokes equation, the motion characteristics of large-scale turbulence was founded. On the other hand, small-scale turbulence motion still applied the above-mentioned model.

It is important that in realistic situations the suitable models are based upon the characteristics of specific issues. The general principle of choice may meet the features such as high precision, simple application, saving calculation time and generalization. Therefore, the first one is commonly used in CFD simulations. The following contents describe several concrete turbulent models which will be applied in the CFD calculations of this chapter.

The Spalart-Allmaras model is a one-equation model that solves a modeled transport equation for the kinematic eddy turbulent viscosity. This model was initially designed for use with aerospace applications (developed specifically for aerodynamic flows such as transonic flow over airfoils) and involved wall-bounded flows. The model has demonstrated good results when used with boundary layers subjected to adverse pressure gradients. Turbo-machinery applications have also grown as an area this model is applied to.

In its original form, the Spalart-Allmaras model is effectively a low Reynolds number (Re) model, requiring the viscosity-affected region of the boundary layer to be properly resolved ($y^+ \approx 1$ mesh). Here, y^+ is $(y\mu_t)/\upsilon$. In the CFD software, the Spalart-Allmaras model has been extended with a y^+-insensitive wall treatment (enhanced wall treatment), which allows the application of the model independent of the near-wall y^+ resolution.

The formulation blends automatically from a viscous sub-layer formulation to a logarithmic formulation based on y^+. On intermediate grids ($1 < y^+ < 30$), the formulation maintains its integrity and provides consistent wall shear stress and heat transfer coefficients. While the y^+

sensitivity is removed, it still should be ensured that the boundary layer is resolved with a minimum resolution of 10–15 cells.

As stated, the Spalart-Allmaras model was initially developed specifically for aerodynamic flows which means that it is not calibrated for general industrial flows and, as a consequence, produces relatively large errors for free shear flows such as plane and round jet flows. Furthermore, the model cannot be relied upon to accurately predict the decay of homogenous, isotropic turbulence. In terms of key limitations, this model under-predicts the separation and decaying of turbulence.

7.2.2.2 *Standard k–ε Model*

The standard *k–ε* model is capable of resolving through the boundary layer and it has grown to become a well-established model which is needed to solve turbulent kinetic energy and dissipation rate equations. Exact equations are needed to derive turbulent kinetic energy transport equations. Dissipative equations, however, use physical inference and mathematical simulation. The standard *k–ε* model is only suitable for a totally turbulent flow model given that it assumes the flow is totally turbulent while the effects of molecular viscosity can be ignored.

7.2.2.3 *Re-Normalization Group k–ε Model*

Derived from the re-normalization group (RNG) mathematical method, the RNG *k–ε* model is an instant Navier-Stokes equation. It is similar to the standard *k–ε* model in terms of the turbulent kinetic energy and dissipation rate.

7.2.2.4 *Realizable k–ε Model*

The third model — realizable *k–ε* — is a relatively recent development that offers an improvement over the standard *k–ε* model while differing in two key ways. Realizable *k–ε* offers a novel formulation for the turbulent viscosity as well as a new transport equation for the dissipation rate — referred to as ε — which is derived from the exact equation for the transport of the mean-square vorticity fluctuation. The model is referred to as realizable *k–ε* due to the fact that "realizable" refers to its capacity to satisfy distinct mathematical constraints on the Reynold stresses in a way that is consistent with the physics of turbulent flows. On the contrary, neither the standard nor the RNG *k–ε* models are realizable in this way. The model introduces a variable $C\mu$ instead of constant. This means that an immediate and key benefit of the realizable *k–ε* model is it can provide improved predictions when it comes to calculating the spreading rate of both planar and round jets. The model also achieves better performance in terms of flows involving rotation, boundary layers under strong adverse pressure gradients, recirculation, and separation. Realizable *k–ε* is capable of capturing a superior mean flow of the complex structures in practically every measure of comparison.

The three –ε models are similar in that they all have the *k* and ε transport models. There are three key differences, the first being that the method to calculate turbulent viscosity is different. The second is that the turbulent Prandtl number is different when used to control turbulent diffusion. Finally, the relationship to the produce item is different. In terms of limitations, the *k–ε* has no-slip walls, adverse pressure gradients, strong curvatures, slowing of jet and difficulty in solving for ε.

7.2.2.5 *Shear Stress Transport k–ω Model*

A two-equation eddy-viscosity model called the shear stress transport (SST) *k–ω* turbulence model by Menter[21] has become very popular in recent years given that it combines the best of two worlds. The SST *k–ω* model can be used as a low-Re turbulence model without any extra damping functions given that a *k–ω* formulation is used in the inner parts of the boundary layer. The SST formulation also switches to a –ε behavior in the freestream in order to avoid the common k–ω problem that the model is too sensitive to the inlet freestream turbulence properties. The model is lauded among its proponents for the good behavior that it exhibits in adverse pressure gradients and separating flow. While this is much less pronounced than in normal k–ε models, the SST k–ω model still produces large turbulence levels in regions with large normal strain such as stagnation regions and those with strong acceleration. Cross-diffusion can be included when away from the wall but not near it due to the inclusion of a blended function based on wall distance. In other words, SST works like k–ε in the far field and k–ω near the target geometry by using wall distance as a switch. While some literature suggests that the model offers superior performance in simulating boundary layers with a diverse pressure gradient when compared to the k–ε model, the performance of SST k–ω is not vastly different to the realizable k–ε two-layer model. The accuracy level at the wind turbine scale is used as the determining factor when choosing between the two.

Turbulent flow is affected by the wall. The closer it is to the wall, the greater the impact of viscous damping in terms of reducing the tangential velocity fluctuation. The wall also plays a role in preventing normal velocity fluctuation. When far away from the wall, turbulence energy is generated and increased immediately which makes the turbulent flow strengthen due to the increase in the average velocity gradient. This means that the processing of the near wall has a clear impact on the simulation results given that the wall is the main source of vortices and turbulence.

The RANS turbulence model implies a constraint is placed on the grid near the vicinity of a wall. The turbulent boundary layer can either by modeled or completely computed. Figure 7.16 demonstrates how the near-wall zone can be divided into three layers, according to the results of experimental studies.[22] The flow is almost laminar in the innermost layer, which is the viscous sub-layer. Momentum, heat, and mass transport are all affected by molecular viscosity. The outer region is fully turbulent and

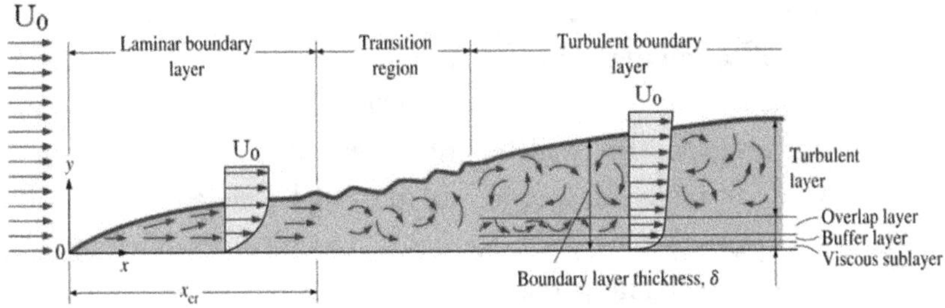

Figure 7.16 Boundary layer.[22]

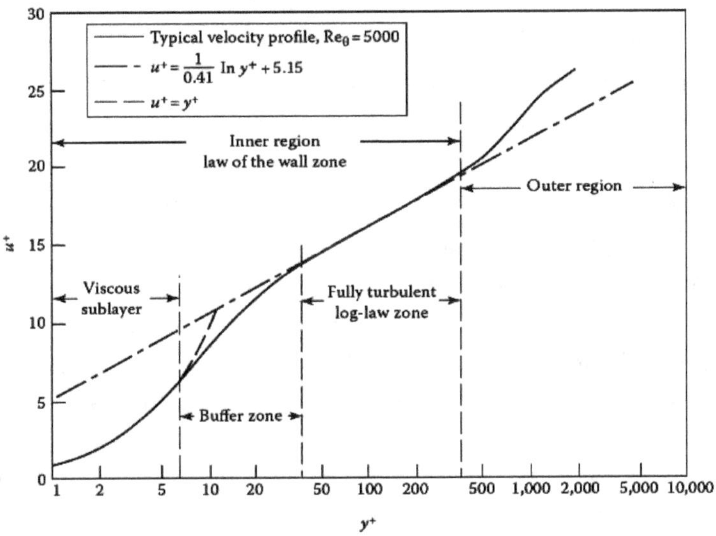

Figure 7.17 Different layers in the near-wall region. Reproduced with permission from Tannehill, J. C., Anderson, D. A., Pletcher, R. H. *Computational Fluid Mechanics and Heat Transfer*, 2nd edition. Taylor & Francis **1997**.

the blending region is found between these two, with molecular viscosity and turbulence both playing considerable functions in this region. The different layers in the near-wall region are shown in Figure 7.17.

Near-wall conditions can be processed using two distinct methods. The first applies a semi-empirical formula which is called wall function in order to correlate the viscous stress at the wall with flow data in the fully turbulent region (referred to as the log layer). This approach helps in avoiding modification to the models and simulates the effect of the wall on the turbulence directly, given that the laminar sub-layer and buffer layer are not resolved. The wall function method can save calculation resources while also offering accuracy for most high Reynolds flow problems. This is referred to as "high-Reynolds" turbulence modeling as is commonly used throughout the industry. However, the wall function method is not suitable when we study low-Reynolds problems as it cannot satisfy the assumption conditions. This means that the model needs to be capable of being solved to the wall. "Low-Reynolds" turbulence modeling is used to modify the turbulent model by resolving the inner layer. In order to account for the wall, given that near the wall the flow is almost laminar, the turbulence models have to be modified with near-wall models. In order to dampen the turbulence adequately near the wall, damping functions must also be introduced into the turbulence equations. The different grids which need to be used for both approaches are outlined in Figure 7.18. Use of the first approach results in lower cost computations and is frequently used for industrial application, but it is only valid for flows when the assumptions for the use of wall functions are valid. If this is not the case, the second approach should be used in circumstances

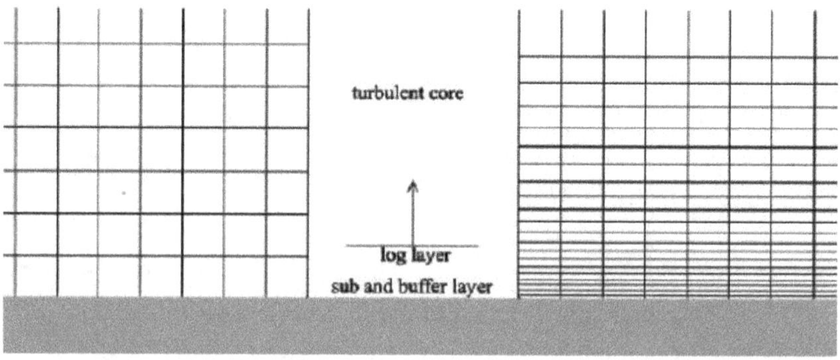

Figure 7.18 Left: Grid for high-Reynolds turbulence modeling. Right: Grid for low-Reynolds (Ansys CFD user guide).

where the near-wall models are appropriate in relation to the type of flow which is being considered.

A good result for the high-Reynolds flow problem will be achieved with standard wall function. Non-equilibrium wall function is also extended from the wall function to the non-equilibrium flow process and pressure gradient. If, however, the flow deviates from the ideal conditions, the wall function will no longer be suitable. Turbulence and average flow have a strong interaction so solving the turbulent problem is dependent on the mesh. Different near-wall treatments result in different grid requirements for the near-wall grid. The distance between the first-layer gird to the wall has to be in the logarithmic zone, with the usual calculation distance being y^+ or y^*.

The right choice from the wall for the first cell becomes obsolete with the introduction of blending functions that appear between the low-Reynolds and high-Reynolds approaches. If the first layer is in the buffer region then we must refrain, so that the y^+ is ideally less than 5 or more than 30.

7.2.2.6 *Discretization Methods*

Analytical solutions of partial differential equations involve closed-form expressions which give the variation of the dependent variables continuously throughout the domain. Numerical solutions, on the other hand, can give answers at only discrete points in the domain.

Figure 7.19, for example, outlines a section of a discrete grid placed in the xy-plane. It will be assumed — for the sake of convenience — that the spacing of the grid points in the x-direction is uniform and given by Δx, while the y-direction spacing points are also uniform and given by Δy.[19]

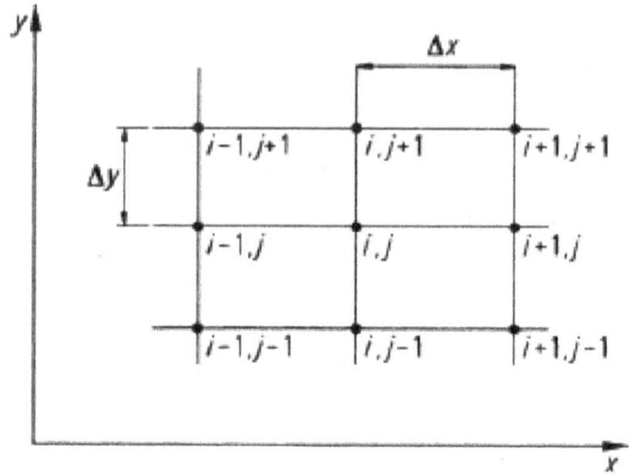

Figure 7.19 Discrete grid points.[20]

Δx and Δy are, in general, different. While it is not necessary that the two are uniform and we could work with totally unequal spacing in both directions, the vast majority of CFD applications involve numerical solutions on a grid which requires uniform spacing given that this makes the programming of the solution much simpler while saving storage space and generally resulting in greater levels of accuracy. The physical *xy* space does not witness this uniform spacing as is frequently done in CFD, but the numerical calculations are performed in a transformed computational space that has uniform spacing in the transformed independent variables, but which relate to the physical plane which has non-uniform spacing. Uniform spacing in each coordinate direction will be assumed throughout this chapter, but not necessarily equal spacing in both directions. That is to say, it will be assumed that Δx and Δy are constants, but that Δx does not have to equal Δy.

CFD frequently uses finite differences widely, and this means that the majority of this chapter will center around matters of finite differences. The philosophy of finite difference methods is to replace the partial derivatives appearing in the governing equations of fluid dynamics with algebraic difference quotients, yielding a system of algebraic equations which can be solved for the flow-field variables at the specific, discrete grid points in the flow. Let us now proceed to derive some of the more common algebraic difference quotients used to discretize the partial differential equations.

Finite difference representations of derivatives are based on Taylor's series expansions. For example, if $u_{i,j}$ denotes the x-component of velocity at point (i, j), then the velocity $u_{i+1,j}$ at point $(i + 1, j)$ can be expressed in terms of a Taylor's series expanded about point (i, j) as follows:

$$u_{i+1,j} = u_{i,j} + \left(\frac{\partial u}{\partial x}\right)_{i,j} \Delta x + \left(\frac{\partial^2 u}{\partial x^2}\right)_{i,j} \frac{(\Delta x)^2}{2} + \left(\frac{\partial^3 u}{\partial x^3}\right)_{i,j} \frac{(\Delta x)^3}{6} \tag{7.40}$$

The above equation is mathematically an exact expression for $u_{i+1,j}$ if:

(a) The number of terms is infinite and the series converges, and/or
(b) $\Delta x \to 0$.

For numerical computations, it is impractical to carry an infinite number of terms in Eq. (7.40). Therefore, this equation is truncated. For example, if terms of order $(\Delta x)^3$ and higher are neglected, Eq. (7.40) reduces to

$$u_{i+1,j} \approx u_{i,j} + \left(\frac{\partial u}{\partial x}\right)_{i,j} \Delta x + \left(\frac{\partial^2 u}{\partial x^2}\right)_{i,j} \frac{(\Delta x)^2}{2} \tag{7.41}$$

We say that Eq. (7.41) is of second-order accuracy, because terms of order $(\Delta x)^3$ and higher have been neglected. If terms of order $(\Delta x)^2$ and higher are neglected, we obtain from Eq. (7.40)

$$u_{i+1,j} \approx u_{i,j} + \left(\frac{\partial u}{\partial x}\right)_{i,j} \Delta x \tag{7.42}$$

where Eq. (7.42) is of first-order accuracy. In Eqs. (7.41) and (7.42) the neglected higher-order terms represent the truncation error in the finite series representation. For example, the truncation error for Eq. (7.41) is

$$\sum_{n=3}^{\infty}\left(\frac{\partial^n y}{\partial x^n}\right)_{i,j}\frac{(\Delta x)^n}{n!} \tag{7.43}$$

and the truncation error for Eq. (7.42) is

$$\sum_{n=2}^{\infty}\left(\frac{\partial^n u}{\partial x^n}\right)_{i,j}\frac{(\Delta x)^n}{n!} \tag{7.44}$$

This chapter uses second-order upwind discretization scheme in the CFD calculation. When second-order accuracy is desired, quantities at cell faces are computed using a multidimensional linear reconstruction approach. In this approach, higher-order accuracy is achieved at cell faces through a Taylor series expansion of the cell-cantered solution about the cell centroid. Overall, second-order upwind discretization scheme is more accurate and stable.

There has been considerable successes in implementing SST k–ω turbulence model of Ansys CFX for the flow around the nacelle of CoolerTop Vestas wind turbines. An accuracy level well below 5% was achieved from experimental results and therefore the CoolerTop concept, design, and implementation were patented.[23,24] This method is valid for applications at the industry level given that CoolerTop is used in almost all of the product families of Vestas turbines. This means that large-eddy simulation and direct numerical simulation are still computationally expensive. This is why the authors have decided to leverage this method for FOWT rotor aerodynamic assessment. Uncertainty quantification is not focused on in this work as SST k–ω turbulence model was validated in wind turbine level applications.

A brief overview of Ansys CFX CFD Solver is illustrated in Figure 7.20. ANSYS CFX consists of three software modules that take a geometry

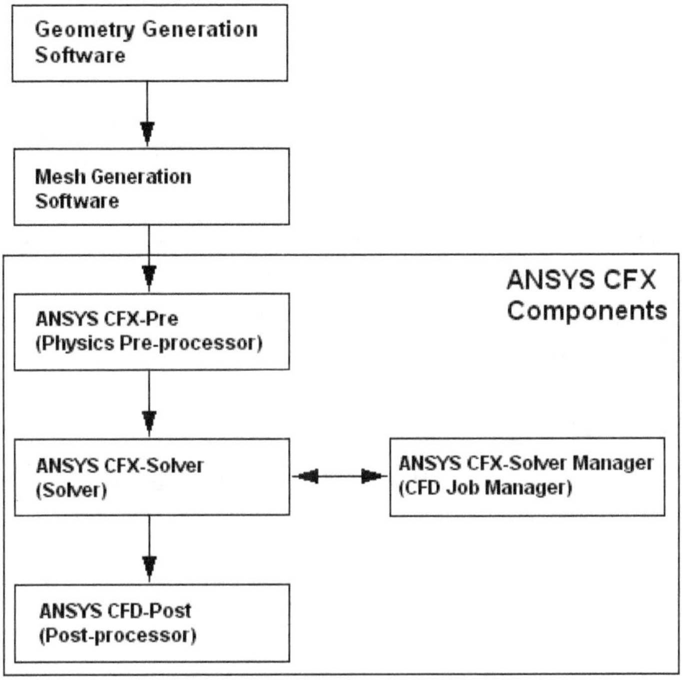

Figure 7.20 General sequence of operations in Ansys CFX analysis.

and mesh and pass the information required to perform a CFD analysis:

7.2.3 *Vortex Lattice Method*

The normal panel method is comparable to the vortex lattice method. This method is very simple to use and it is capable of offering insight into wing aerodynamics and component interaction given that it is based upon the idea of a vortex singularity as the solution of Laplace's equation. This method was used as one of the earliest on computers in order to assist aerodynamicists when estimating aircraft aerodynamics. This method is not considered in this research work, however, given that it is not quite attractive in terms of industrial applications and it lacks the valid compatibility assessment to be integrated with aero-servo-elastic control codes.

7.3 Scaled Rotor Design and Unsteady Experimentation

Floating turbines remain a nascent technology in the early prototype stage of development. This means that when it comes to validating numerical models for the aerodynamics of these machines, there is a very limited amount of data available. Accurate numerical modeling tools are essential to conduct useful design work. The validation of numerical models through the application of full-scale prototype data would be ideal, but the few full-scale floating tests which have been conducted are private and the data is not available to the research community. In order to remedy this complex situation, a number of scaled experiments have been performed in wave tanks around the world. A literature review will follow, critically discussing these experiments while offering comparisons.

Scaling laws must be adhered to in order to establish functional scale relationships. The modeling of floating offshore wind turbines follows these scaling relationships:

a. From prototype to scale model the Froude number similitude is employed. The Froude number and geometric similarity are typically used to scale offshore platform wave basin tests. Despite the fact that the Froude model does not scale all parameters properly the dominant factor in the wave mechanics problem — inertia — is properly scaled.[25] This means that when it comes to a floating wind turbine, the key properties of interest which influence the global dynamic response time of the system are covered, apart from the aerodynamic wind forces. It is impractical for a floating body subjected to wave forcing to employ a Re scaling scheme which is a common model for aerodynamic experiments. This means that Froude scaling is optimal when it comes to model testing floating wind turbines. The Froude number for a free surface wave is:

$$Fr_{wave} = \frac{C}{\sqrt{gL}} \tag{7.45}$$

where C is the wave celerity, or propagation speed, g is the local acceleration due to gravity and L is a characteristic length. The scaling relationship maintained from model scale to the full-scale prototype is given as

$$Fr_p = Fr_m \tag{7.46}$$

where p and m stand for prototype and model, respectively. Forces such as airfoil lift and drag are reliant on Re.

b. During basin model testing Froude scaled wind is employed. If aerodynamic turbine features are insensitive to Re, then the wind force to wave force ratio from prototype to model scale is maintained by utilizing Froude scaled wind and can be shown as

$$Fr_{wind} = \frac{U}{\sqrt{gL}} \tag{7.47}$$

An alternative, yet consistent, way to represent Froude scaled wind is by maintaining the ratio of wind speed to wave celerity from model to full scale. This ratio is identified by the variable Q and defined as

$$Q = \frac{U}{C} \tag{7.48}$$

where U is the wind inflow velocity.

c. The wind turbine tip-speed ratio, TSR, is to be maintained from prototype to scale model. TSR is computed as

$$TSR = \frac{\Omega R}{U} \tag{7.49}$$

where Ω is the rotor rotational speed and R is the blade tip radius. Maintaining TSR ensures that the turbine rotational speed as well as any system excitation frequencies resulting from rotor imbalance or aerodynamic interaction with the tower will scale properly. In addition, maintaining TSR will yield properly scaled turbine thrust forces and rotor torque in conjunction with a Froude scaled wind environment, assuming a low dependence on Re for the wind turbine airfoil section lift and drag coefficients. Maintaining TSR between the prototype and model is given as

$$TSR_p = TSR_m \tag{7.50}$$

The details of seven FOWT scaled experiments are discussed in this section. The tests are outlined in Table 7.3 but it is important to note that no comprehensive unsteady aerodynamic experiments have yet been conducted at scaled rotor level.

Researchers from Kyoto University in Japan performed a 1/22.5 scale experiment using a model spar buoy — as outlined in Figure 7.21(a) — in

Table 7.3 Scaled model test details.

Name of Experiment	Scale Ratio	Test Location	Platform	Aerodynamic Setup
Spar at NRMI (2009)	1/22.5	NMRI	Spar buoy	Steady force
WindFloat (2010)	1/105	UC Berkeley	Semi-submersible	Actuator disk + rotating mass
DeepCWind (2011)	1/50	MARIN	Semi-submersible, spar buoy and tension-leg platform	Cylinder
DeepCWind, continued (2013)	1/50	MARIN	Semi-submersible	Full rotor (Froude-scaled)
Tension-Leg Bouy (2011)	1/100	MARINTEK	Tension-leg and spar buoy	None
Tension-Leg Bouy (2014)	1/40	IFREMER	Tension-leg buoy N	None
Concrete Star (2014)	1/40	ECN	Braceless semi-submersible	Ducted fan
MARINTEK Braceless (2015)	1/30	MARINTEK	Braceless semi-submersible	Novel actuator
INNWIND.eu Model Test (2015)	1/60	ECN	10MW semi-submersible	Ducted fan and Froude-scaled rotor

2009. The free decay, regular wave, and irregular wave tests were conducted at the NMRI in Tokyo, Japan.[26] In addition, a series of tests were executed which combined regular waves with the application of a constant force to the top of the tower in a bid to replicate a steady thrust force. Aside from these tests, no other aerodynamic forces or interactions were considered.

Principal Power is a company from Oregon and it installed a full-scale prototype of its WindFloat platform just off the coast of Portugal. A small-scale experiment (1/105) which used this platform (shown in Figure 7.21(b)) was conducted in 2010 by Roddier *et al.*[27] During the experiment an actuator disk was used to replicate wind thrust, while a scaled spinning mass was used to generate gyroscopic forces as if a genuine rotor was present. The principal objective of this experiment was to test the performance of the platform in a 100-year wave case, but a series of regular wave cases were also conducted in order to achieve a baseline response from the platform.

The DeepCWind Consortium is led by the University of Maine (UMaine) and it conducted two sets of experiments. The first series was performed in a wave pool at MARIN in the Netherlands in 2011,[26] with the second

(a) NRMI Spar (b) WindFloat Semi-Submersible

Figure 7.21 Spar and semi-sub type porotypes.[26,27]

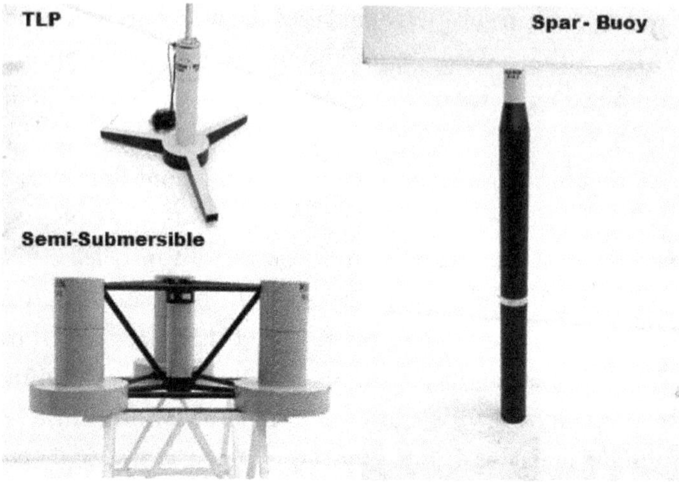

Figure 7.22 DeepCWind platforms.[26]

being performed in the same location in 2013 by Martin[28] and Koo *et al.*[29] The 2011 test used a semi-submersible and tension-leg platform designed by UMaine, in addition to a spar buoy based on the OC3 spar buoy[30] (Figure 7.22). These models were tested at a 1/50 scale in a wide variety of conditions such as free decay, regular waves, irregular waves, and wind. In the first experiment a Froude-scaled rotor based on the NREL 5MW blade design was used. A greater wind speed was called upon in order to replicate the full-scale aerodynamic/hydrodynamic force balance, due to the Reynolds mismatch when Froude scaling. The aerodynamic performance fell below expectations at the lower Reynolds numbers of the test, due to the fact that the blades were direct geometric scales of the NREL 5MW blades. The second round of tests executed in 2013 used a new rotor designed to be equal in performance to the full-scale NREL 5MW rotor and the semi-submersible platform.

A tension-leg buoy platform shown in Figure 7.23(a) was developed and tested in a partnership between researchers from the Norwegian University of Life Sciences (UMB) and the Institute for Technology

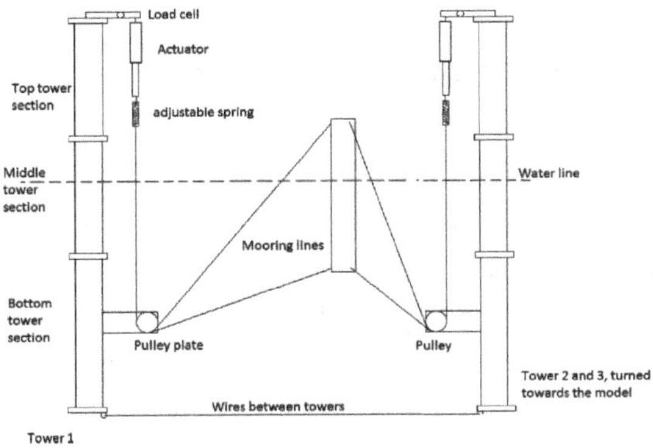

(a) UMB/IFE Tension-Leg Buoy

(b) Concrete Star Wind Floater

(c) MARINTEK Braceless Semi-Submersible

Figure 7.23 Semi-submersible platforms.[31–33,34,36]

(IFE).[31–33] The 1/100 scale platform was analyzed in a series of tests in 2011 with a particular focus paid to comparing a spar buoy with more conventional catenary mooring lines in a MARINTEK wave tank. In 2014, a 1/40 scale tension-leg buoy was also developed in testing at IFREMER. These tests did not include a rotor or other actuator to stimulate aerodynamics given that the tests revolved purely about hydrodynamics. Both testing campaigns ran a series of free decay, regular wave, and irregular wave tests.

A braceless semi-submersible called the "Concrete Star Wind Floater" (shown in Figure 7.23(b)) was tested in 2013 in the wave tank at the Ecole Centrale de Nantes (ECN). This experiment used a feedback-controlled ducted fan which made it possible to simulate aerodynamic forces.[34] The semi-submersible was designed by Olav Olsen AS and it is designed to use concrete at full scale. The Reynolds mismatch for aerodynamics when using Froude scaling we resolved by the use of the ducted fan approach. NREL's design code FAST was leveraged in order to calculate the aerodynamic forces which existed due to motion from the measured platform and a simulated turbulent or steady wind was created by the ducted fan. Free decay and regular/irregular wave tests were performed both with and without the aerodynamic forces being simulated by the ducted system.

Another experiment was recently conducted for the INNWIND.eu project. This was performed in order to test a scaled version of the OC4 DeepCWind semi-submersible.[35] Modifications were made to the platform so that it could support a 10MW wind turbine before being scaled to 1/60 scale for the test which used a smaller model. A rotor with high chord blades was also designed for the experiment, the blades being implemented in order to match the rotor thrust despite the Reynolds numbers which mismatched between the model and full-scale sizes. In a similar fashion to the Concrete Star experiments, a feedback-controlled ducted fan was used. The test parameters for this series included free decay, regular waves, irregular waves, and extreme wave conditions.

A 2015 test at MARINTEK offered another approach in order to generate the realistic aerodynamic forces with feedback.[36] The experiment was based around a braceless semi-submersible (shown in Figure 7.23(c)), but differed from previous experiments in that a series of tensioned wires connected to actuators generated the simulated forces rather than a ducted fan. FAST was once again used in order to calculate aerodynamic forces in real time while the tests were being conducted.

7.3.1 *Floating Offshore Wind Turbine Scaled Model Evaluation*

Froude scaling was used in each of the experimental campaigns above but they varied widely in terms of the scale factors which were used: the smallest was on a 1/105 scale while the largest was on a 1/22.5 scale. These differences were present predominantly due to the size of the wave tank facilities which were used as part of the test. The aerodynamic loading methods were one of the main differences between the experiments: the UMB/IFE tension-leg buoy project maintained an exclusive focus on hydrodynamics, whereas the NRMI spar buoy experiment simulated a steady thrust force with a simple constant force. The WindFloat, DeepCWind, and INNWIND experiments used fans in order to generate aerodynamic forces but differed slightly: WindFloat used an actuator disk while the DeepCWind and INNWIND tests used an actual spinning rotor. Novel actuators were used to simulate wind forces in the Concrete Star Wind Floater, MARINTEK, and INNWIND experiments which accommodated for correctly scaled wind forces as well as a realistic interplay between motions and aerodynamic forces. Despite the mechanical differences, the quality of the experiments is determined by the simulation tool which was used to create the wind forces, which in both cases was the AeroDyn module of NREL's FAST program. Among the research community there is a prevalent concern that BEM codes like AeroDyn fail to adequately accommodate for important aerodynamic phenomena such as dynamic stall, which could be more prevalent for floating platforms as a consequence of the increased rotor motion.

This means that while the simulated aerodynamic force actuators were indeed successful in providing both dynamic and correct scaling, the experiments may have been limited by the aerodynamic simulator which means the results are most useful for studying the hydrodynamics of floaters. Model validation is a key concern in each of these experiments, and will continue to be so for future experiments of this type. The DeepCWind data is publicly available and it has been used for validation in the Offshore Code Collaboration and Comparison, Continued (OC4) project and its continuation OC5.

A portion of the IRP Wind program is making an attempt to resolve this lack of available data by executing a project which will see it focus on compiling a database for researchers. This data will be made available to assist researchers in benchmarking design codes and it will include data from the MARINTEK braceless semi-submersible in addition to the UMB/IFE tension-leg buoy. In the future there are plans for these data to be published online in an accessible format, as well as a series of preliminary benchmark activities for a variety of design codes. By providing a basis for comparison of not only real experimental data but design code results benchmarked in the past, the online database will become an important resource for future developers of wind turbine design codes. The University of Strathclyde worked with the support of Lloyd's Register in an attempt to offer a scaled rotor for more detailed study of unsteady aerodynamic behavior.

7.3.2 *Floating Offshore Wind Turbine Rotor-Scaling Methodology and Application*

In order to complete a valid and reliable scale model test, proper scaling of both a model and environmental conditions must be executed effectively. Established modeling techniques are pending for floating wind turbine wind/wave testing and this chapter will explore the scaling relationships, factors, and modeling techniques.

When it comes to understanding the complex behavior exhibited by mechanical systems in a variety of environmental and operation conditions, scale model tests are an invaluable tool.[37] Where fluid dynamics are involved this is particularly true, especially when coupled with a floating wind turbine's response to aerodynamic and hydrodynamic factors. The response and action of the small-scale system needs to be as accurate as possible and this means small-scale models must be structured and scaled in such a way that the data is representative of the full-scale arrangement.[38] When it comes to a floating wind turbine, the Froude number should be maintained between the full and model scales in order to adequately scale the hydrodynamic forcing and response. This technique is considered best practice and is regularly used when it comes to the study of offshore structures, whether fixed or floating.

Designing a model-scale wind turbine rotor to measure global loads in a controlled environment which are similar to full-scale references is an incredibly complex task.[39] This is due to a major challenge in satisfying the three key scaling laws: geometric, kinematic, and dynamic similarity. If aero-elastic effects are ignored these still need to be present in order to create an aerodynamically similar rotor.[40] By maintaining a constant length scale factor and TSR across scales, geometric and kinematic similarity is easily achieved. Dynamic similarity, however, requires that the experiment maintains the blade chord Re on an entirely different scale, which is complicated in the case of a floating wind turbine given that the Re is limited to values which are markedly lower than at full scale as this is a function of the wind velocity at the blade and the chord length.[41]

A number of "performance matching" solutions have been proposed in order to try and address the aerodynamic response mismatch in the Froude scaled wind environment.[38,39,42,43] One such solution involves designing a model-scale rotor which has a similar non-dimensional aerodynamic response as the full-scale reference rotor in the Froude scaled environment. This is achieved by easing the typically strict geometric criteria so that the cross-sectional profile of the blade allows for attached flow conditions during the operation of the model. It is possible to simultaneously maintain the Froude scaled environment and first-order aerodynamic response if the blade dimensions are chosen in such a way that the global non-dimensional response of the model rotor emulates that of the full-scale reference.

The MARIN Stock Wind Turbine (MSWT), for example, is designed in such a way that it has a similar thrust response to the NREL 5MW wind turbine, meaning that various floating foundation concepts can be analyzed at the Wind/Wave Basin at MARIN.[27]

Martin *et al.*[37] offers two potential performance-matching scaling methodologies for floating wind turbine motors. The DAR (Direct Aerofoil Replacement) methodology has been developed to such an extent that the aerodynamic similarities of the rotor to full-scale references have been maximized, whether steady or unsteady. This leap means that direct measurements of the global rotor loading can be measured and legitimate conclusions can be drawn. GFRD (Geometrically Free Rotor Design) is the second and is applicable in environments similar to the MSWT but the similarity of the rotor torque is a design variable.

In almost all scenarios the blade chord Re is achievable in the experimental analysis of wind turbine rotors but is deemed to be insufficient to be considered aerodynamically similar to a full-scale reference machine. If the model is designed in such a way, however, that the attached flow occurs over the blade then the comparison of both the measured and numerically predicated performances can determine the validity of the prediction code.[44]

$$Re = \frac{\rho W_r c}{\mu} \tag{7.51}$$

While a variety of techniques have been detailed in the literature in terms of approaches to small-scale wind turbine model design, it is generally accepted that in order to maximize the blade chord Re, the design incorporates a series of geometric simplifications. To achieve attached flow over the blade simple aerofill profiles are used to define the cross-sectional area of the blade. These profiles are suited to the experimental Re and there are various publicly available measures of performance which exist. The literature contains examples of both constant and tapered chord distributions, as well as both twisted and non-twisted blades.[45] Following these methods will, naturally, result in a model rotor which does not bear geometric resemblance to the full-scale wind turbine rotor upon which it is based. The basic aerodynamic principles under which it operates, however, are the same and therefore it can be safely assumed that similar mathematical descriptions can be outlined which apply to both. The validated numerical modeling tools which result from this exercise are used in order to ascertain the performance of full-scale wind turbine rotors. Vermeer *et al.*[45] outlined a range of rotor designs in the literature prior to 2003, and each of these generally follows the described approach.

It is important that the rotor scaling methodology is defined in such a way that uncertainties around the degree of similarity of the unsteady aerodynamic response of the model rotor to the reference is minimized. This is made difficult by the complications that arise when attempting to numerically model the anticipated response of the rotor in the scaled floating environment, due to the generally poor level of understanding when it comes to unsteady aerodynamic behavior of wind turbine rotors.[44] It is proposed that, in the first instance, the measure of similarity between the model and reference rotors be the steady non-dimensional thrust ($C_T = T/0.5\rho\pi R^2 V^2$) and torque ($C_Q = Q/0.5\rho\pi R^3 V^2$) at a range of operating points as defined in Eq. (7.52). This follows the design criteria described by Martin *et al.*[37] and de Ridder *et al.*[39]

$$\begin{aligned} C_Q(\lambda)|_{full} &= C_Q(\lambda)|_{model} \qquad \forall\lambda \in [\lambda_{min}, \lambda_{max}] \\ C_T(\lambda)|_{full} &= C_T(\lambda)|_{model} \qquad \forall\lambda \in [\lambda_{min}, \lambda_{max}] \end{aligned} \tag{7.52}$$

Secondly, a measure of similarity should be developed which takes into consideration the unsteady response of the rotor. If this is not possible, attempts should be made to understand the degree of dissimilitude to the reference so that an estimate can be made of the influence any potential aerodynamic simplification has on the data which is collected. The full-scale reference turbine for this project will be the NREL 5MW Baseline Wind Turbine. Despite the fact that this is a conceptual machine, it is used because the specification is available publicly and it is extensively used in the literature.[47]

7.3.2.1 *Direct Aerofoil Replacement Methodology*

The blade chord Re deficiency — defined in Eq. (7.51) — at experimental scale in comparison to full is the key driver in the global load dissimilitude of a geometrically scaled rotor to the reference. Given that at model scale it is not possible to achieve the full scale Re, it is suggested that the rotor blade be redesigned in an effort to ensure the model-scale aerodynamic response is similar to that of the full scale. This can be achieved by reducing the strict adherence to geometric similarity criteria for the cross-section of the blade so that each distinct aerofoil profile is redesigned by ensuring that the two-dimensional performance at the blade level is similar. A benefit from the DAR Methodology when attempting to minimize the unsteady loading dissimilitude is that it allows for the maintenance of the non-dimensional chord and twist distributions at model scale.

7.3.2.2 *Geometrically Free Rotor Design Methodology*

Given that a model rotor designed using the DAR methodology can be expected to demonstrate a non-dimensional torque deficiency in the potentially complex three-dimensional geometry, a second procedure is proposed in which the primary design objectives are the performance match of both the non-dimensional thrust and torque and a simple rotor arrangement. Other than the geometric constraints imbued by the manufacturer of the rotor and the maintenance of design simplicity, no others are imposed. This means that any consideration of the unsteady aerodynamic response of the model motor is negated. In the case of a wind/wave basin assessment of a floating wind turbine model where the full system motion is the measurement of interest, however, only the matching of the steady rotor response at multiple operating points will account for the scaled rotor loading with accuracy.

7.3.2.3 *Scaling Methodology Evaluation*

In order to produce example-scale rotor designs for use in the University of Strathclyde's Kelvin Hydrodynamics Laboratory 76.0 m × 4.6 m × 2.5 m towing tank, the DAR and GFRD scaling methodologies are applied. A larger blade chord Re is achieved when assessing the aerodynamic response of the rotor hydrodynamically, at an estimated 1.9×10^5, which would be the case if operated in a wind tunnel. This is beneficial when attempts are being made to design a model rotor that has a similar aerodynamic performance to the NREL 5MW baseline wind turbine, the full-scale reference.[17] To reduce the effects of blockage a geometric scale factor of 126.0 will be used, which results in a model scale rotor diameter of 1.0 m. In order to scale the boundary layer conditions and improve the stability of the aerofoil performance, the flow will be tripped at the leading edge in both cases on the upper and lower surfaces at $x/c =$ 0.1. Forcing the transition of the flow in this manner was also proposed by Martin *et al.*[37] Evaluating the aerodynamic response of a wind turbine hydrodynamically in a towing tank requires that the impact of the testing environment be considered in more detail than would be the case in a wind tunnel. The effect of the free surface on the measured data must be considered, however. Bahaj *et al.*[47] noted that this was minimal for an actuator disk. Cavitation must be avoided as this phenomenon is not representative of the aerodynamic response of a wind turbine rotor. The mitigation of this effect can be incorporated as an absolute constraint into the aerofoil design optimization algorithm by comparing the XFOIL pressure distribution simulation to the cavitation number.[48]

In summary, the scaling methodologies outlined by Martin *et al.*[37] make an effort to directly confront the difficulties which are linked to the experimental aerodynamic assessment of a model wind turbine rotor model which is scaled down from its full-scale reference. These methodologies have elaborated upon the performance-matching techniques used by de Ridder *et al.*[39] and Campagnolo[42] by placing an emphasis upon the unsteady aerodynamic response of the model rotor from the very moment that the design process begins. This is achieved primarily through the proposal of surrogate similarity criteria in a bid to maximize the similarity of the unsteady response when there is an absence of accurate, efficient computational resources. The DAR scaling methodology is capable of producing a highly accurate performance match in terms of the rotor thrust, and though its torque results are less accurate it is still positive and stable between operator points of TSRs below 9. The analysis which has been presented demonstrates that there is a similarity in terms of the way in which the rotor loads are produced at model scale to the reference, as demonstrated by showing a favorable comparison between the axial and tangential induction onto the rotor and the local blade angle of attack. A more detailed flow analysis, however, needs to be conducted to arrive at a conclusive resolution.

The DAR rotor — as a consequence of this and similar twist and chord distributions — will respond to an unpredictable aerodynamic event in a manner which is comparable to that of the full-scale reference. Because there are potentially complex manufacturing requirements of the DAR scale rotor, an additional performance-matching methodology is offered which aims to produce model rotor designs which are less complex due to minimal geometries. The extent of torque match is similar to that of the DAR rotor, but results in a generally poorer performance given that the thrust is significantly higher for the GFRD rotor than the reference. The produced design is, however, easily manufacturable and the aerodynamic efficiency of the rotor is comparable to that of a large generic wind turbine. To achieve detailed experimentation and validation, the scaled rotor is manufactured based on the above and the various numerical codes are discussed in the next chapter.

7.4 Remaining Useful Life Prediction of Floating Offshore Wind Turbine Power Converter

After the FOWT aerodynamic numerical models are validated using scaled experiments, certification bodies are able to use them with more confidence for the certification of FOWT. Going beyond the certification itself, the numerical tool can also be used for FOWT wind farm level operation and maintenance (O&M) under the digital twin framework of the certification body's advisory business model. There are about 4–5 major components which make up more than 80% of O&M costs as per the certification body.[49,50] The power and frequency converter is one such major cost, which is why this section will explore the power converter's remaining useful life (RUL) prediction in more depth, given that it is critical for an effective and predictive O&M maintenance strategy. When compared to fixed-bottom wind turbines, the power converters found in FOWT will be more heavily affected by wind, wave, and blade control.

7.4.1 *Wind Farm Operation and Maintenance*

A decrease in the cost of parts, fabrication of structures, construction, more reliable site characterization, alongside a reduced cost of finance, and reduced operating costs mean that wind farms are being offered at lower and lower costs. Operating costs are currently fairly high given the inadequate access to failure data from service companies, many of which also act as the original equipment manufacturers. This failure data is valuable in that it indicates diagnosis (with techniques such as root cause analysis), the timing and descriptions of any maintenance interventions, details of which parts were used, and whether the parts are new or refurbished (if the parts are indeed refurbished, information will be included about previous service, too). The superior maintenance strategy depends upon two key factors: the specifics of the equipment and the commercial environment in which it operates. An O&M cost model can factor in all of these variables to provide a deterministic "best" scenario.

As wind farm operating companies gather more experience it is becoming more common for them to propose nascent policies for consistent assessment and management of maintenance across all of the assets that they have under their care. The success of these policies relies upon establishing novel ways of measuring reliability and the risk of failure faced by particular elements at particular times in the lifetime of a wind farm. In order to succeed this method of managing a wind farm requires the integration and assimilation of all relevant information sources and digital data streams available to the operator. A replicable methodology is also key for breaking down the physical assets into sub-assemblies and functions which may need to be assessed according to their vulnerabilities.

The cost of maintenance for wind farms is generally high due to the fact that accessibility is subject to a range of favorable weather conditions (tidal currents, daylight, temperatures, wind, water levels, waves). Each of these factors has the potential to increase lost electricity production onto the turbine, which increases the odds a technician may be abandoned at the location due to extreme or unfavorable weather conditions. The technicians also run the risk of having their work disrupted due to sea sickness.

7.4.2 *Why Predictive Over Condition-Based Maintenance?*

Condition-based maintenance (CBM) is a maintenance method applied to those wind turbine components which are considered to present the highest level of risk. The most common examples include gearboxes and blades. Gearboxes are particularly problematic given they can fail before the end of their design life and result in large repair and replacement costs in addition to the heavy machinery which is required in order to execute this complex maintenance task. Naturally, these costs are higher for offshore wind farms when compared to their onshore counterparts. Blades are single components which have a number of failure modes. This means that each of these modes is not often completely understood and subsequent replacement can be time consuming, hazardous, and expensive. Over the past 25 years — since the Vindeby wind farm was connected to the grid — a great deal has changed in terms of the way wind turbines are designed. Unfortunately, new designs for components were implemented in large numbers before being comprehensively checked for reliability, leading to serial failures which threatened the businesses of established and well-regarded suppliers of components and wind turbines. Many wind turbines use condition-monitoring systems which regularly and silently measure vibrations, oil temperatures, accelerations, oil impurities, and other such factors in a big to achieve early detection of issues. This early identification of potential issues allows for the scheduling of repair and replacement without the cost and issues caused by disruption to production.

This system is not without its issues: by the time an issue is flagged damage has already been done and this means that the available options may simply be limited to a replacement of the component or even the whole sub-assembly, leading to significant costs. Offshore wind farms have further issues given that maintenance intervention can only be scheduled in optimal weather conditions and major replacements may need multiple days of relatively calm weather. This means that even the simplest maintenance task could require a minimum period of four hours to allow technicians to transfer to and from the turbine and to perform their task.

Bearing this in mind, it is clear that predictive maintenance could allow for more sophisticated planning. With this system orders can be placed with an appropriate amount of time for delivery, while favorable prices can be secured through negotiations on parts, labor, tools, and more. A forecast energy yield can also be analyzed in order to minimize any impact on electricity production. Forecast weather conditions, too, can be used in order to select appropriate maintenance strategies ahead of time.

7.4.3 *Power Converter*

One of the major wind turbine components is the power or frequency converter.[49,50] An overview of the power converter and its technical aspects related to O&M will be presented in this section.

Table 7.4 and Figure 7.24 demonstrate the share of grid-connected turbines belonging to wind turbine manufacturers between 2014 and the first half of 2015 (in units).

The statistics show that the market is clearly controlled by Siemens and Vestas, the two powerhouses reaching a combined total of 86% of the

Table 7.4 2014–2015 fixed-bottom wind turbines installed in total.

Manufacturer	Units Connected, 2014	Units Connected, 2015	Total Number of Turbines, 2014–2015
Siemens	349	347	696
Vestas	47	111	158
Senvion	2	35	37
Areva	9		9
Adwen		91	91

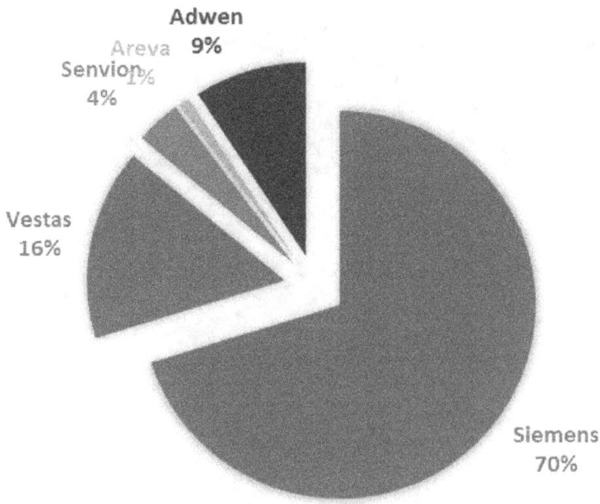

Figure 7.24 2014–2015 fixed-bottom wind turbines installed in total — in percentage share.

Table 7.5 Type of wind turbine generator used.

Type of Generator	Number of Wind Farms	Number of Turbines
Asynchronous	36	1803
DFIG	31	828
PMSG	10	198

offshore wind turbines installed in Europe. Table 7.5 and Figure 7.25 show the total number of turbines installed which use different types of generators. Currently, practically every turbine is installed with a power converter — the type of power converter is also presented in Table 7.6.

While the majority of existing offshore wind farms are installed by coasts, new projects aim to be placed far offshore such as German Bight and the UK Dogger Bank.[51] Accessibility is becoming the critical difference

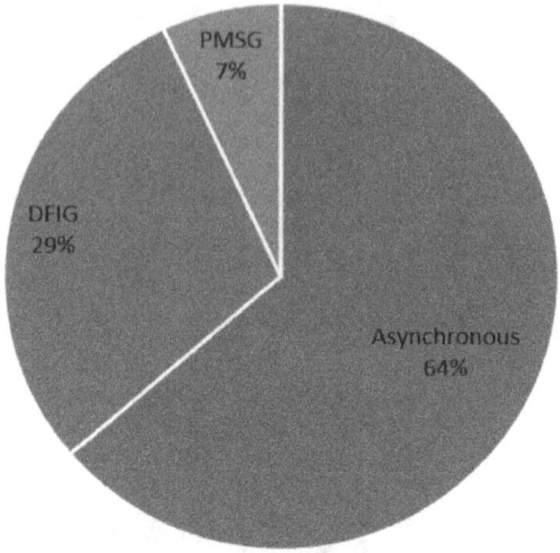

Figure 7.25 Type of wind turbine generator used — in percentage share.

Table 7.6 Type of wind turbine power converter used.

Type of Generator	Power Converter Type	Number of Turbines
Asynchronous	Full scale	1664
	None	137
	Partial scale	2
	Total	**1803**
DFIG	Full scale	38
	Partial scale	790
	Total	**828**
PMSG	Full scale	198

between offshore and onshore wind farms due to this fact, creating a necessity for systems to be particularly reliable. This means that the design of electronic components needs to pay particular attention to not requiring regular repair and maintenance.[51] This also stands when it comes to the power converters of offshore wind turbines too.

This is due to the fact that the power electronic converter has traditionally demonstrated high failure rates. This phenomenon is no longer acceptable due to accessibility challenges and a strong negative impact on the power grid. In fact, improvements to power converters have been singled out as one of the most important causes of failure on variable speed offshore wind turbines.[52] This means that the reliability of this component is critical and and there is a strong need for offshore wind farm operators. Advances will improve the lifetime of wind turbines which also leads to a reduction in energy costs.[53]

There is an integration between wind turbines and wind farm control systems. Both turbines and farms alike deal with environment factors: the turbines face wind share turbulence, waves, and sea current while the wind farm deals with electrical losses and wake effects.[51] See Figure 7.27.

Figure 7.28 outlines the loads which are associated with an offshore wind turbine which uses a monopile foundation. Due to the wave on the floater causing a variety of complex movements to the drive train components and structures within the wind turbine, there are additional loads within floating wind turbines. This fact has an impact on the load fluctuations within the power converter too. This is why when compared to a

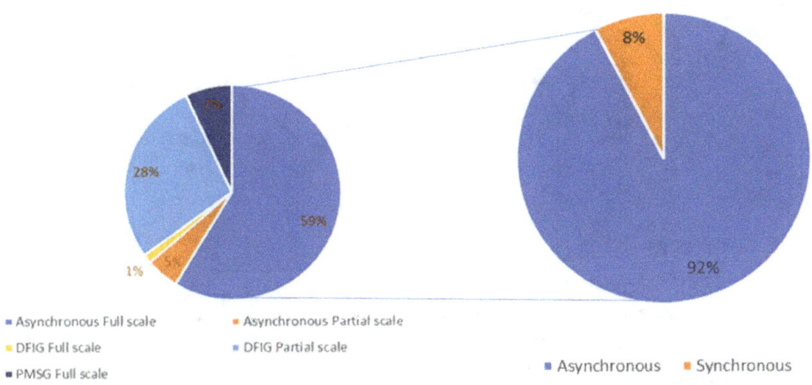

Figure 7.26 Combination of percentage share with type of power converter used.

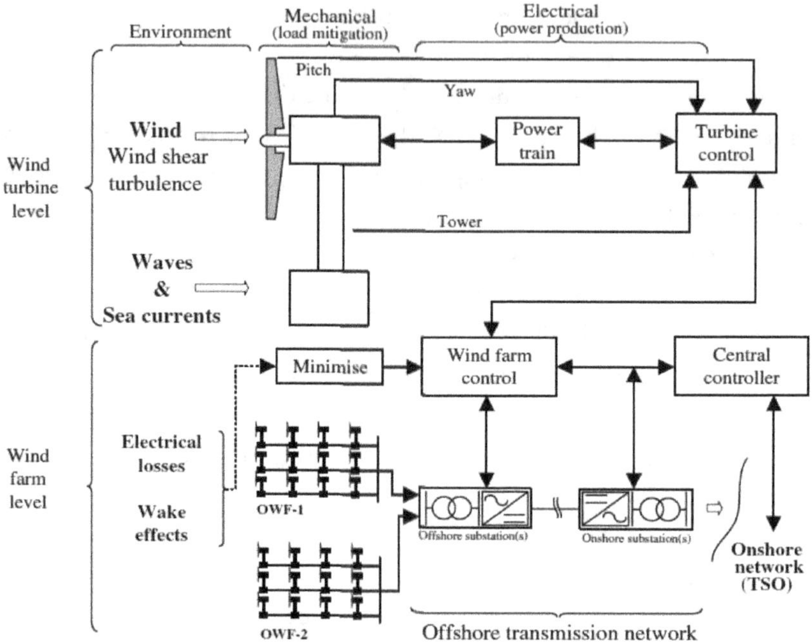

Figure 7.27 Offshore wind turbine and wind farm control network — overview.

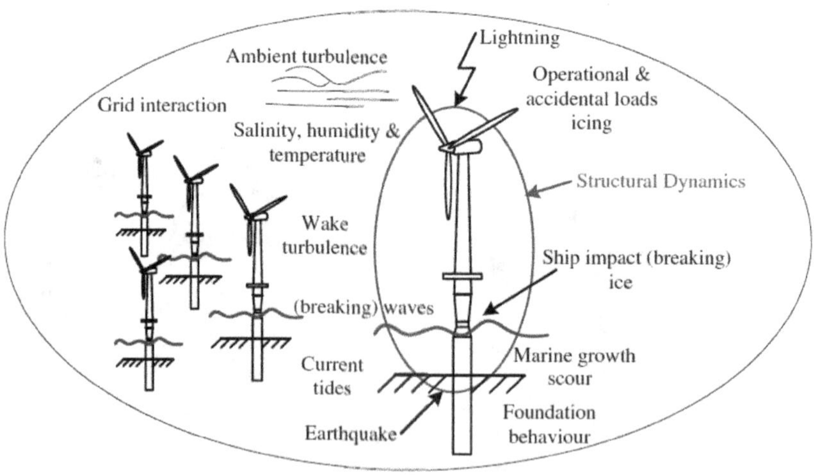

Figure 7.28 Offshore wind farm — external factors to be considered.

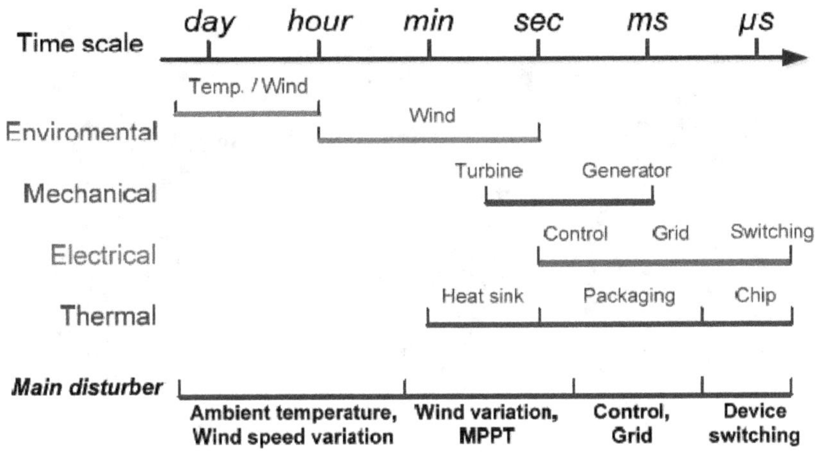

Figure 7.29 Thermal behaviors at different timescales.[53]

fixed-bottom turbine, the RUL of an FOWT power converter is entirely different.

Electrical parts, environmental conditions, and mechanical parts all make up the complete thermal behavior which is exhibited in power converters. Figure 7.29 outlines how different timescales are applied to the loading conditions within a power converter. On a scale from microseconds to years, the environmental, mechanical, electrical, and thermal sub-systems have a range of different timescales according to shifts in ambient temperature. When it comes to wind turbines, three model levels are proposed[53] in order to accommodate for variances in terms of power converter behaviors, reflecting varying causes of thermal dynamics or cycling at a range of different sampling rates as in Figure 7.30.

Over the course of the last two to three decades, a range of converter technologies have been deployed, each of them exhibiting a variety of different advantages and disadvantages. The most common power converters used within a commercial wind turbine generator system include

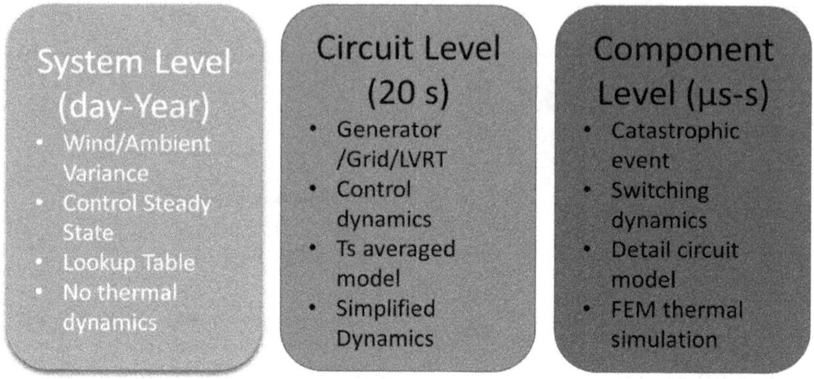

Figure 7.30 System level to component level — examples.[53]

back-to-back bidirectional converters, full-rated converters, and diode rectified-based unidirectional converters.[54]

Variable frequency and variable magnitude make up the diode-based rectified converters. A diode rectifier circuit first converts AC power to DC power, before being converted back to AC power with a different voltage level and frequency. The generator to grid, diode uncontrolled rectifier transfers power in just a single direction. This type of converter is usually used within a wound rotor synchronous generator (WRSG) or a permanent magnet synchronous generator (PMSG) and rather than being based upon asynchronous or induction generators, it is based upon a wind power generation system. There are serious disadvantages with this converter, though, despite the fact that it has cheap manufacturing costs and is relatively simple to implement. Key disadvantages include unidirectional power handling capability and large of harmonics production.[54] See also Figure 7.31.

The back-to-back converter is a controlled rectifier and controlled inverter-based converter. As illustrated in Figure 7.32, it is made up of two conventional pulse width-modulated voltage source inverters.

This study will analyze the fully enclosed asynchronous/induction generator with a squirrel-cage rotor. Through a full-scale power converter, its stator is connected to the 690 V side of the transformer. This configuration has a range of benefits, allowing the generator to operate at variable speed, frequency, and voltage at the same time that the power converter provides power at constant frequency and voltage to the transformer with fault ride-through capabilities.[55] Composed of insulated gate bipolar transistors (IGBTs), the rectifier and converter are controlled by space vector pulse width modulation. The schematics of the full AC-AC converter are represented in Figures 7.33 and 7.34.

IGBTs modules are used by modern wind turbines in two varying packaging types depending upon the voltage level that is demonstrated. As in Figure 7.35, the conventional module and classical presspack are

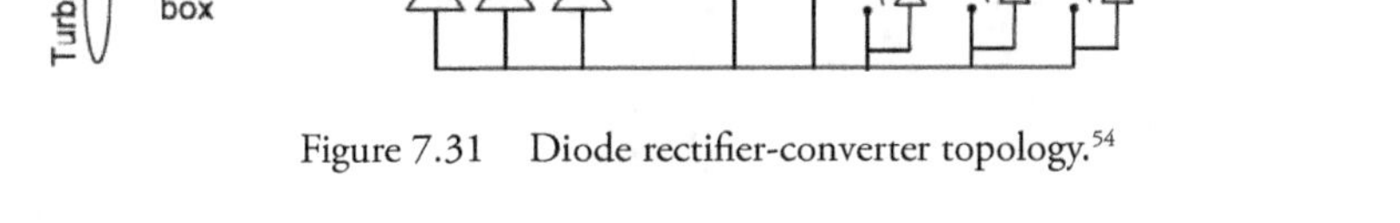

Figure 7.31 Diode rectifier-converter topology.[54]

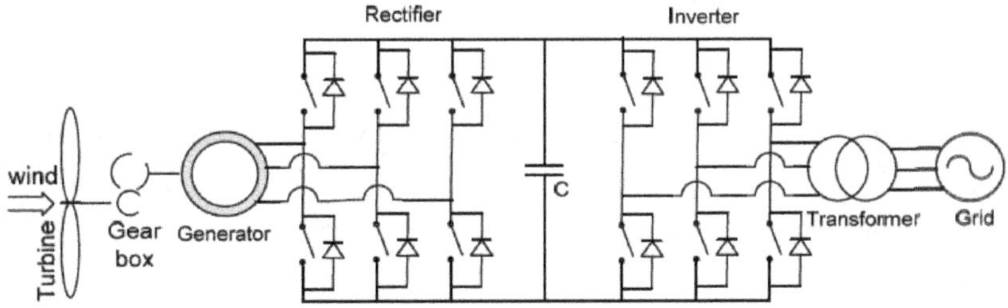

Figure 7.32 Back-to-back converter.[54]

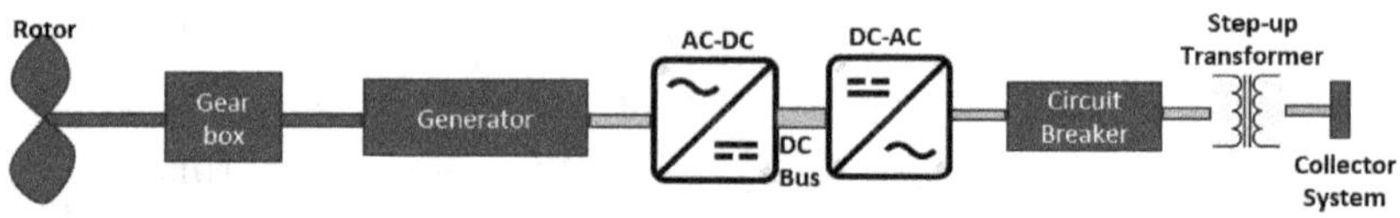

Figure 7.33 Full AC-AC converter.

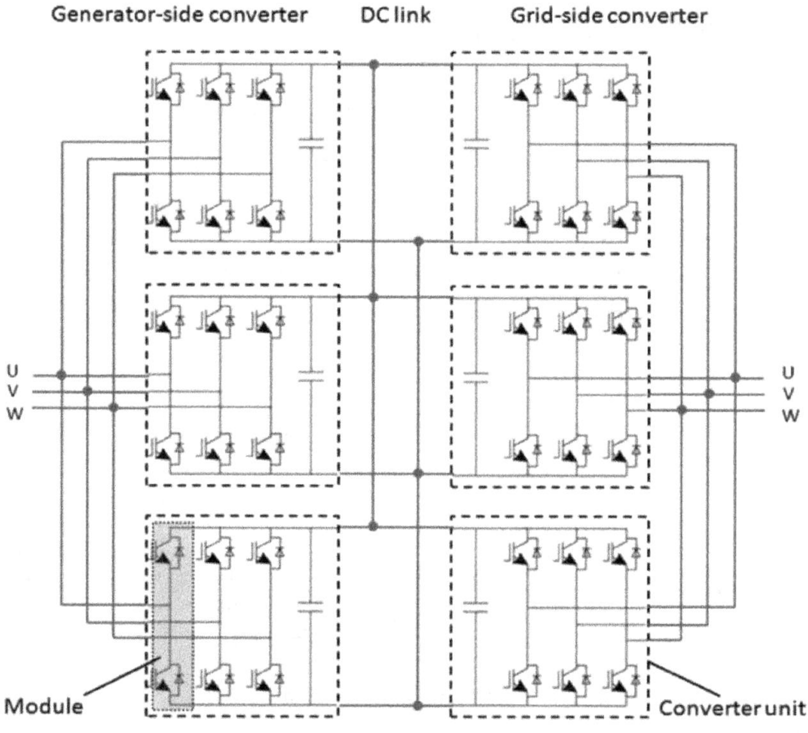

Figure 7.34 Full-scale power converter schematics.[54]

outlined. When the voltage is less than 1000 volts the conventional module is used for application and it is assumed both IGBT modules have varying failure modes.

The squirrel cage induction generator is the generator which is associated with the fully rated power converter. See Figure 7.36. The only part of the

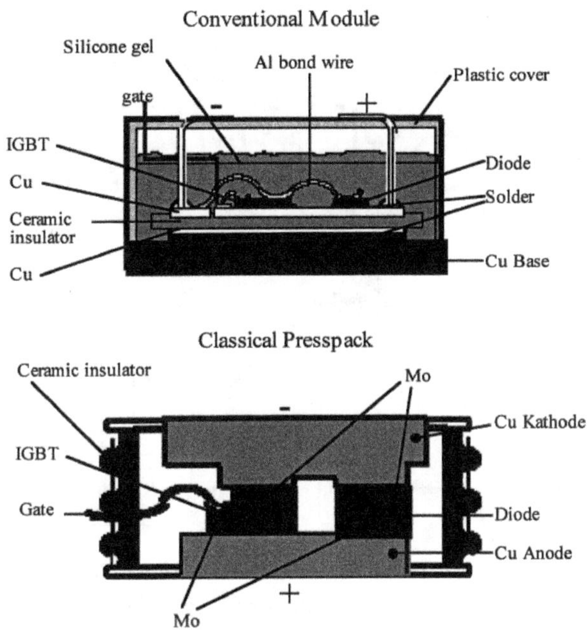

Figure 7.35 Conventional module and classical presspack IGBT modules. Reproduced with permission from Zehringer, R., Stuck, A., Lang, T. Material requirements for high voltage, high power IGBT devices. *Solid-State Electronics* **1998**, *42*(12), 2139–2151.

generator which is connected to the grid is the stator, given that the rotor windings are connected in short circuit.[56] Due to its construction simplicity, reasonable price, low operating costs and the fact that it is brushless with no commutators, it is a very attractive solution for offshore wind turbines.

7.4.4 *Investigation on Converter Failures*

One of the most common sources of failure in offshore wind turbines is the power converter system which can lead to high maintenance costs and significant downtime. Due to this fact, there is a desire to further study the reliability of power modules in order to identify appropriate maintenance intervals while improving replacement costs. Half of the

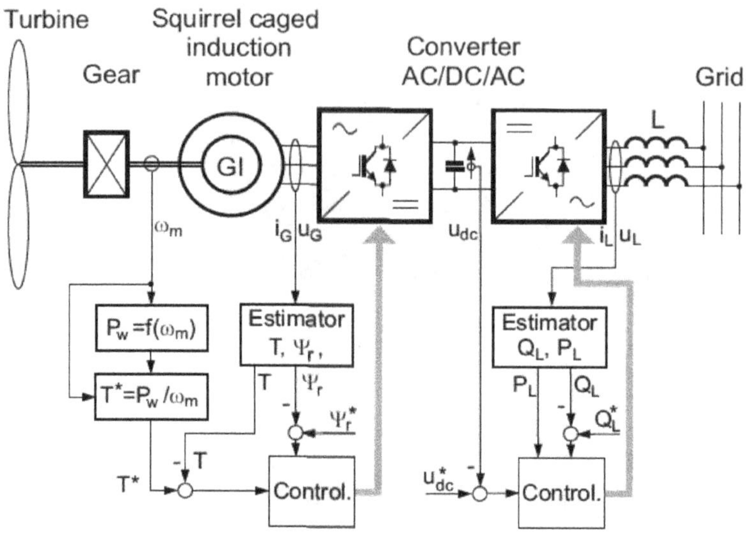

Figure 7.36 Control diagram of the power converter. Reproduced with permission from Sikorski, A., *et al.* Cooperation of induction squirrel-cage generator with grid connected AC/DC/AC converter. ***Bialystok Technical University*** **2009**.

reported failures can be exceeded by power devices and capacitors. System transients, overload conditions, and environmental factors typically cause these failures.[57]

According to the data analysis, there is a correlation between temperature, frequency converter failure on both sides (generator side and grid side), generator failure, and grid failure. In order to properly determine failure causes of the cable and converter modules, further literature review analysis is required.

The Elforsk Report from 2012 offered a detailed analysis of alarm data and converter-specific failure. The study concluded by asserting that failure rate of frequency converters does not vary between onshore and offshore. The report also suggested that the fatigue-related phenomenon of bond wire lift-off from the IGBT chips is not in fact among the most dominant failure mechanisms. Thermal cycling, however, could represent problems for IGBTs during operations close to synchronous speed.[58]

A range of other failure causes and failure modes are discussed below.

Lightning strikes and converter failure. Electrical overstress can be caused to converter modules in this situation. The results demonstrate that if a lightning strike hits the surrounding structure a discharge path through the converter module can be found.[58] Galvanic coupling occurs between the power circuit and the cooling fin, demonstrated by the electric measurements outlined in the previous section. See Figure 7.37.

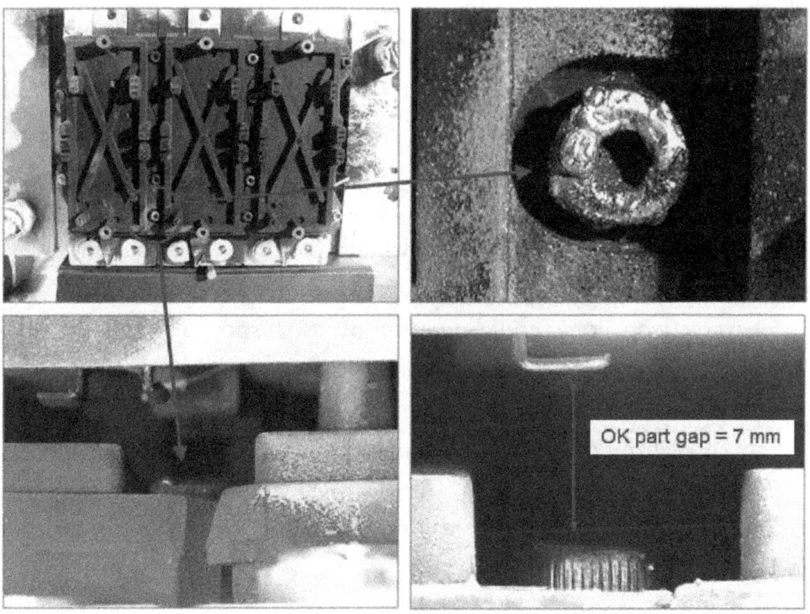

Figure 7.37 Discharge due to lightning strike path over the converter module.

Converter cabinets where dead insects are found. While it is unclear how this could be a failure root cause from a failure point of view, it is important that countermeasures are taken in offshore sites.

Long periods of wind turbine stoppage leading to condensation. It is important that pre-heating techniques are deployed and regularly checked.

Corrosion due to salt. Should salt enter the converter modules during the transportation, installation, and operation phases, corrosion could occur so preventative measures are recommended.

Wang *et al.*[59] presents a range of lifetime models which outline the impact that temperature and temperature cycling have on converters. Capacitors are identified in this study as being the most fragile of the components which appear in converter systems, second only to printed circuit boards.

The following points list the failure mechanisms related to the chip identified by Yang *et al.*:[60]

- Electrical overstress: related to overvoltage and overcurrent conditions; heating effects under high-voltage conditions can be significant
- Latch-up and triggering of parasitic structures
- Charge effects, ionic contamination or hot carrier injection
- Electro-migration, contact- and stress-induced migration
- Thermal activation
- False triggering due to cosmic radiation

The following failure mechanisms have been identified by Fischer *et al.*[55] and Yang *et al.*[60]:

- Bond wire lift-off
- Solder fatigue
- Degradation of thermal grease

- Fretting corrosion at pressure contacts
- Tin whiskers

Fischer *et al.*[58] classified the failures of the converter (1269 years of operation data) according to the following categories:

- Phase module (including IGBT modules and corresponding driver boards, DC-link capacitors, busbars)
- Converter control board
- Crowbar (doubly fed induction generator (DFIG) only)
- Cooling system
- Semiconductor fuse
- Main circuit breaker
- Grid-coupling contactor
- Other converter failures

Figure 7.38 shows the average failure rates for each category described above for a DFIG, while Figure 7.39 conveys the percentage failure distribution and repair cost per converter category.

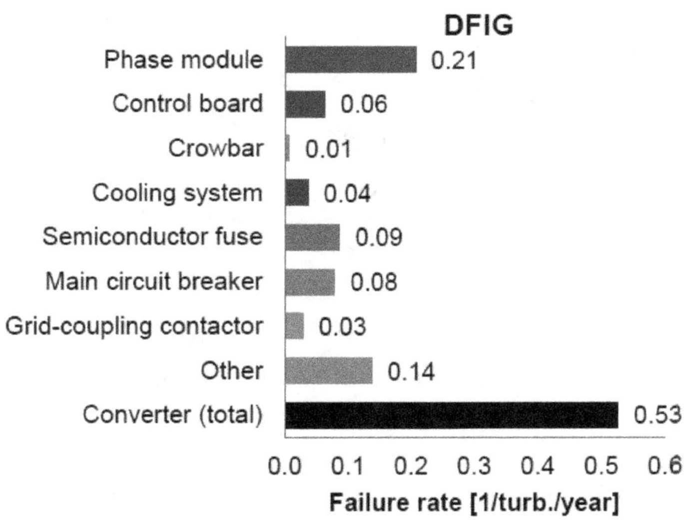

Figure 7.38 Failure rates (average) for converter components.

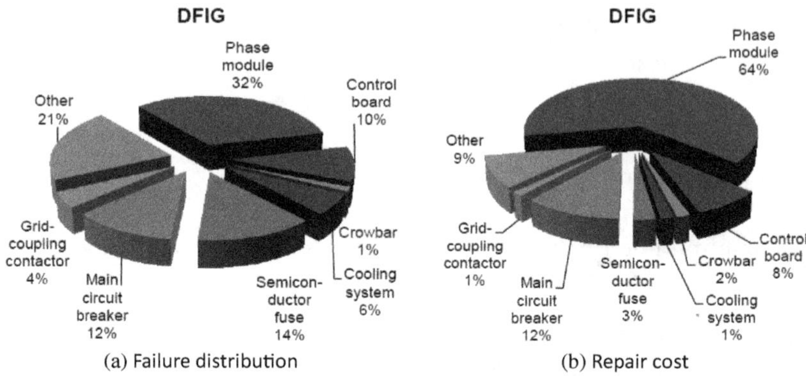

Figure 7.39 Failure distributions in percentage and their repair costs per converter category.

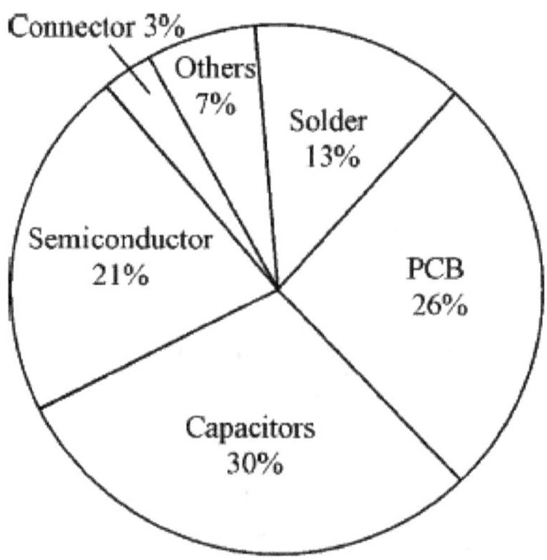

Figure 7.40 Failure distributions.[55]

Figure 7.40 illustrates the failure distribution per converter components which is described by Yang *et al.*[60]

As shown in Figure 7.41, several mechanical layers are used in the construction of power modules.

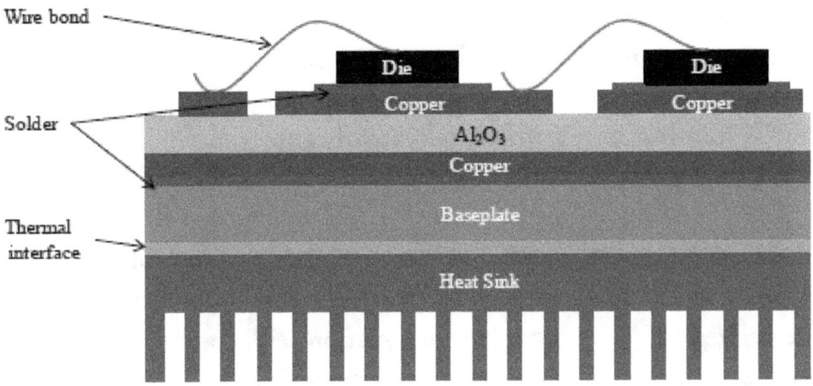

Figure 7.41 Power module used in a power electronic converter.[57]

(a) (b)

Figure 7.42 (a) Bond wire crack. (b) Solder delamination.[57]

When it comes to the sources of wear, two principal failure mechanisms have been identified: solder fatigue with consequential solder delamination and bond wire lift-off.[57] As outlined in Figure 7.42(a), bond wire is a cost-effective process when it comes to forging an interconnection with the source on top of the die. The edges of the interface are where degradation begins and appears as crack growth as shown in Figure 7.42(b). The contact area is decreased given that the cracks grow inwards toward the center of the interconnection. This electrical resistance increases in a manner which is measurable as a growing of a forward voltage drop. This means that the power losses can rise to a higher junction temperature during power-on time. Higher thermos-mechanical stress will ultimately lead to a bond wire lift-off. The degradation process will be accelerated by a single lift-off given that the current per bond is increased by the full load current for the remaining wires which leads to higher temperature at the interconnection.

The second most common failure mechanism in power modules is solder delamination. Cracks are formed by stress inside the solder interfaces. Degrenne *et al.*[57] suggests that the junction temperature of the die is constantly raised due to the increase of thermal impedance.

Thermal cycling and wear-out fatigue in the power semiconductor is suggested to be one of the major causes of wind power converter system failures.[53] The number of cycles or the time to a certain probable failure are used in order to quantify the strength, which is defined as the ability to withstand specified stress levels on electronic components.[53]

According to the study from Givaki,[52] there are a variety of failure modes which vary depending on the packaging which is used. While presspack IGBTs are designed in such a way to fail with short circuits, conventional modules are designed in such a way that they fail in both open and short circuits. Failure modules as chip-related and package-related failures are defined in this study. The most frequent failures in the power electronic devices are said to be packaging related. This failure mechanism is primarily due to the difference in the thermal expansion coefficient.[52] The study identified the main failure mode:

- Solder fatigue as a consequence of thermal cycling brought about by alternating currents which also produce pulsating loss. Alternating currents (converter load) are due to the wind speed variations which lead to shifts in temperature due to power losses.
- Wang *et al.*[59] presents a case of 2.3 MW wind power converter with a focus on critical IGBT modules. Figure 7.43 outlines the results.

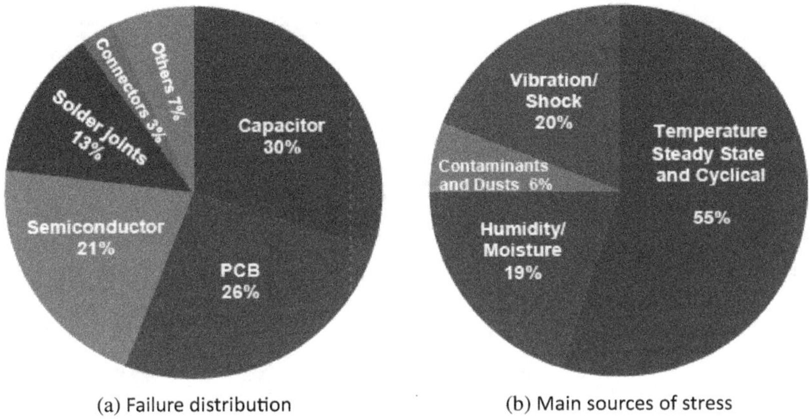

Figure 7.43 Failure analysis.[59]

The main causes of stress which impact the reliability of power electronic components have been identified as temperature and temperature cycling.

7.4.5 *Studies Pertaining to Temperature-Related Failures in Power Converters*

In order to further explore the effect that temperature has on component degradation, physical models have been developed as part of several studies in the literature.

Normally considered to be the most important indicators in a reliability assessment, the mean junction temperature and junction temperature fluctuation are analyzed to estimate the RUL in the study.[62] Conduction loss and switching loss are mostly responsible for loss dissipation of a power switching device. The conduction loss can be calculated by accumulation of every switching pattern within the fundamental period of the loading current. The DC-link voltage is a benchmark against which the switching loss can be considered proportional. While the junction temperature fluctuation will not be disturbed by the thermal model of the cooling method, the mean junction temperature will be.

The following equations are given to junction temperature T_{jm} and junction temperature fluctuation dT_j:

$$T_{jm_T/D} = P.\sum_{i=1}^{4} R_{thjc\_T/D(i)} + P.\sum_{j=1}^{3} R_{thca\_(j)} + T_a \tag{7.53}$$

$$dT_{j\_T/D} = P.\sum_{i=1}^{4} R_{thjc\_T/D(i)} \cdot \frac{\left(1 - e^{-\frac{t_{on}}{\tau_{thjc\_T/D(i)}}}\right)^2}{1 - e^{-\frac{t_p}{\tau_{thjc\_T/D(i)}}}} \tag{7.54}$$

The Coffin-Mansion model — in which the mean junction temperature, junction temperature swing, and the on-state time t_{on} are considered — can be used in order to roughly estimate the power cycles to failure N_f. The consumed lifetime per year (CL) for each averaged wind speed n can be calculated by

$$CL_n = \frac{N_n}{N_{fn}} = \frac{t_n \cdot f_{rn}}{N_{fn}} = \frac{D_n.365.24.3600.f_{r\_n}}{N_{f\_n}} \tag{7.55}$$

where CL_n is the percentage which experienced thermal cycles at wind speed of n and consumed the total lifetime of the device. N_{fn} means the number of cycles to failure or total lifetime of device under a certain stress level. N_n and t_n represent experienced number of thermal cycles and total time, respectively, at wind speed of n. D_n denotes the percentage of time for each wind speed accounting for the total time of one year, and f_{rn} denotes the frequency of rotor current under wind speed of n. Assuming that the fatigue damage is linearly accumulated, the total consumed lifetime (TCL) can then be estimated by the decomposition of wind speed.

$$TCL = \sum_{n=4}^{25} CL_n \tag{7.56}$$

The model for lifetime estimation can be represented using the following structure for the FOWT application.[52] See Figure 7.44.

In Chapter 10 the turbine model, generator model, converter model, and lifetime estimation model are described in terms specific to FOWT.

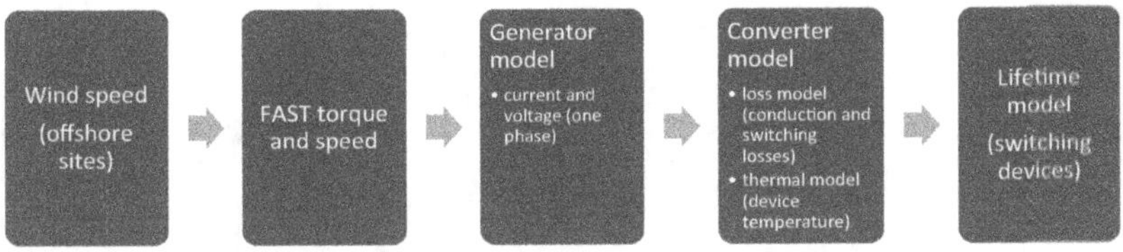

Figure 7.44 Lifetime estimation structure (example).

Two methods can be used in order to develop the lifetime model: either physical or analytical. Analytical models are used in order to describe the dependence of the number of cycles to failure on temperature cycle parameters including minimum and maximum temperature, duration, frequency, mean value, amplitude, and dwell time. Physical models, on the other hand, offer modeling of the physical processes which lead to failure.[52] There is a further exploration of those physical models used in order to assess the impact of temperature and temperature cycling.

In Wang *et al.*,[59] two models are outlined which can explore the impacts of temperature and temperature cycling. The first degradation model illustrates the degradation of a material from an initial state which is stable. Free energy difference between E_1 and E_2 — defined as ΔE, where E_1 is free energy of initial stable state and E_2 is the free energy of a degraded state — is the driving force behind this degradation. The energy to move from one state to the other is provided by the heat which comes from power losses in power electronic components. The net reaction rate is described in the equation below:

$$k_{net} = k_{forward} - k_{reverse} = k_0 e^{-\left(\frac{E_a}{E_B T}\right)}\left[1 - e^{-\left(\frac{\Delta E}{E_B T}\right)}\right] \approx k_0 e^{-\left(\frac{E_a}{E_B T}\right)} \qquad (7.57)$$

where $k_{forward}$, $k_{reverse}$ and k_{net} are the degradation rate, recovery rate and net reaction rate, respectively.

$$k_0 = k_B \left[\frac{\partial In\,(k_{net})}{\partial\left(\frac{1}{T}\right)} \right] \tag{7.58}$$

k_B is the Boltzmann's constant ($8.62 \times 10^{-5} eV/K$), T is the temperature in Kelvin and k_0 is a material-/device-specific constant.

The activation energy ΔE is dependent on additional applied stress ε such as electrical stress. The parameters a and b are based on data of degradation induced by stress. The final equation is given by:

$$k_{net} = k_0 \xi^n e^{-\left(\frac{E_a}{E_B T}\right)} \tag{7.59}$$

Secondly, the lifetime model to predict the temperature cycling effect of loading variations and commutation of power switching devices on the converter is given as

$$N = k(\Delta T - \Delta T_0)^{-m} \tag{7.60}$$

where a and k are empirically determined constants and N_f is the number of cycles to failure. ΔT is temperature cycle range and ΔT_0 is the portion of ΔT in the elastic strain range. Regarding this study, if ΔT_0 is negligible compared to ΔT, it can be dropped out from the above equation, which becomes the Coffin-Manson model.

An electro-thermal model is presented in the study by Lei *et al.*[61] in order to provide a means by which to calculate the power loss of power electronics components. The following equation outlines the conduction losses:

$$P_{con}\left(I_L, T_j\right) = \left(a_0 + a_1 T_j\right) I_L + \left(b_0 + b_1 T_j\right) I_L^2 \tag{7.61}$$

Load current, temperature, and manufacturer data are used as inputs. Curve fitting is used to obtain parameters *a* and *b*. Switching losses described as turn-on and turn-off losses are given by the formula:

$$E_{SW} = E_{SW}^{ref} \cdot \left(\frac{I_L}{I_L^{ref}}\right)^{Ki} \left(\frac{V_{CC}}{V_{CC}^{ref}}\right)^{Kv} (1 + TC_{Esw} \cdot (T_j - T_J^{ref})) \tag{7.62}$$

where the blocking voltage V_{cc} is equivalent to the DC-link voltage, I_L is the load current and T_j the device junction temperature. K_i and K_v are current- and voltage-dependent exponents; TC is the temperature coefficient.[61]

Methods relevant to lifetime prediction of power modules have been presented in the study by Degrenne *et al.*,[57] covering the failure mechanisms which are present in typical IGBT-based power modules.

Aging of devices is brought about by variance in terms of thermal coefficients of adjacent layers. The varying expansion of adjacent layers causes a sheer stress to be applied along the contact surface during both heat-up and cool-down phases. The causes of degradation are quantified by stress-based methods. Power and thermal models are used in order to estimate the stress. In order to determine the RUL of a device the time history of the state of health (SoH) is required. See Figures 7.45 and 7.46. It can be linearly extrapolated using the following equation:

$$RUL_n = \frac{n * SoHn}{100 - SoHn} \tag{7.63}$$

where n is time variable and SoH is given in percentage.

The study also describes a condition-monitoring prognostics method. “Failure precursors” is used to refer to the Damage-Sensitive Electrical

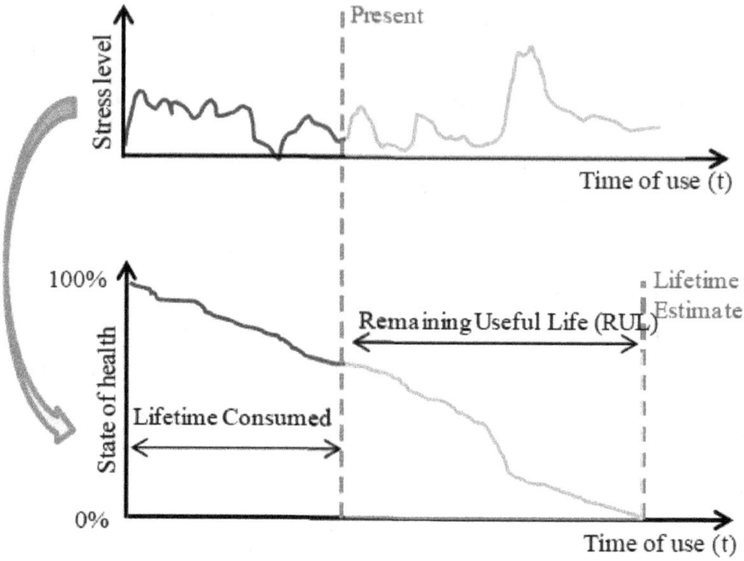

Figure 7.45 Remaining useful life of power converters.

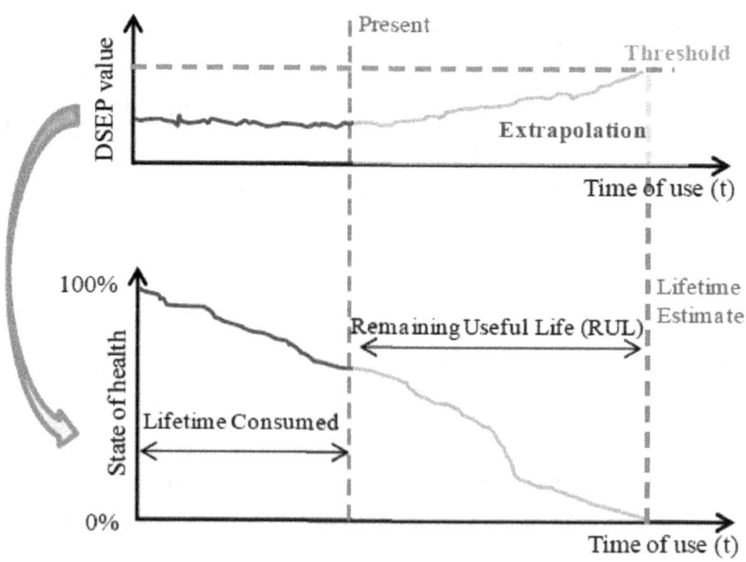

Figure 7.46 RUL estimation based on condition monitoring variables.

Parameters (DSEPs) and the evolution of one of more of these is studied. Two DSEPs are described: R_{TH}-based methods and voltage at high current.

7.4.6 *Proposal Summary: Remaining Useful Life Estimation Model of Power Converter*

For technology demonstration purposes, the 5MW NREL FOWT is chosen. Figure 7.47 outlines the schematics of a physics-based approach

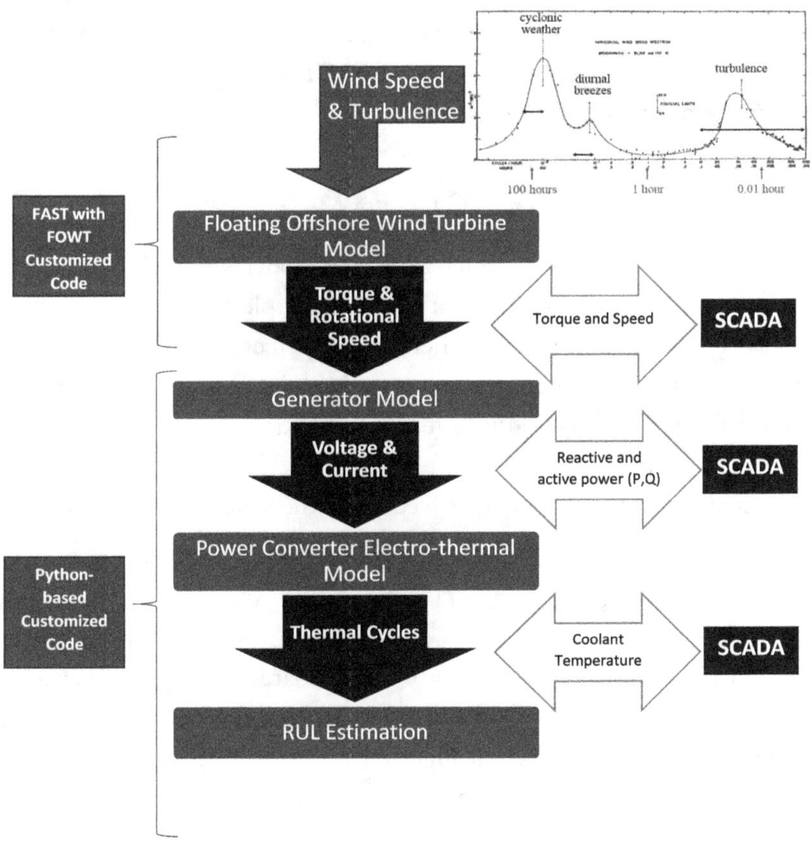

Figure 7.47 RUL model structure.

for RUL estimation. Prognostic and diagnostic approaches are guided by the figure and the concept is adopted in this chapter. RUL estimation can be an effective O&M tool for all the major components which are integrated with the digital twin approach. Chapter 10 describes this process in further details for the power converter, while details around FAST models are outlined in the following section. Given that the data is coming from supervisory control and data acquisition (SCADA) to estimate the consumed lifetime, it can be used for diagnostic purposes. The series of events can subsequently be classified as per the DLCs for sequencing futuristic scenarios which are based upon the wind profile from the SCADA data. The FAST model, based upon a sequence of futuristic scenarios, can be used to estimate the RUL through prognostic analysis. The general capabilities of FAST for RUL prediction are described in more details below.

7.4.6.1 *FAST Capabilities — Generator Model in FAST*

The GenModel determines which generator model will be chosen. There are just four parameters used by the FAST simple induction generator model (GenModel = 1): rated generator slip percentage (SIG_SlPc), the synchronous (zerotorque) generator speed (SIG_SySp), the rated torque (SIG_RtTq), and the pullout ratio (SIG_PORt).

The rated rotor speed, Ω_R, is derived from the synchronous speed and the slip percent.

$$\Omega_R = SIG_{SySp}.(1 + 0.01.SIG_{SlPc}) \tag{7.64}$$

FAST uses a more accurate generator model that is made up of Thevenin Equivalent Circuit equations for a three-phase induction generator when GenModel = 2. Rather than using the peer-unit values (normalized by base values), the values are input in engineering units often found in generator specification sheets. FAST's Thevenin equivalent equations assume a three-phase generator configuration, which is Y-connected. It is essential to divide impedances and the voltage by three to convert the values to a Y-connected configuration in the instance of a delta-connected configuration. A bespoke generator model can also be created by modifying subroutine UserGen() and compiling the modified file to link it with the rest of the code. FAST will call UserGen()set GenModel = 3.

Generator losses can also be simulated by setting generator efficiency. FAST is capable of determining electrical generator power by multiplying the mechanical generator power by its efficiency. The Thevenin model does not use generator efficiency since it incorporates a more complex expression for electrical power based on input circuit resistances. The electrical generator torque and the instantaneous electrical generator power are the outputs for the three options of the generator model.

The instantaneous power is the product of instantaneous voltage and current, $P = VI$, where

$$V = V_m \sin \omega t \tag{7.65}$$

$$I = I_m \sin(\omega t - \varnothing) \tag{7.66}$$

with reference to Figure 7.48.

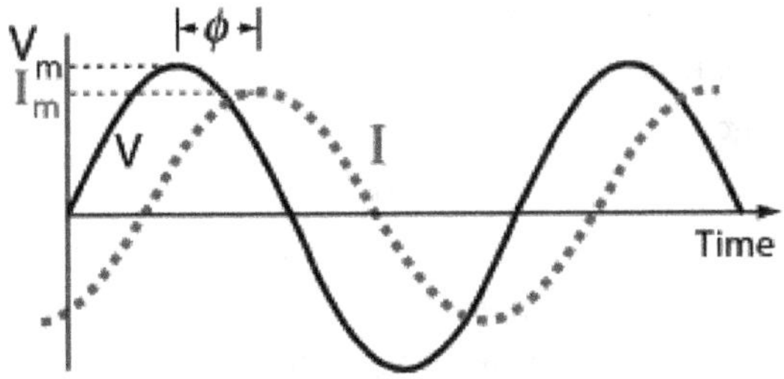

Figure 7.48 Instantaneous voltage and current.

The instantaneous power at any time t can be expressed as

$$P_{instantaneous} = V_m I_m \sin \omega t \sin(\omega t - \varnothing) \tag{7.67}$$

Applying

$$\sin(\omega t - \varnothing) = \sin \omega t \cos \varnothing - \cos \omega t \sin \varnothing \tag{7.68}$$

$$P_{instantaneous} = V_m I_m \cos \varnothing \sin^2 \omega t - V_m I_m \sin \varnothing \sin \omega t \cos \omega t \tag{7.69}$$

FAST does not model the details of the electrical drive, but rather it focuses on getting the torque-speed curve correct (which has an impact on turbine loads). The most sophisticated built-in model available in FAST in terms of the induction machine is the Thevenin Equivalent Circuit model. Some FAST users have implemented detailed induction gear models using Simulink, however, which have been interfaced to FAST. The details of the electrical drive, in these instances, can be extracted using that Simulink model.

Python and Scipy:

- A free, open-source software which is widely used and boasts an experienced community and advanced level of support.
- A range of rich modules and libraries.
- Capable of producing readable and well-structured code due to being such a sophisticated language.
- Capable of handling a range of complex numbers and equations.
- Performance can be improved to reach a level comparable to compiled languages.
- GUI application can be developed.
- SciPy is an open-source library of scientific tools for use in Python. Capable of gathering a variety of high-level science and engineering modules together as a single package, it provides an excellent tool to perform engineering modeling.

7.4.6.2 *Squirrel Cage Induction Generator*

A Squirrel Cage Induction Generator is a three-phase induction machine which is composed of three windings in the stator as well as three more windings in the rotor. The same set of equations which describe motors can also be used to describe generators. The following hypothesis is used in order to simplify the equations:

- A single winding rotor and constant gap on a symmetric and balanced three-phase induction machine.
- The iron saturation is discarded which makes it possible to assume the material is linear.
- The magnetic flux density is radial to the gap, which means that iron magnetic permeability is assumed to reach infinity in front of the air permeability.
- Losses within the iron are neglected.
- A sinusoidal magnetic field distribution is generated in the gap due to the stator windows and the rotor windings which represent distributed windings.

7.4.6.3 *Wind Turbine Specification*

The specification of the chosen digital wind turbine for the FOWT application is outlined in Table 7.7. The RUL prediction is based on physics under digital twin framework and this is expanded upon in Chapter 10.

7.4.6.4 *Power Converter as Integrated Power Modules*

An appropriate thermal model needs to be defined in order to accurately and appropriately describe the thermal behavior of IPMs or power semiconductors. Losses made in the IPM are the combination of switching and conduction losses. The current is divided assuming that each IPM

Table 7.7 Specification of FOWT wind turbine for digital twin framework.

Turbine type	NREL 5MW: three-bladed, horizontal axis, upwind, spar buoy floater.
Nominal rotor speed	5–13 rpm.
Power regulation	Pith regulation with variable speed.
Blades	As per NREL 5MW reference turbine.
Gearbox	As per NREL 5MW reference turbine and OC3 Phase IV specification.
Generator	As per NREL 5MW reference turbine and OC3 Phase IV specification.
Converter	Fully rated, modular arrangement or integrated power modules (IPM), water cooled.
Operational information	Self-starting when the wind speed reaches 3 to 5 m/s avg. The power is regulated at rated power 12 to 13 m/s. The wind turbine is shut down by feathering the blades when wind reaches 25 m/s avg.

consumes the same power with parallel IPMs. IPMs work as rectifiers given that generator side power converter is considered. This means, therefore, that diodes can be expected to conduct a great current when compared to IGBTs. The thermal model of the single IPM can be defined through use of the manufacturer information (SEMIKRON).[52]

7.4.6.5 *Thermal Analysis of Power Semiconductor Converters*

If the junction temperature reaches a point of runaway or melting, then the power devices could potentially fail. A thermal exchange system — in this case a water-cooling system — need to be used to remove the heat which is generated within the IPMs. The heat loss of the case of heat sink is mostly carried out through conduction, which can be compared or accurately described using an electrical analogy. Thermal capacitance represents the junction temperature fluctuation while thermal resistance represents the steady-state mean junction temperature. The case to heat sink thermal impedance (R and C) is modeled as a simple thermal resistor. There are a range of time constraints regarding the thermal behaviors of power converters. It is essential that the thermal model takes wind speed, ambient temperature, and the behavior of electrical and mechanical parts into consideration, each of which involves different time constraints. This lifetime estimation method therefore proposes to generate three loadings for long-term, medium-term and short-term time constraints.[57]

The worldwide organization for standardization is the International Electrotechnical Commission (IEC). For offshore wind turbines, the International Standard IEC 61400-3 has been prepared by IEC technical committee 88: Wind turbines — Part 3: Design requirements.

It is a fact that offshore wind turbines constantly endure environmental and electrical conditions that can have a range of potential impacts on their loading, durability, and operation. These environmental conditions are divided into three groups: wind conditions, sea conditions (waves, sea currents, water level, sea ice, marine growth, seabed movement, scour), and other environmental conditions. The network conditions are referred to as part of the electrical conditions. See Figure 7.49.

Normal and extreme categories are used in order to subdivide the external conditions. The normal external conditions, for example, are mostly

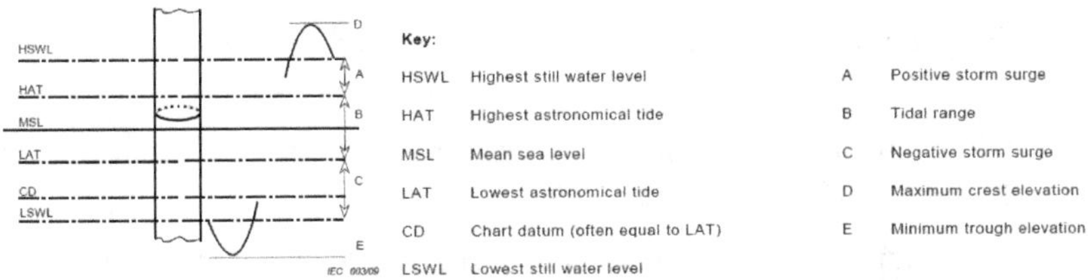

Figure 7.49 Sea state-related water level and its nomenclature as per IEC 61400-3.

concerned with the recurrent structural loading conditions, while the extreme conditions refer to those rare external design conditions. The DLCs shall consist of potentially critical combinations of these external conditions with wind turbine operational modes and other design situations.

The intended site or site type for an offshore wind turbine installation will play a key role in determining which external design choices need to be made. IEC 61400-1 wind turbine classes, for example, are defined in terms of wind speed and turbulence parameters.

The wind regimes are divided into normal and extreme wind conditions. These affect the load and safety considerations. Normal wind conditions will occur more than once per year during normal operation, while extreme wind conditions are defined as having a 1-year or 50-year recurrence period.

The life of a FOWT can be represented using a set of design situations which cover the most significant conditions that an offshore wind turbine may experience. This is performed for design purposes. The load cases are identified via a combination of operational modes and other design situations such as specific assembly, erection, or maintenance conditions within the external conditions. If the relevant load cases are considered to have a reasonable probability of occurrence they will be considered alongside the behavior of the control and protection systems which are in place. The DLCs which are used in order to identify the structural integrity of an offshore wind turbine are calculated by bringing together:

- Transportation, installation, and maintenance design situations as well as the appropriate external conditions;
- Fault design situations and the appropriate external conditions;
- Normal design situations as well as the appropriate normal or extreme external conditions.

References

1. https://gwec.net/global-wind-report-2017/.
2. Ishihara, T.; Phuc, P. V.,; Sukegawa, H.; Shimada, K.; Ohyama, T. A study on the dynamic response of a semi-submersible floating offshore wind turbine system Part 1: A water tank test. *ICWE12 CAIRNS* **2007**.
3. Wang, C. M.; Utsunomiya, T.; Wee S. C.; Choo, Y. S. Research on floating wind turbines: a literature survey. *The IES Journal Part A: Civil & Structural Engineering* **2010**, *3*, 267–277.
4. Stewart, J. F.; Shuck T. L. Flight-testing of the self repairing flight control systems using the F-15 highly integrated digital electronic control flight research facility. *Technical Memorandum* **1990**, NASA-TM-101725.
5. Jonkman, J.; Musial, W. Offshore code comparison collaboration (OC3) for IEA Task 23 offshore wind technology and deployment. *Technical Report* **2010**, NREL/TP-5000-48191.
6. Betz, A. Das Maximum der theoretisch möglichen Ausnützung des Windes durch Windmotoren. *Zeitschrift für das gesamte Turbinenwesen* **1920**, *26*, 307–309.
7. Glauert, H. Airplane Propellers. In: Durand, W. F., Ed., *Aerodynamic Theory*, Springer **1935**, pp. 169–360.
8. Hansen, M. O. *Aerodynamics of Wind Turbines*, 2nd edition. Earthscan **2008**.
9. Snel, H.; Schepers, J. G. *Joint Investigation of Dynamic Inflow Effects and Implementation of an Engineering Method.* Energy Research Centre of the Netherlands **1995**.
10. Burton, T.; Jenkins, N.; Sharpe, D.; Bossanyi, E. *Wind Energy Handbook.* John Wiley & Sons **2001**.
11. Moriarty, P. J.; Hansen, A. C. AeroDyn theory manual. *Technical Report* **2005**, NREL/TP-500-36881.
12. Xu, G.; Sankar, L. N. Application of a viscous flow methodology to the NREL phase VI rotor. *ASME Wind Energy Symposium*, **2002**, Paper No: WIND2002-30, pp. 83–93.
13. Hand, M. M.; Simms, D. A.; Fingersh, L. J.; Jager, D. W.; Cotrell, J. R.; Schreck, S.; Larwood, S. M. Unsteady aerodynamics experiment phase VI: wind tunnel test configurations and available data campaigns. *Technical Report* **2001**, NREL/TP-500-29955.

14. Glauert, H. A general theory of the autogyro. *Reports and Memoranda* **1926**, *1111*, 559–593.
15. Buhl, M. L., Jr. A new empirical relationship between thrust coefficient and induction factor for the turbulent windmill state. *Technical Report* **2004**, NREL/TP-500-36834.
16. Manwell, J. F.; McGowan, J. G.; Rogers, A. L. *Wind Energy Explained: Theory, Design and Application.* John Wiley & Sons **2002**.
17. Jonkman, J.; Butterfield, S.; Musial, W.; Scott, G. Definition of a 5-MW reference wind turbine for offshore system development. *Technical Report* **2009**, NREL/TP-500-38060.
18. Kooijman, H. J. T.; Lindenburg, C.; Winkelaar, D.; van der Hooft, E. L. DOWEC 6 MW Pre-Design: Aero-elastic modeling of the DOWEC 6 MW pre-design in PHATAS. *Technical Report* **2003**, ECN, Petten, Netherlands.
19. Bazilevs, Y.; Hsu, M. C.; Akkerman, I.; Wright, S.; Takizawa, K.; Henicke, B.; Spielman, T.; Tezduyar, T. E. 3D simulation of wind turbine rotors at full scale. Part I: Geometry modeling and aerodynamics. *International Journal of Numerical Methods in Fluids* **2010**, *65*, 207–235.
20. Anderson, J. D. *Computational Fluid Dynamics.* McGraw-Hill Education **1995**.
21. Menter, F. R. Zonal two equation k-w turbulence models for aerodynamic flows. *23rd Fluid Dynamics, Plasmadynamics, and Lasers Conference.* Orlando **1993**.
22. Çengel, Y. A.; Cimbala, J. M. *Fluid Mechanics — Fundamentals and Applications.* McGraw-Hill **2006**.
23. Sivalingam, K.; Bahuguni, A.; Kandasamy, R.; Narasimalu, S.; Grevsen, J. K.; Nyvad, J.; Tietze, P. T. *Wind Turbine Nacelle With Cooler Top* (U.S. Patent number: 9039369-2015). United States Patent and Trademark Office -USPTO-May 26, **2015**.
24. Sivalingam, K.; Bahuguni, A.; Kandasamy, R.; Narasimalu, S.; Grevsen, J. K.; Nyvad, J.; Tietze, P. T. *Wind Turbine Nacelle With Cooler Top* (U.S Patent number: 9074582-2015). United States Patent and Trademark Office -USPTO-May 26, **2015**.
25. Chakrabarti, S K. *Fluid Structure Interaction in Offshore Engineering.* Computational Mechanics Publications **1994**.

26. Utsunomiya, T.; Matsukuma, H.; Minoura, S.; Ko, K.; Hamamura, H.; Kobayashi, O.; Sato, I.; Nomoto, Y.; Yasui, K. On sea experiment of a hybrid SPAR for floating offshore wind turbine using 1/10 scale model. *29th International Conference on Ocean, Offshore and Arctic Engineering.* Shanghai **2020**.
27. Roddier, D.; Cermelli, C.; Aubault, A.; Weinstein, A. Windfloat: A floating foundation for offshore wind turbines. *Journal of Renewable and Sustainable Energy* **2010**, *2*, 033104.
28. Martin, H. R. *Development of a Scale Model Wind Turbine for Testing of Offshore Floating Wind Turbine Systems* (PhD thesis). Maine Maritime Academy **2011**.
29. Koo, B. J.; Goupee, A. J.; Kimball, R. W.; Lambrakos, K. F. Model tests for a floating wind turbine on three different floaters. *Journal of Offshore Mechanics and Arctic Engineering* **2014**, *136*, 020907.
30. Jonkman, J.; Musial W. Offshore code comparison collaboration within IEA Wind Task 23: Phase IV results regarding floating wind turbine modeling. *Technical Report* **2010**, EWEC, NREL/CP-500-47534.
31. Myhr, A.; Maus, K. J.; Nygaard, T. A. Experimental and computational comparisons of the OC3-HYWIND and tension-leg-buoy (TLB) floating wind turbine conceptual designs. *ISOPE International Society of Offshore and Polar Engineering Conference* **2011**, paper number ISOPE-I-11-560.
32. Myhr, A.; Nygaard, T. A. Experimental results for tension-leg-buoy offshore wind turbine platforms. *Journal of Ocean and Wind Energy* **2014**, *1*, 217–224.
33. Myhr, A.; Nygaard, T. A. Comparison of experimental results and computations for tension-leg-buoy offshore wind turbines. *Journal of Ocean and Wind Energy* **2015**, *2*, 12–20.
34. Azcona, J.; Bouchotrouch, F.; Gonzalez, M.; Garciandia, J.; Munduate, X.; Kelberlau, F.; Nygaard, T. A. Aerodynamic thrust modeling in wave tank tests of offshore floating wind turbines using a ducted fan. *Journal of Physics: Conference Series* **2014**, *524*, 012089.
35. https://www.sintef.no/globalassets/project/eera-deepwind-2015/presentations/e/e2_sandner_univ-stuttgart.pdf.
36. Sauder, T.; Bachynski, E. MARINTEK Ocean energy review: experimental modeling of wind loads on offshore wind turbines in wave tanks. *Tech-*

nical Report **2014**, Norwegian Marine Technology Research Institute, MARINTEK.

37. Martin, S; Day, S.; Gilmour, C. B. Rotor scaling methodologies for small scale testing of floating wind turbine systems. *34th International Conference on Ocean, Offshore and Arctic Engineering* **2015**, paper number OMAE2015-41599.
38. Martin, H. R.; Kimball, R. W.; Viselli, A. M.; Goupee, A. J. Methodology for wind/wave basin testing of floating offshore wind turbines. *Journal of Offshore Mechanics and Arctic Engineering* **2014**, *136*, 021902.
39. de Ridder, E.-J.; Otto, W.; Zondervan, G.-J., Huijs, F.; Vaz, G. Development of a scaled-down floating wind turbine for offshore basin testing. *33rd International Conference on Ocean, Offshore and Arctic Engineering* **2014**, paper number OMAE2014-23441.
40. Haans, W. *Wind Turbine Aerodynamics in Yaw: Unravelling the Measured Rotor Wake* (PhD thesis). Technische Universiteit Delft **2011**.
41. Jain, A.; Robertson, A. N.; Jonkman, J. M.; Goupee, A. J.; Kimball, R. W.; Swift, A. H. P. FAST code verification of scaling laws for deepcwind floating wind system tests. *22nd International Offshore and Polar Engineering Conference.* Greece **2012**.
42. Campagnolo, F. C. *Wind Tunnel Testing of Scale Wind Turbine Models: Aerodynamics and Beyond* (PhD thesis). Poletecnico di Milano, **2013**.
43. Whelan, J. I.; Stallard, T. Arguments for modifying the geometry of a scale model rotor. *Proceedings of 9th European Wave and Tidal Energy Conference.* Southampton **2011**.
44. Sant, T. *Improving BEM-based Aerodynamic Models in Wind Turbine Design Codes* (PhD thesis). Technische Universiteit Delft **2007**.
45. Vermeer, L. J.; Sorensen, J. N.; Crespo, A. Wind turbine wake aerodynamics. *Progress in Aerospace Sciences* **2003**, *39*, 467–510.
46. Leishman, J. G. Challenges in modeling the unsteady aerodynamics of wind turbines. *Wind Energy* **2002**, *5*, 85–132.
47. Bahaj, A. S.; Myers, L. E.; Rawlinson-Smith, R. I.; Thomson, M. The effect of boundary proximity upon the wake structure of horizontal axis marine current turbines. *Journal of Offshore Mechanics and Arctic Engineering* **2012**, *134*, 021104.

48. Molland A. F.; Bahaj, A. S.; Caplin, J. R., Batten, W. M. J. Measurement and predictions of forces, pressures and cavitation on 2-D sections suitable for marine current turbines. *Journal of Enginering for the Maritime Environment* **2004**, *218*, 127–138.
49. Sivalingam, K.; Martin, S.; Singapore Wala, A. A. Numerical validation of floating offshore wind turbine scaled rotors for surge motion. *Energies* **2018**, *11*, 2578.
50. Sivalingam, K.; Sepulveda, M..; Spring, M.; Davies, P. A review and methodology development for remaining useful life prediction of offshore fixed and floating wind turbine power converter with digital twin technology perspective. *2nd International Conference on Green Energy and Applications.* Singapore **2018**, 197–204.
51. Anara-Lara, O.; Campos-Gaona, D.; Moreno-Goytia, E.; Adam, G. *Offshore Wind Energy Generation: Control, Protection and Integration to Electrical Systems.* John Wiley & Sons **2014**.
52. Givaki, K.; Parker, M.; Jamieson, P. Estimation of the power electronic converter lifetime in fully rated converter wind turbine for onshore and offshore wind farms. Wind Energy Systems. *7th IET International Conference on Power Electronics, Machines and Drives.* Manchester **2014**.
53. Ma, K.; Zhou, D.; Blaabjerg, F. Evaluation and design tools for the reliability of wind power converter system. *Journal of Power Electronics* **2015**, *5*, 1149–1157.
54. Islam, M. R.; Guo, Y.; Zhu, J. Power Converters for Wind Turbines: Current and Future Development. In Mendez-Vilas, A., Ed., *Materials and Processes for Energy: Communicating Current Research and Technological Development*, Formatex Research Center **2013**.
55. Fischer, K.; Stalin, T.; Ramberg, H.; Thiringer, T.; Wenske, J.; Karlsson, R. Investigation of converter failure in wind turbines. *Elforsk Report* **2012**.
56. Dominguez, J.; Luis, J. *Modeling and Control of Squirrel Cage Induction Generator with Full Power Converter Applied to Windmills* (Master's thesis). University of Oulu **2009**.
57. Degrenne, N.; Ewanchuk, J.; David, E.; Boldyrjew, R., Mollov, S. A review of prognostics and health management for power semiconductor

modules. *Annual Conference of the Prognostics and Health Management Society*. Beijing **2015**.

58. Fischer, K.; Wenske, J. Towards reliable power converters for wind turbines: Field-data based identification of weak points and cost drivers. *Proceedings of EWEA Conference*. Paris **2015**.
59. Yang, S.; Xiang, D.; Bryant, A, Mawby, P.; Ran, L.; Tavner, P. Condition monitoring for device reliability in power electronic converters: a review. *IEEE Transactions on Power Electronics* **2010**, *25*, 2734–2752.
60. Wang, H.; Zhou, D.; Blaabjerg, F. A reliability-oriented design method for power electronic converters. *2013 Twenty-Eighth Annual IEEE Applied Power Electronics Conference and Exposition (APEC)*. Long Beach **2013**, 2921–2928.
61. Lei, T.; Barnes, M., Smith, S.; Hur, S. H.; Stock, A.; Leithead, W. E. Using improved power electronics modeling and turbine control to improve wind turbine reliability. *IEEE Transactions on Energy Conversion* **2014**, *30*, 1043–1051.
62. Ma, K.; Liserre, M.; Blaabjerg, F. Lifetime estimation for the power semiconductors considering mission profiles in wind power converter. *2013 IEEE Energy Conversion Congress and Exposition*. Colorado **2013**, 2962–2971.

Chapter 8

Aerodynamic Analysis of Floating Offshore Wind Turbine

8.1 NREL 5MW Wind Turbine Details

The NREL 5MW wind turbine was chosen for the current study in order to compare the aerodynamic effects of this FOWT with CFD and BEM. The chosen turbine was widely studied with sufficient validated data available in the public domain. It has a shaft tilt of 5°, and a rotor and hub radius of 63 m and 1.5 m, respectively. The blade aerodynamic model uses 17 blade elements for integration of the aerodynamic and structural forces. The inboard three and outboard three elements are two-thirds the size of the 11 equally spaced midspan elements in order to better capture the large structural gradients at the blade root and the large aerodynamic gradients at the blade tip. The geometric properties at the blade nodes, which are located at the center of the blade elements, are given in Table 8.1. The blade node locations, labelled as RNodes in the table, are along the radial direction.

Table 8.1 Distributed aerodynamic properties.[1]

Node (–)	RNodes (m)	AeroTwst (deg)	DRNodes (m)	Chord (m)	Airfoil Table (–)
1	2.8667	13.308	2.7333	3.542	Cylinder1.dat
2	5.6000	13.308	2.7333	3.854	Cylinder1.dat
3	8.3333	13.308	2.7333	4.167	Cylinder1.dat
4	11.7500	13.308	4.1000	4.557	DU40_A17.dat
5	15.8500	11.480	4.1000	4.652	DU35_A17.dat
6	19.9500	10.162	4.1000	4.458	DU35_A17.dat
7	24.0500	9.011	4.1000	4.249	DU30_A17.dat
8	28.1500	7.795	4.1000	4.007	DU25_A17.dat
9	32.2500	6.544	4.1000	3.748	DU25_A17.dat
10	36.3500	5.361	4.1000	3.502	DU21_A17.dat
11	40.4500	4.188	4.1000	3.256	DU21_A17.dat
12	44.5500	3.125	4.1000	3.010	DU64_A17.dat
13	48.6500	2.319	4.1000	2.764	DU64_A17.dat
14	52.7500	1.526	4.1000	2.518	DU64_A17.dat
15	53.1667	0.863	2.7333	2.313	DU64_A17.dat
16	58.9000	0.370	2.7333	2.086	DU64_A17.dat
17	61.6333	0.106	2.7333	1.419	DU64_A17.dat

As indicated in Table 8.1, the NREL 5MW baseline model incorporates eight unique airfoil data tables. The two innermost airfoil tables represent cylinders with drag coefficients of 0.50 (Cylinder1.dat) and 0.35 (Cylinder2.dat) and no lift. The remaining six airfoil tables were created by incorporating 3D corrections to the 2D data coefficients of the six airfoils.

As the drive train, nacelle and tower details are not the focus in this study, their details have not been included. The performance curves for the NREL 5MW turbine are shown in Figure 8.1. The rated operating wind speed for the turbine is 11.2 m/s. The wind speeds and corresponding rotor RPM, blade pitch, rotor power, etc., for the CFD study are extracted from these curves. Additional details are available in the literature.[1]

8.2 General Aerodynamic Analysis of Floating Offshore Wind Turbines

This section characterizes the aerodynamic behaviors of FOWT for the OC3 phase IV case 5.1 scenario. Unlike onshore turbines, additional degrees of freedom (DoFs) (surge, sway, heave, roll, pitch, and yaw) as shown in Figure 8.2 would considerably cause the time-varying geometric angle of attack to be along the blade span.[2] This in turn would alter the flow behaviors along the blade span. Platform motion and variation in the geometric angle of attack vary at the same frequencies. It is therefore vital to analyze the impact of the platform motions on the unsteady aerodynamic behavior of the rotor in an effort to determine the most influential platform DoFs. Similar methodology will be adopted in the subsequent sections in predicting the aerodynamic behavior of FOWT.

Reduced frequency, k, is used to characterize the degree of unsteadiness of an aerodynamic system because of disturbance with some frequency, ω. Reduced frequency is defined as

$$k = 0.5\,\omega c / V \tag{8.1}$$

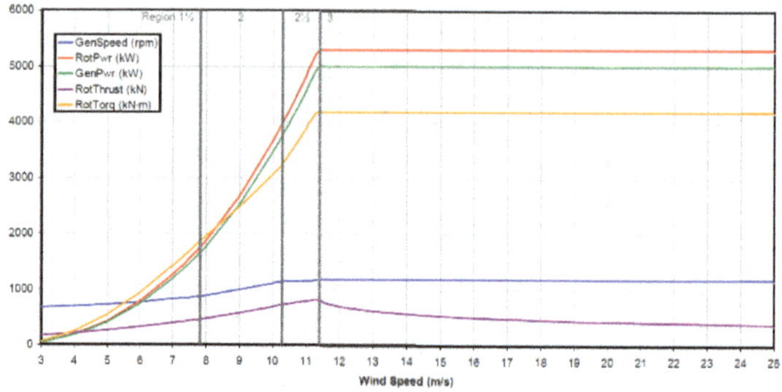

(a)

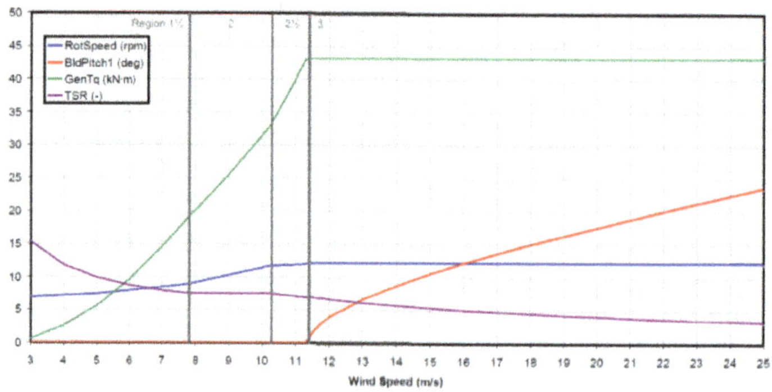

(b)

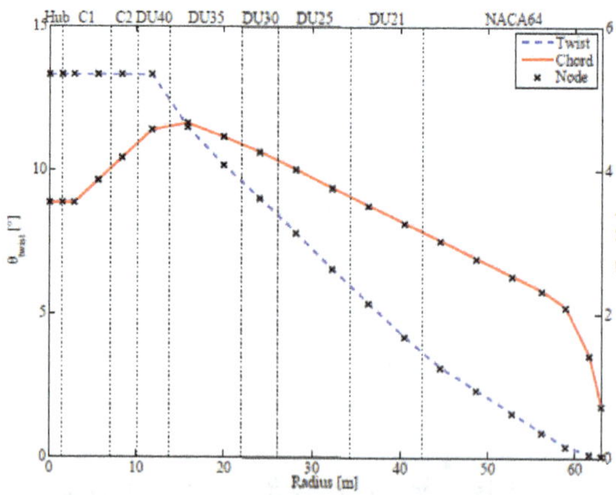

(c)

Figure 8.1 (a) Wind speed vs generator speed and rotor power. (b) Wind speed vs rotor speed and blade pitch. (c) Blade radius vs blade twist angle for various nodes.

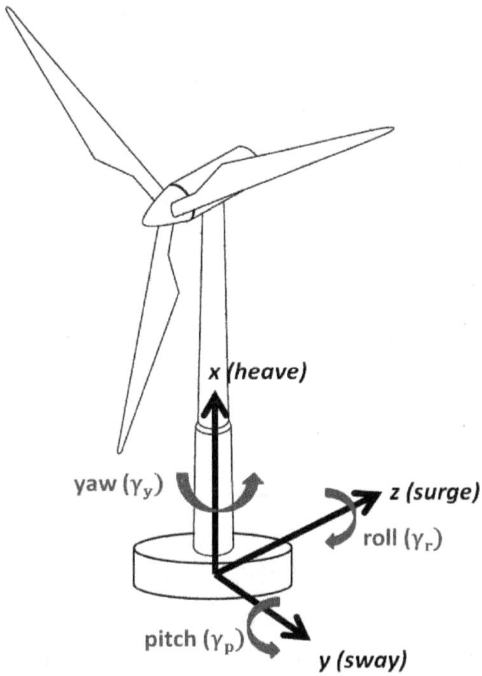

Figure 8.2 Representation of FOWT platform motion details.

Table 8.2 Degree of flow unsteadiness, as determined by reduced frequency.

Range	Flow Type
$k = 0$	Steady
$0 < k < 0.05$	Quasi-steady
$0.05 < k < 0.2$	Unsteady
$k > 0.2$	Highly unsteady

The degree of unsteadiness, as determined by reduced frequency, is typically categorized as shown in Table 8.2.

The local velocity of a wind turbine blade section has a rotational velocity component, given in terms of the rotor speed Ω, and a wind velocity component, U_∞.

The overall magnitude of the resultant or kinematic velocity is defined as

$$V = \sqrt{U\infty^2 + (r\Omega)^2} \tag{8.2}$$

where r is the local blade radius. Note that for a wind turbine, wind velocity and the rotor speed are not independent; as the wind speed increases, the rotor speed also increases up to a design-limited rated speed, after which the rotor speed is held constant. Eq. (8.2) may be substituted into Eq. (8.1) to give

$$k = \frac{\omega c(r)}{2\sqrt{U\infty^2 + (r\Omega)^2}} \tag{8.3}$$

where c is the local blade chord and is a function of r. It is meant to provide useful insight regarding the necessity of unsteady aerodynamic models when analyzing the aerodynamics of a wind turbine. Therefore, induced velocity and structural motions are neglected. A demarcation frequency, ω_d, above which flow can be unsteady, may be obtained by rearranging the above equation, substituting in the unsteady inequality from Table 8.2 ($k = 0.05$), as given by the equation below.

$$\omega_d = \frac{0.1\sqrt{U\infty^2 + (r\Omega)^2}}{c} \tag{8.4}$$

Eq. (8.4) may be rewritten in terms of ordinary rather than angular frequency, resulting in Eq. (8.5).

$$f_d = \frac{0.05\sqrt{U\infty^2 + (r\Omega)^2}}{\neq c} \tag{8.5}$$

The observed geometric angle of attack frequencies are expected to be driven in part by the platform modes for a floating turbine. Therefore, platform modal frequencies that are greater than the demarcation frequency are expected to cause unsteady aerodynamic loading on the blades. Blade geometries and steady-state operating behavior (recall that Ω is a function of U_∞) for a wind turbine may be substituted into Eq. (8.5) to generate the plot in Figure 8.3. This plot illustrates the cut-in, 8 m/s (for OC3 phase IV case 5.1 and when it is subjected to 0.1 Hz pitching motion as well), rated, and cut-out demarcation frequency curves for the NREL 5MW wind turbine. The area between the cut-in and cut-out curves represents all operating conditions for the turbine. The demarcation frequencies increase with increasing wind speed; at higher wind speeds, higher frequency platform motions are necessary to cause unsteady aerodynamic loading. The effect of span-wise chord variation on demarcation frequency curves shows up more clearly with increasing wind speeds.

Oscillation frequencies that fall to the right of a demarcation frequency curve, associated with a particular wind speed, may cause unsteady loading. As shown in Figure 8.3, as *r* increases for a given operating point, the demarcation frequency increases. Using the NREL 5MW turbine

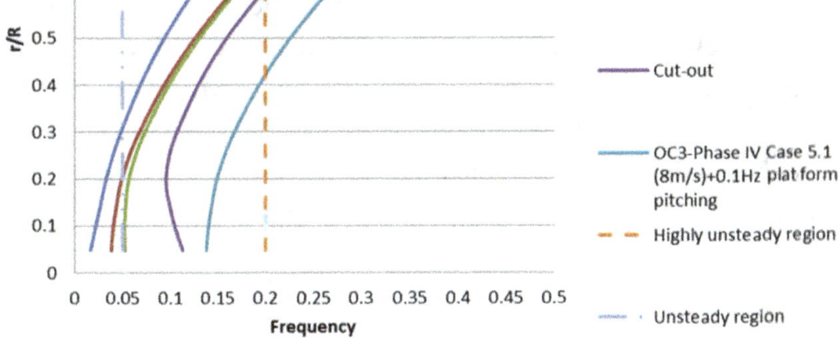

Figure 8.3 Demarcation curves for the NREL 5MW turbine for OC3 phase IV case 5.1.

with OC3 phase IV case 5.1 conditions as an example, an oscillation (pitching) of 0.1 Hz would yield unsteady loading for up to 40% of the rotor and remaining sections would be in highly unsteady condition. It is evident from the above fact that the unsteady behavior is inevitable in floating wind turbine applications.

8.3 OC3 Phase IV Case 5.1 — with Normal Sea State

8.3.1 *Benchmark Simulation Scenarios for FAST*

To compare and benchmark the results, OC3 phase IV cases (with small modifications) were considered.[9] A 3D blade model was created with sufficient details from NREL 5MW turbine aerofoil data. In this research work, the rotor is considered as a rigid body and rotating in a fixed rotor plane similar to an onshore turbine in order to establish CFD method against BEM. These simplifications will enable us to shed light on the aerodynamics of the blades, and to characterize the induced velocities of FOWT rotor due to pitching motion. As is known, CFD tools cannot replace the FAST-like tools as CFD tools are expensive in both cost and turnaround time for the simulations involved in wind turbine blade design and design certifications. CFD helps to study the finer details of certain scenarios which are not possible with softwares like FAST.

The basic version of OC3 phase IV FAST model was obtained from NREL, USA. The model is modified to include the requirements of the simulations, such as mooring line and aero-hydrodynamic models to simulate time series of the turbine and platform responses to environmental operating conditions. Table 8.3 summarizes the FAST-simulated load cases for the NREL 5MW spar buoy OC3 phase IV.

FAST model settings:

- The tower fore-aft and side-side DOFs were switched off to make rigid body.

Table 8.3 FAST simulation scenarios.

Uniform Wind Speed (m/s)	Sea State	Rigid Body Assumption	Rotor Speed (rpm)
8	6 m wave height and 10 s wave period	Yes	9.16
10	6 m wave height and 10 s wave period	Yes	11.45
11.2	6 m wave height and 10 s wave period	Yes	12.05

- First and second flapwise blade DoF were switched off and edgewise DoF was also switched off to make rigid body.
- An equilibrium BEM inflow model was used along with Prandtl tip and hub loss factors (according to *Aerodyn Theory Manual*,[3] GDW is better for when the wind speed is equal to or greater than 8 m/s, and it is easy to implement correction terms with equilibrium BEM for FOWT applications)
- Blade pitch and generator torque controllers were switched off (for 11.2 m/s and below, blade pitch control is not needed as per the design curve given in Figure 8.2. Moreover, all the scenarios are with fixed and uniform wind speed).
- Freestream wind was defined as constant, unidirectional, and without shear.
- The six platform DOFs were enabled.
- Wind and wave data from OC3 phase IV case 5.1 were used to define the sea state in terms of significant wave height, Hs, and spectral period, Tp. The simulated time series was created with these parameters.

Time domain simulations were performed for each of the cases described in Table 8.3, with each simulation lasting 1860 seconds. The outputs generated by the initial 1800 seconds of each simulation were omitted in the analysis.

The initial run was performed to extract the motion of the platform for the above specified OC3 case. Figure 8.4 represents the roll, pitch and yaw motions of the platform.

As seen in Figure 8.4, the platform pitching dominates when compared to yaw and roll. A close-up of the pitching motion is shown in Figure 8.5, from which it can be concluded that the blade plane motion can be

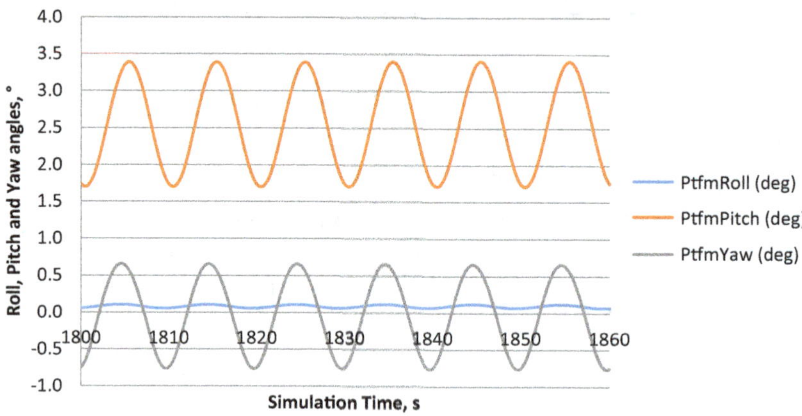

Figure 8.4 Platform motions of the simulation scenario of Table 8.3.

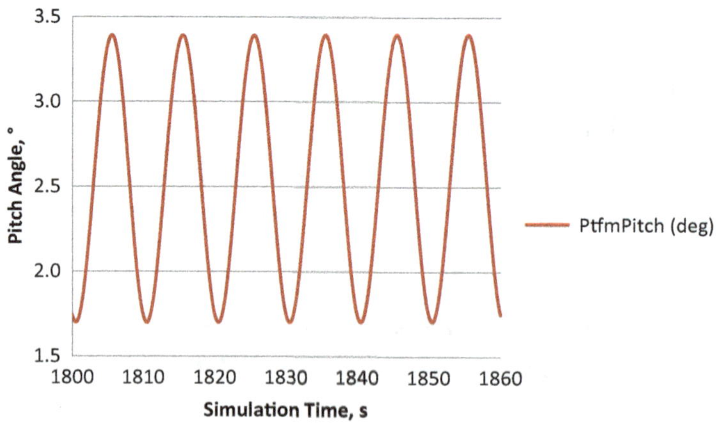

Figure 8.5 Platform motions of the first simulation scenario of Table 8.3.

characterized as a mean pitch angle of 2.54° with 1.6° range from the mean.

Platform translational movement details are given in Figure 8.6. Rotor plane movements were derived from the platform pitching motion as shown in Figure 8.7.

From Figure 8.6, it is evident that surge motion plays a crucial role in the translational directional changes. As the objective is to understand

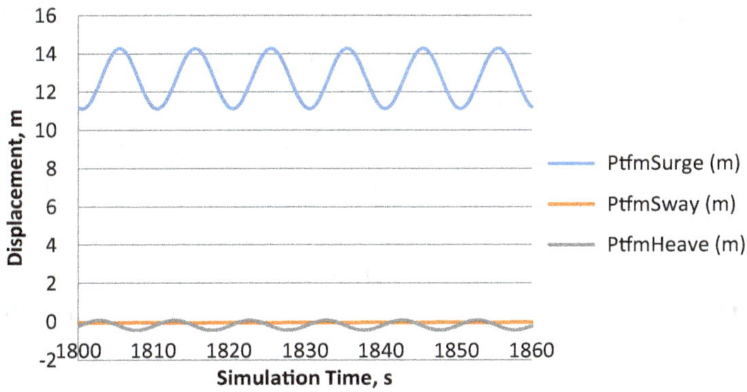

Figure 8.6 Platform surge, heave and sway movements.

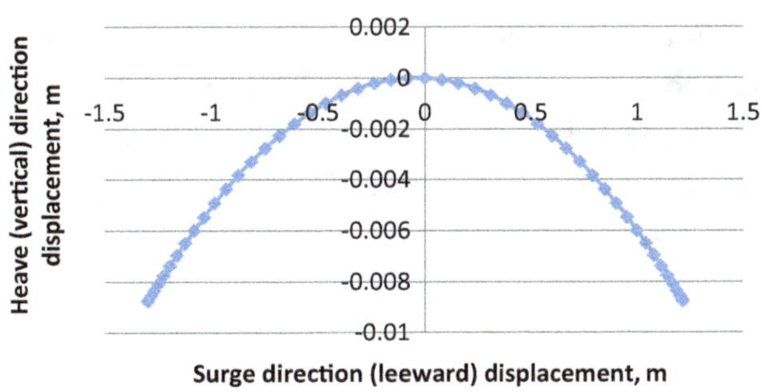

Figure 8.7 Rotor plane displacement due to platform pitch.

the effects of platform pitching motion on the rotor plane aerodynamics, hub height has been accounted for to determine the displacement of the rotor plane. In this particular study, the AeroDyn output of FAST results were taken into consideration as the applied load onto the rotor structure alone can be compared with CFD results. Blade and generator control were not considered as the steady wind is applied.

8.3.2 *Methodology*

Induction factors are mainly used in the BEM equations to calculate the load on the rotor. Hence, much attention has been dedicated to calculate the induction factors with the data obtained from complete CFD simulations as CFD can predict the load more accurately in prescribed conditions than BEM. The CFD simulations were set up based on chosen case scenarios from FAST. They were carried out with known steady state cases and the rotor power was compared with the value provided by the design curve to gain confidence on the results. Once the simulation set-up including the mesh parameters and boundary conditions was finalized, the calculations were carried out for the chosen turbine motion scenarios.

8.3.3 *Induction Factors*

The axial and tangential induction factors are the two most important quantities that a BEM code calculates iteratively. The amount of axial flow through the rotor is decided by the axial induction factor, a, and the rotation of the wake behind the turbine is determined by the tangential induction factor, a', expressed as

$$a = 1 - \frac{U}{U_{\circ}} \tag{8.6}$$

$$a' = \frac{\omega}{2\Omega} \tag{8.7}$$

The aerodynamic forces are then calculated with these parameters along with the empirical coefficients for the blade sections and additional loss/correction factors.

The tip and hub losses are accounted for using various loss functions. In this paper, the Prandtl tip loss formulation has been included to account for the tip losses. The Prandtl tip loss is given by $F_t = (2/\pi)\cos^{-1}(-f_t)$ where $f_t = B(R - r)/(2r \sin Ø)$, with B being the number of blades, R the rotor radius, and Ø the local flow angle. The hub losses are calculated as $F_h = (2/\pi)\cos^{-1}(-f_h)$ where $f_h = B(r - R_{hub})/(2r \sin Ø)$. The total loss factor is calculated as

$$F = F_t \times F_h \tag{8.8}$$

Various other corrections like Glauert correction, Skewed wake correction, etc., can be incorporated to capture the physics of the problem more accurately.

8.3.4 *Elemental Torque and Thrust*

The BEM code solves a set of equations iteratively to calculate the induction factors and hence the forces on the blade elements.[3] Torque and thrust on the blade elements are given by Eqs. (8.9) and (8.10), respectively.

$$dT = 4\pi r^3 dr\, \rho\, U_\infty\, F\,(1 - a)a'\, \Omega \tag{8.9}$$

$$dF = 4\pi r\, dr\, \rho\, U^2_\infty\, F\,(1 - a)a \tag{8.10}$$

where dT is the elemental torque, dF is the elemental thrust, F is the combination of hub loss and Prandtl tip loss factors, a is the axial induction factor and a' is the tangential induction factor.

These equations will be compared against the results from the CFD studies to obtain a new set of induction factors. These are then compared with the ones obtained by the BEM code.

8.3.5 Computational Fluid Dynamics- and FAST-Based Induction Factors

The CFD-based approach to calculate the induction factors is a first step towards the verification of these values when complex turbine motions are involved. The approach involves the calculation of the forces on the wind turbine blades with high-fidelity CFD simulations and extracts the elemental force values. These values are compared with the forces from the BEM equations and the induction factors are derived. The two equations used for the above mentioned approach are the thrust and the torque equations.

Both axial and tangential induction factors can be evaluated using the two equations. The axial induction factors were calculated by equating the thrust equations with the elemental thrust values obtained with CFD, as given in Eq. (8.11). There are two solutions possible for the induction factors with the thrust equation. The choice of the solution for the induction factor can be concluded from the thrust equation as one root is <0.5 and the other is >0.5. All the roots <0.5 were chosen in compliance with FAST. The axial induction factors thus obtained were used with the elemental torque equation and equated with the elemental torque values obtained from CFD data to calculate the tangential induction factors as given by Matha *et al.*[4]

$$a(1-a)=\frac{dF_{CFD}}{4\pi r\ dr\rho U_{\infty}^{2}F} \tag{8.11}$$

$$a'=\frac{dT_{CFD}}{4\pi r^{3}\ dr\rho U_{\infty}F(1-a)\Omega} \tag{8.12}$$

8.3.6 Computational Fluid Dynamics Simulation Scenarios

The CFD simulations were set up for FAST OC3 phase IV 5.1 case with rigid body assumption. A 6 m significant wave height is used for hydrodynamics calculations. The resulting turbine motions provided by FAST were included in the CFD simulations. The details of the case output values used to set up the CFD simulation are given in Table 8.4.

The surge motion is not accounted for in this section on steady state CFD simulation scenarios as the displacement is constant for the given wind speed. Three operating points were selected from the design curve of the turbine such that the rotor performance encompasses cut-in and rated wind speed. The details of the wind speeds chosen along with other parameters used in the study are also shown in Table 8.4.

The blade geometrical model can be generated in ICEM CFD by sweeping various cross-sections of the NREL 5MW blade. The 3D 5MW NREL wind turbine rotor model can constructed using the following aerofoil sections.

1. NACA64
2. DU40
3. DU35
4. DU30
5. DU25
6. DU21

Aerofoil data can be obtained from the NREL website as "DOWEC-NREL 5MW blade airfoil data-v2 (1)". This data can then be dimensionalized to the Cartesian coordinate system to create 19 sets of aerofoil data from blade root to the tip. All 19 sets of data were exported to chord

Table 8.4 Output from OC3 phase IV, used to build CFD simulation.

Wind Speed (m/s)	Platform Pitching Amplitude (°)	Platform Mean Pitch Angle (°)	Mean Rotor Speed (rpm)	Average Power (MW)	Blade Pitch (°)
8	0.85	2.54	9.16	1.84	0
10	0.85	4.23	11.45	3.67	0
11.2	0.85	4.90	12.05	4.96	0

center/aerodynamic center in ANSYS ICEM CFD to construct the blade geometry. Roots are made of four varying cross-section cylinders. Lofting and mirror features are used in modeling the blades.

The complete rotor was modeled for the calculations, excluding the hub. The roots of the blades were extended to meet at the center of the rotor. The inlet boundary was located at ~3 rotor diameters in front of the rotor and the domain extends up to ~6 rotor diameters behind the rotor as shown in Figure 8.8. A multiblock meshing strategy was used to create hexahedral mesh in the whole domain. 'O-grid'-type mesh was created in the blade domains. The geometry and mesh are shown in Figures 8.9 and 8.10,

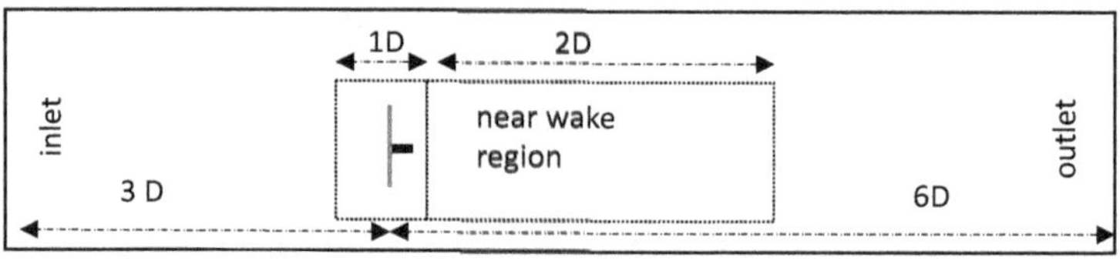

Figure 8.8 Computational domain for simulations with inlet and outlet are respectively 3D and 6D away from the turbine to minimize the effect of boundary conditions.

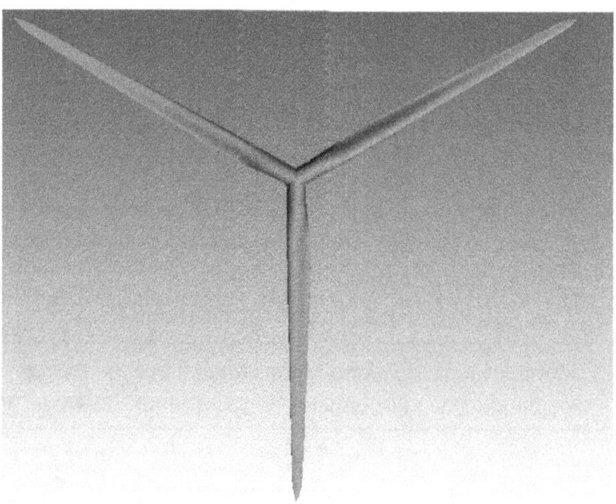

Figure 8.9 NREL 5MW rotor with the blade roots extended to meet at the center.

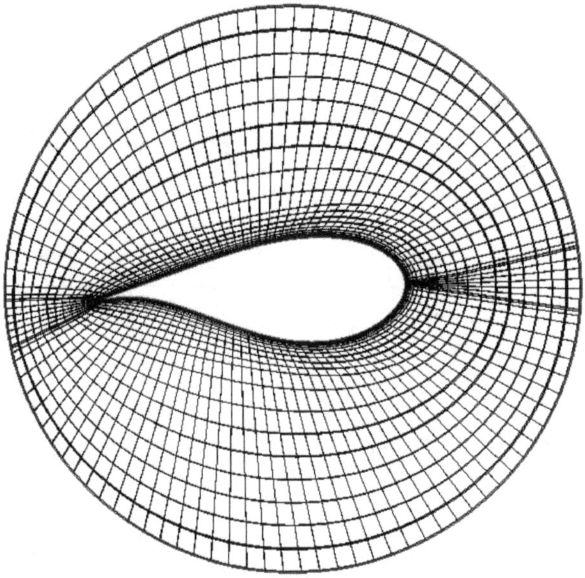

Figure 8.10 Structured mesh on the blade with the O-grid section of the blade domain.

respectively. The blade was meshed to yield a y^+ of ~1 near the tip and less than 1 for the rest.

As a first step of CFD model validation against 5MW NREL wind turbine, mesh sensitivity and domain size studies are performed for domain- and mesh-independent solutions. The 5MW NREL wind turbine's designed power curve by Chow and van Dam[5] is shown in Figure 8.11.

From the domain-independent CFD study, it is found that 6 rotor diameter spacing in the leeward direction and 3 rotor diameter spacing in the windward direction are adequate. Mesh-independent study was carried out with the above mentioned domain size. The overall CFD computational domain, enlarged view of inner domain and rotor region are given in Figures 8.12 to 8.14.

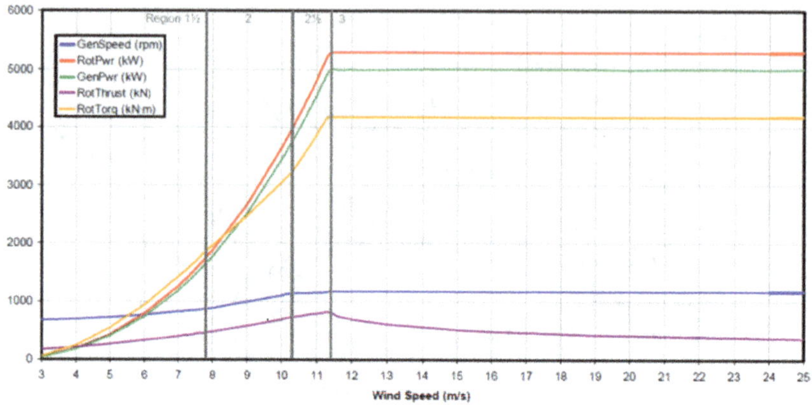

Figure 8.11 5MW NREL wind turbine-designed power curve.

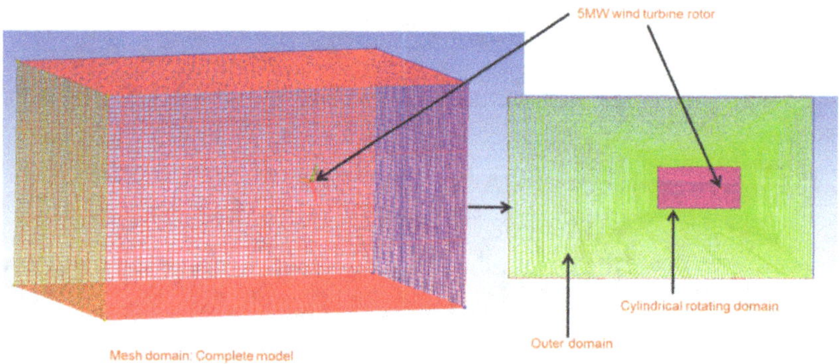

Figure 8.12 CFD model of 5MW NREL wind turbine (outer domain).

As the prediction of applied force on the rotor relies on the mesh clustering around it, four different meshes are used around the rotor in the CFD model for mesh-independent study. Figure 8.15 shows the four mesh category types. The chosen wind speed is 8 m/s with 0° rotor pitch angle.

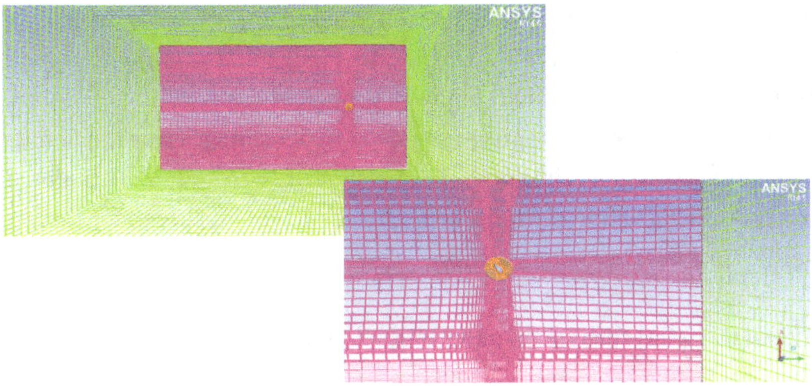

Figure 8.13 CFD model of 5MW NREL wind turbine (zoomed-in view — cylindrical and rotor domains).

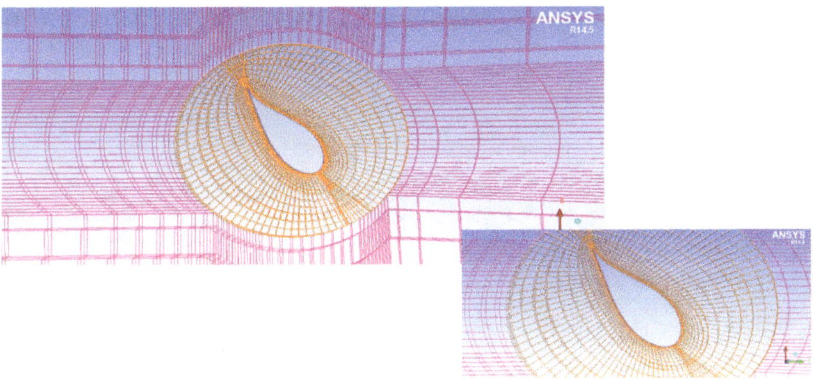

Figure 8.14 CFD model of 5MW NREL wind turbine (zoomed-in view of rotor).

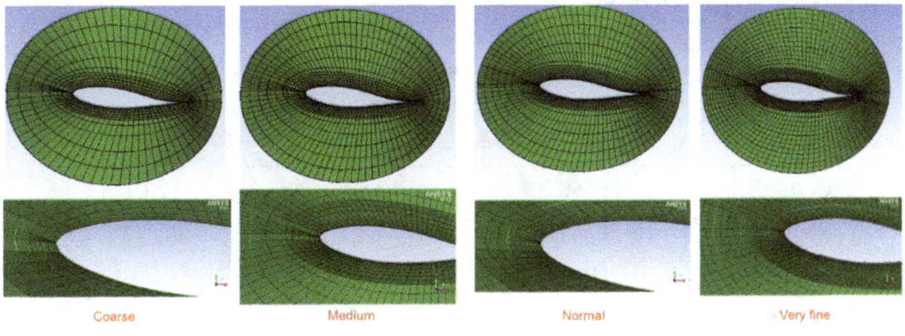

Figure 8.15 Mesh-independent study — rotor CFD mesh models.

The four mesh category results are compared against the designed power curve of 5MW NREL wind turbine in Figure 8.16. As per designed power curve, 1876 kW is the power generated for 8 m/s wind speed. Table 8.5 shows the comparison of power in percentage error for all four category mesh sizes against power curve value of 1876 kW. As highlighted in yellow in Table 8.5, the normal mesh size category is chosen for further study, i.e., power curve validation.

When the wind turbine operates in the low wind regime, the change in the axial induction factor will go beyond 0.5. Hence, it is decided to use the wind speed of region 1 ½, 2 and 2 ½ (up to 11.4 m/s) in the designed power curve. In an effort to validate the CFD model and its parameters against the designed power curve, three chosen wind speeds are used to compute the power. The designed power curve and the CFD-computed power curve are compared as in Figure 8.17 and Table 8.6.

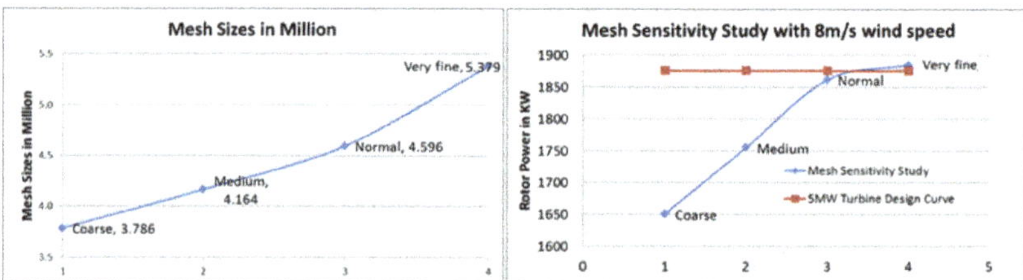

Figure 8.16 Mesh sizes (cells in millions) and mesh-independent study.

Table 8.5 Percentage error and chosen mesh size (normal mesh size) for designed power curve validation.

Mesh Size	CFD Predicted Power (kW)	Designed Power (kW)	% Difference
Coarse	1650.7	1876	12.01
Medium	1755.6	1876	6.42
Normal	1862.2	1876	0.74
Very fine	1885.7	1876	−0.52

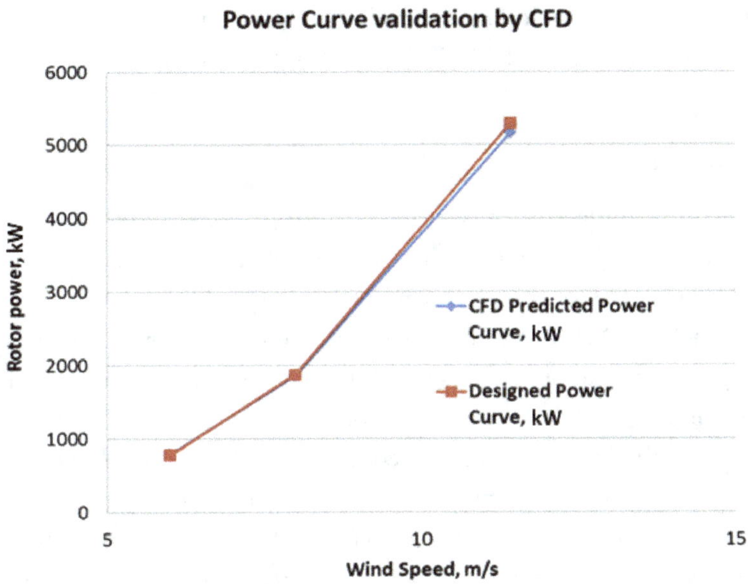

Figure 8.17 5MW NREL wind turbine power curve validation (for the regions 1 ½, 2 and 2 ½ in the power curve).

Table 8.6 5MW NREL wind turbine power curve validation (for the regions 1 ½, 2 and 2 ½ in the power curve — with % error).

Wind Speed (m/s)	CFD Predicted Power (kW)	Designed Power Curve (kW)	% Difference
6	790.3	785	–0.68
8	1862.2	1876	0.74
11.4	5164.5	5294	2.45

By selecting the suitable mesh size and domain size, the error between designed power curve and CFD-predicted power is well below 2.5% (as in Table 8.6), which suffices as per industry practices.

Hence, the CFD model has been validated against the designed power curve (up to the region 2 ½) of the 5MW NREL wind turbine. Further steady and transient simulation studies will be performed through this validated CFD model of 5MW NREL wind turbine in the rest of this study.

The CFD simulations were set up to capture the dynamic effects due to the pitching motion of the NREL 5MW turbine. The steady state MRF simulations were carried out with the rotor tilted by a mean pitch angle in addition to 5° base tilt. The mean rotational speed and the wind speeds chosen for CFD simulations are shown in Table 8.4. The results of the steady state calculations were used to initialize the solutions for a complete transient case involving the pitching motion of the turbine. The pitching motion methodology was developed in the CFD software. The domain was structured in a way that the rotor domain was made as a rotating domain and the mesh motion applied for pitching had an extremely high stiffness in this domain and relaxed in the outer domain. This is to ensure that the mesh in the rotor domain had almost no relative nodal displacement, as the mesh on the blade had its first node in the order of microns to yield a $y^+ \sim 1$, where y^+ is the local non-dimensional wall distance for wall-bounded flow.

The full range of pitching motion was carried out according to OC3 phase IV data. The results were averaged for the mean thrust, torque and hence the induction factors. Hence, the results dealt with for the rest of this study correspond to the averaged values for the above mentioned parameters.

In order to understand the dynamic flow behavior around the rotor and near wake field due to the platform pitching motion, the transient pitching motion study is carried out at a uniform wind speed of 8 m/s as in Table 8.4. In this study, the transient simulation is initialized with steady state results and simulated for 60 seconds, which is 6 wave period time.

The transient simulations as mentioned were carried out at a wind speed of 8 m/s. The rotor pitching motion starts with the mean rotor tilt of 7.54°. As the platform and tower are not modeled in the study, the angular motion at the platform is converted into translational and axis tilting motion at the rotor. Figures 8.18 and 8.19 shows the angular motion cycle at the platform (provided by FAST) and the corresponding rotor motion. It can be noted that, for the chosen pitching motion case, the

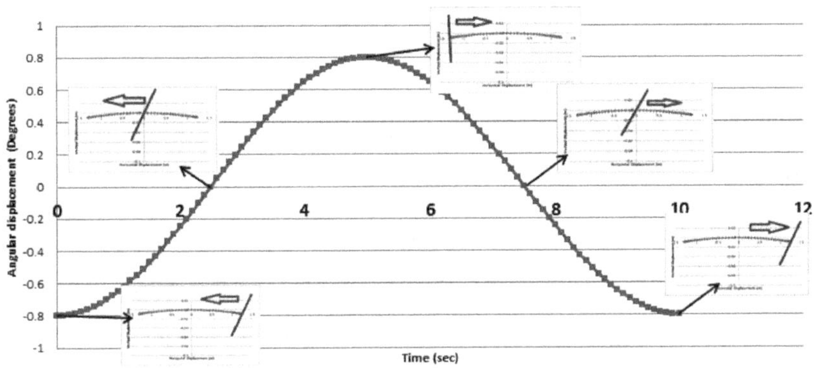

Figure 8.18 Platform pitching motion: angular displacement.

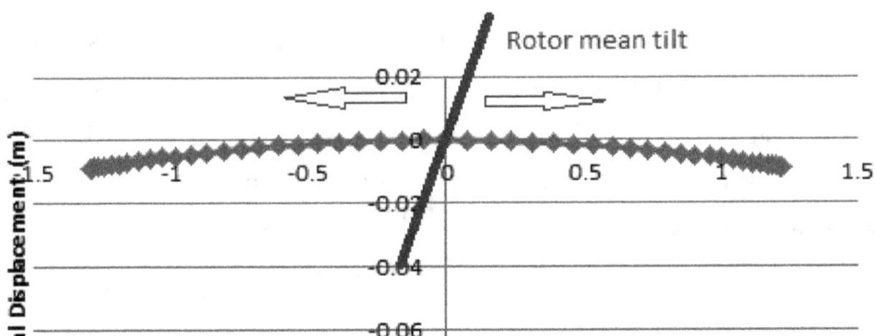

Figure 8.19 Translational displacements at the rotor.

vertical displacement is diminutive when compared to horizontal displacement and hence it is negligible. For the rest of this study, only horizontal displacement is accounted for, which is computed as 1.29 m.

As the wind speed approaching the rotor is not constant due to the platform motion, an oscillating wind speed has to be calculated to supply for the equations that require wind speed for computation. The oscillating wind speed is calculated using the following equations.

$$\theta = A.\omega.t.\cos(\omega t) \tag{8.13}$$

and

$$\text{wind speed} = V_x + V_{tang} + 8 \cdot \cos\left(7.54 \cdot \frac{\neq}{180}\right) \tag{8.14}$$

where

$$V_{tang} = r \cdot \frac{\theta}{t} \cdot \cos\,\theta$$

$$V_{hub} = \frac{\theta}{t} \cdot 90 \text{ (tower height)}$$

$$V_x = V_{hub}\cos(A\sin\omega t)$$

in which A is angular displacement at the platform, r is the radial distance to the blade element from the hub, V_{hub} is the velocity at the hub height, V_x is the horizontal displacement of the rotor, and V_{tang} is the tangential velocity due to rotor pitching motion. The wind speeds for various elements along the blade span calculated with the above formulation are shown in Figure 8.20. The wind speeds thus obtained are used for calculating the axial and tangential induction factors.

The CFD calculations were carried out at constant wind. Atmospheric boundary layer effects are not accounted for. The *k*-ω SST solver was used for the simulations. The air properties were set as incompressible at 15°C. A high-order advection scheme and first-order numerical method were employed for turbulence solutions.

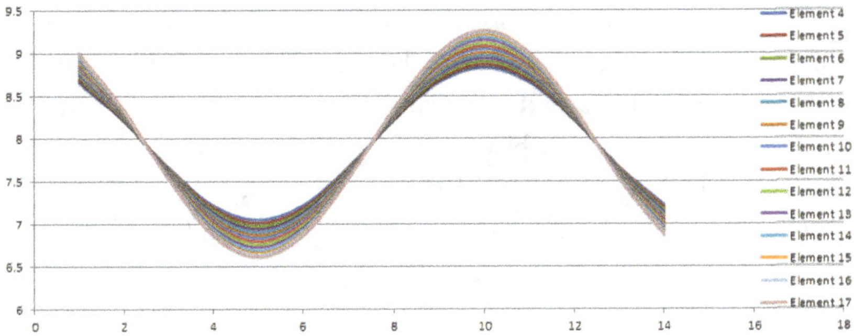

Figure 8.20 Oscillating wind speed for various blade elements (x axis is in time and y axis is in wind speed).

8.3.7 *Results and Discussion*

The results from the CFD simulation are categorized into two sets. One set of results provides the overall aerodynamic behavior obtained by averaging the pitching motion and another set of results provides an in-depth profile of aerodynamic behavior due to transient pitching motion.

Velocity contour, thrust and power details of the rotor and its elements are presented for rotor mean position state. Moreover, axial and tangential induction factors are also derived.

The average pitching motion data is obtained from steady state CFD calculations at specified mean rotor angles, which is the sum of rotor tilt angle and the mean pitching angle, obtained from the FAST software for the specified load case. The rotor speed, wind speed and the blade pitch angles are obtained from the design curve and verified by comparing with the FAST output data. The parameters thus obtained are used in the steady state calculations of CFD.

Various aspects of wind turbine flow were analyzed and compared with the existing literature to establish confidence on the CFD simulations output. The y^+ values on the blades are maintained ~1. It is necessary to maintain low y^+ values to resolve the boundary layers effectively on the blades. Figure 8.21 shows the typical y^+ contours in the current study. The flow patterns near the blade roots are compared with the patterns available in the NREL findings. The comparisons lead to an interesting finding that the flow separation zone exists up to 20 m from the center of the hub. Figure 8.22 shows the flow separations on the blades near the hub.

In addition, the wake pattern behind the rotor was captured with a fine mesh oriented along the wake direction. Figure 8.23 shows the velocity contours and the wake propagation behind the rotor. The above mentioned verification procedures ensure the mesh quality used on the rotor blades and in the wake are sufficiently fine to determine the aerodynamic loads with good accuracy.

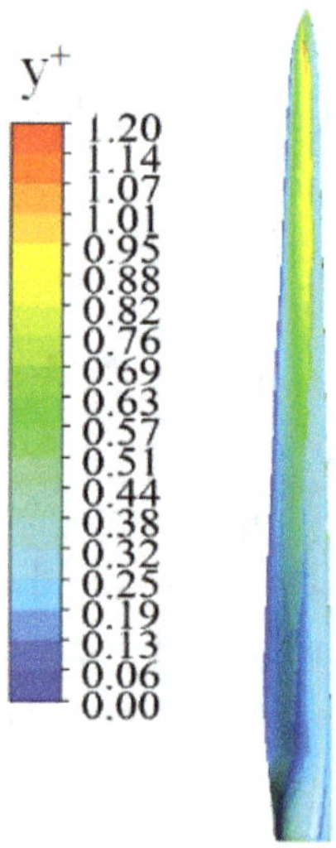

Figure 8.21 Blade y^+ contours at 8 m/s wind speed.

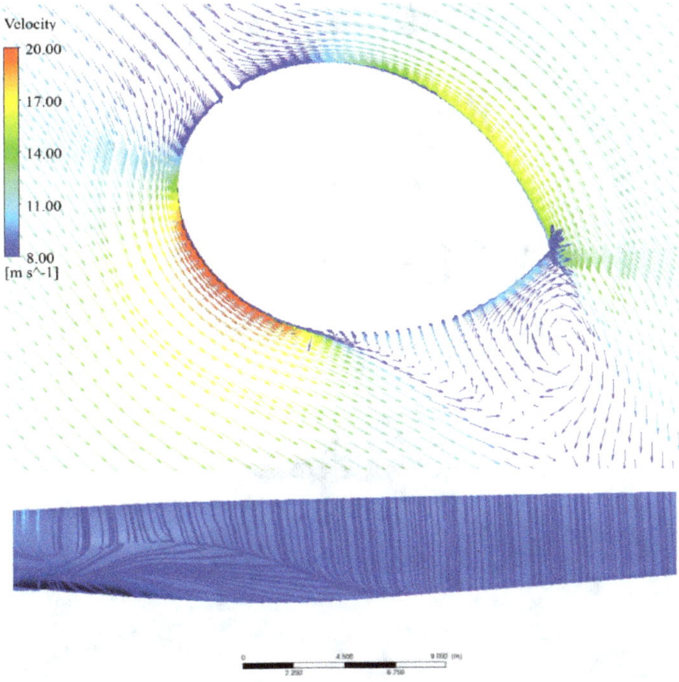

Figure 8.22 Vector plot showing the flow separation at 8 m from the hub (top) and streamline plot showing the extent of the flow separation zone on the blade (bottom).

The elemental power and thrust values are obtained from CFD post-processing. The elemental distances are chosen as per the FAST software listing so that the comparison with the BEM results can be made one-to-one with the CFD results. Also, the elemental results start from element #4 (radial distance 11.75 m) as there is no power generated till element #3 since the blade sections are cylindrical. Figure 8.24 provides the comparison of CFD-based elemental power with those obtained with BEM calculations for three different wind speeds as detailed in Table 8.4. It can be noted that the CFD power thus obtained correlates well with the BEM results.

(a)

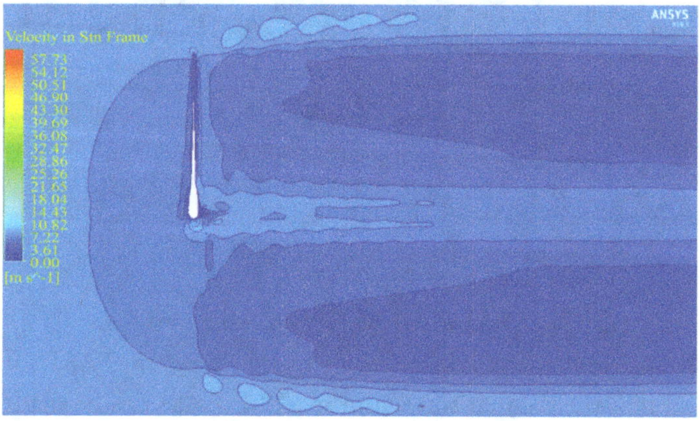

(b)

Figure 8.23 (a) Propagation of tip and root vortices and (b) the flow contours at 8 m/s wind: to show the wake and tip vorticity pattern for the normal case.

The elemental thrust values obtained from CFD and BEM are shown in Figure 8.25. The difference in the thrust values between CFD and BEM is notable for the elemental section near the blade tip, which can be attributed to the tip loss factors included in the BEM model.

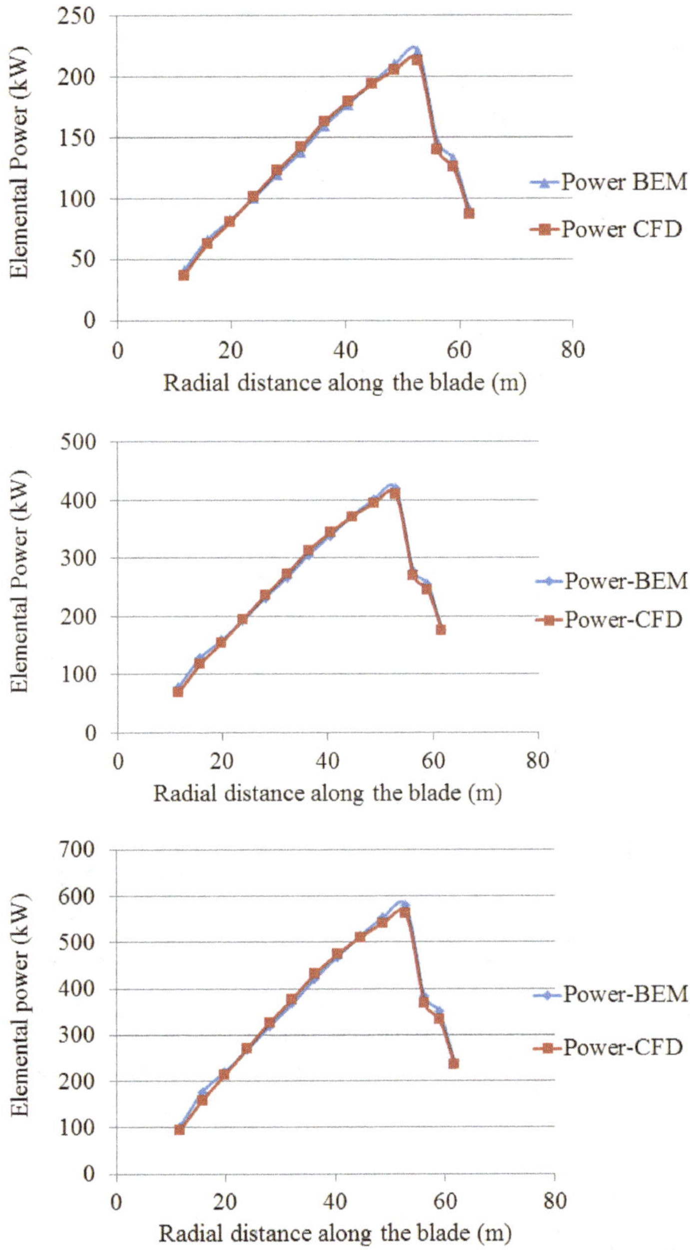

Figure 8.24 Comparison of elemental power obtained with CFD to those obtained with BEM for 8 m/s wind (top), 10 m/s wind (center) and 11.2 m/s wind (bottom).

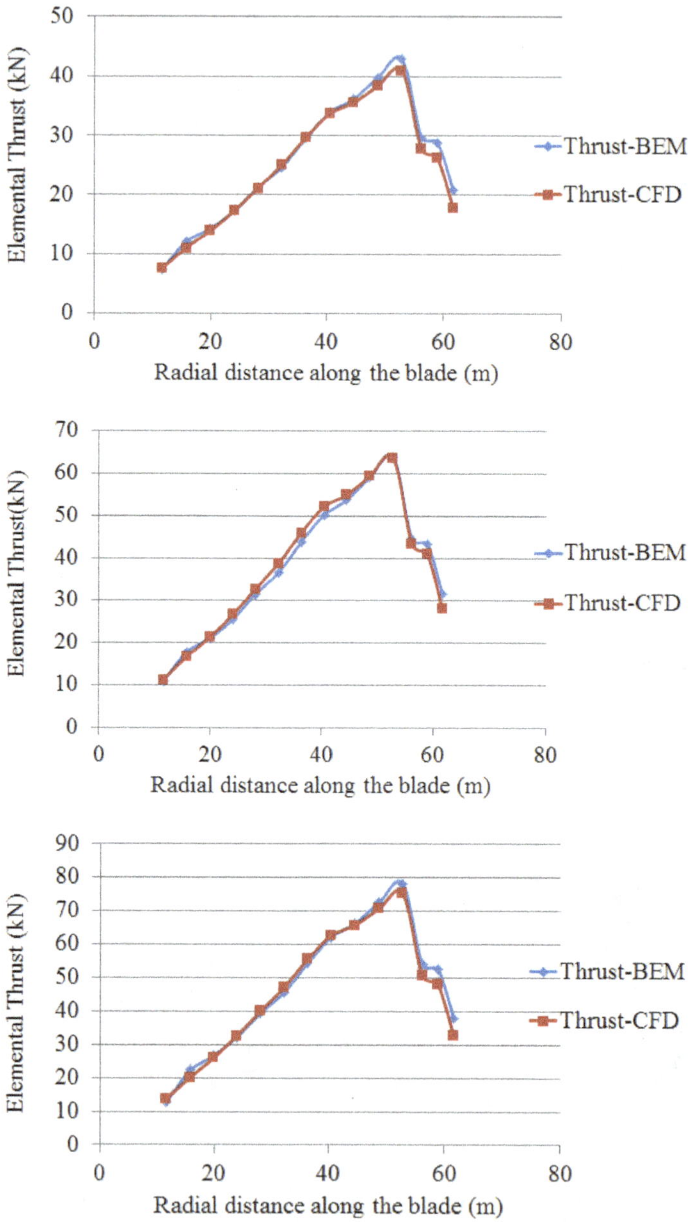

Figure 8.25 Comparison of elemental thrust values from CFD and BEM for 8 m/s wind (top), 10 m/s wind (center) and 11.2 m/s wind (bottom).

8.3.7.1 *Comparison of Axial and Tangential Induction Factors*

The axial and tangential induction factors are calculated using the elemental power and thrust values from CFD and BEM codes as mentioned in the previous section. The axial induction factor derived from CFD is generally lower than BEM prediction due to tip loss corrections accounted for in BEM. The tangential induction factors for CFD and BEM predictions are almost similar. The axial and tangential induction factors are shown in Figures 8.26 and 8.27 respectively.

The transient pitching motion study provides a detailed set of results for every time step for one pitching motion cycle of 8 m/s. In this study, the aerodynamic data were extracted from CFD for 10 equally spaced time steps in one cycle. This study provides an insight into the transient wake effects on the aerodynamic performance of the rotor. In addition, the transient simulation is able to shed light on the flow separations and streamwise flows on the blade sections with the help of pressure coefficient plot (Cp) for every element. Also, the elemental power, elemental thrust and the induction factors are compared between the CFD and BEM results for every time step as in the case of averaged pitching motion study. Figure 8.28 shows the velocity contour at the 60th second during transient pitching motion by CFD.

The elemental power and the thrust values are extracted from the CFD and BEM results for one pitching cycle. The comparison is furnished in Figures 8.29 and 8.30.

The elemental axial and tangential induction factors for all the elements in one complete pitching cycle are provided in Figures 8.31 and 8.32. The axial induction factor obtained with the elemental thrust values attains its peak at 0.5. Hence, the axial induction factors at 0.5 represent a saturation value.

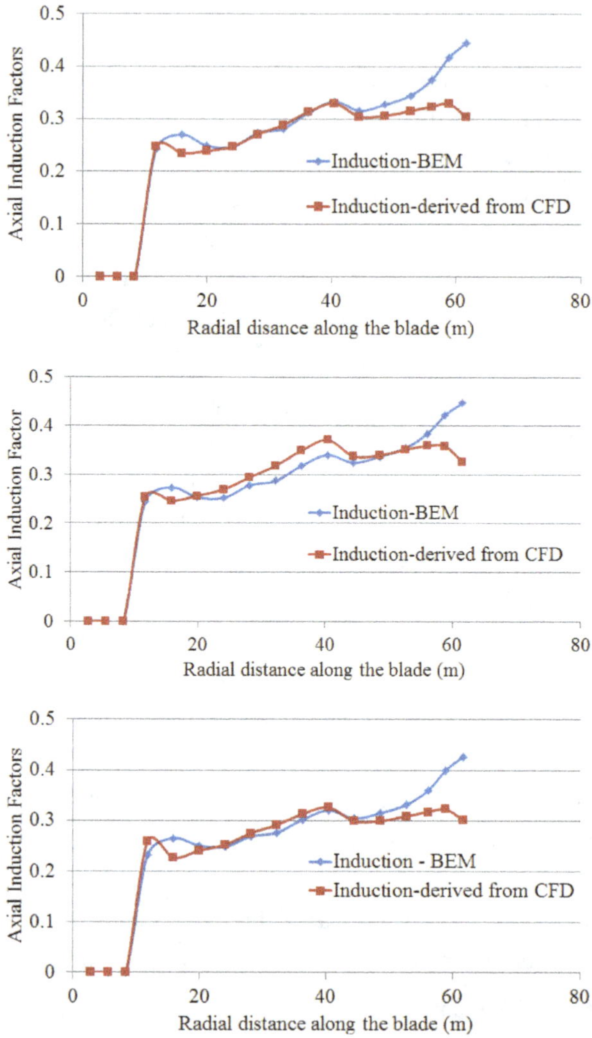

Figure 8.26 Comparison of axial induction factors for 8 m/s wind (top), 10 m/s wind (center) and 11.2 m/s wind (bottom).

As discussed in the earlier sections, FOWT design caters to unique operating conditions, as aerodynamic, hydrodynamic and mooring line forces have an impact on the wind turbine and they induce motions in six DoFs. The associated motions potentially amplify the unsteady

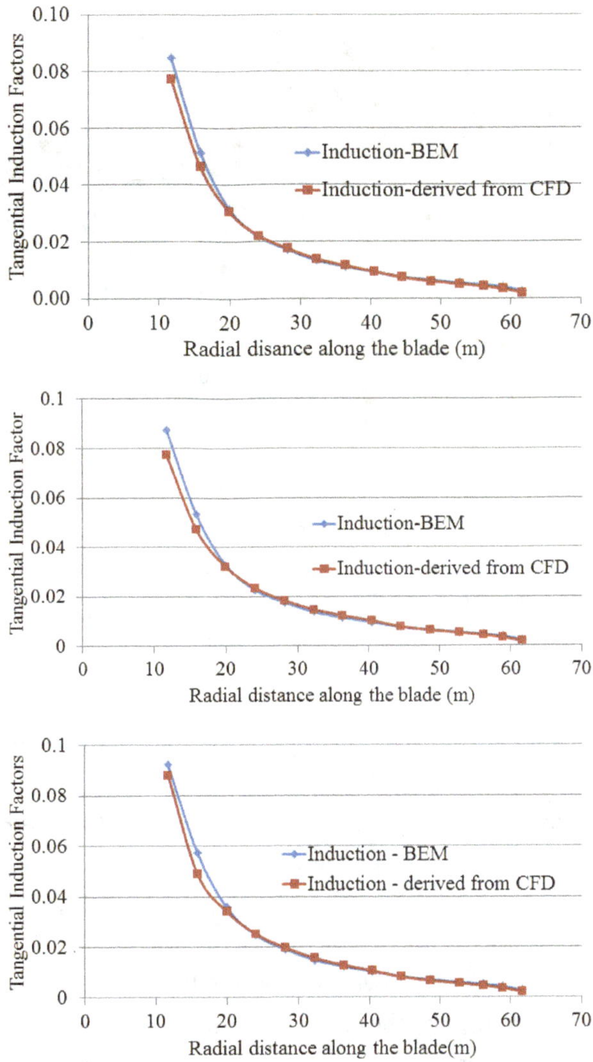

Figure 8.27 Comparison of tangential induction factors for 8 m/s wind (top), 10 m/s wind (center) and 11.2 m/s wind (bottom).

aerodynamic effects (e.g., blade-vortex interaction, dynamic stall, skewed inflow), which cannot be accurately modeled by available methods based on the BEM theory including correction factors. With OC3 phase IV

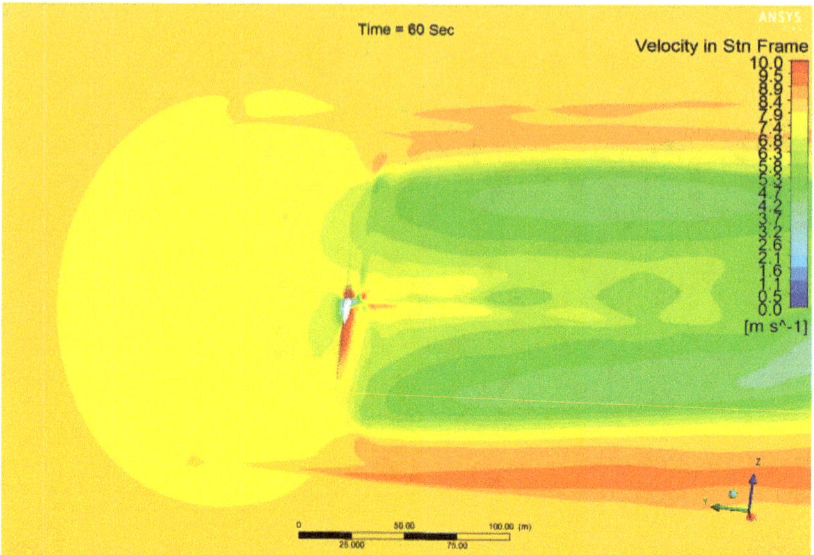

Figure 8.28 Velocity contour at 60th second during transient pitching motion — CFD.

case 5.1 chosen in this study, no significant blade-vortex interaction effects are observed as the wind speed, wave height and wave period generate minimal rotor displacement. Hence, some of the aerodynamic phenomena like dynamic stall and skewed wake flow are also not observed.

The AeroDyn software provides the elemental output values for a single blade. Figure 8.33 displays the comparison between CFD and AeroDyn (BEM) output power for a single blade. The fluctuating wind speed experienced by the rotor due to platform pitching motion is plotted with a second axis. With the blade power data provided by AeroDyn BEM, the effect of tower passage is reflected as periodic asymmetric fluctuations with a period of 120°. Comparatively, CFD power curve is smooth sinusoidal as the effect of tower is not modeled.

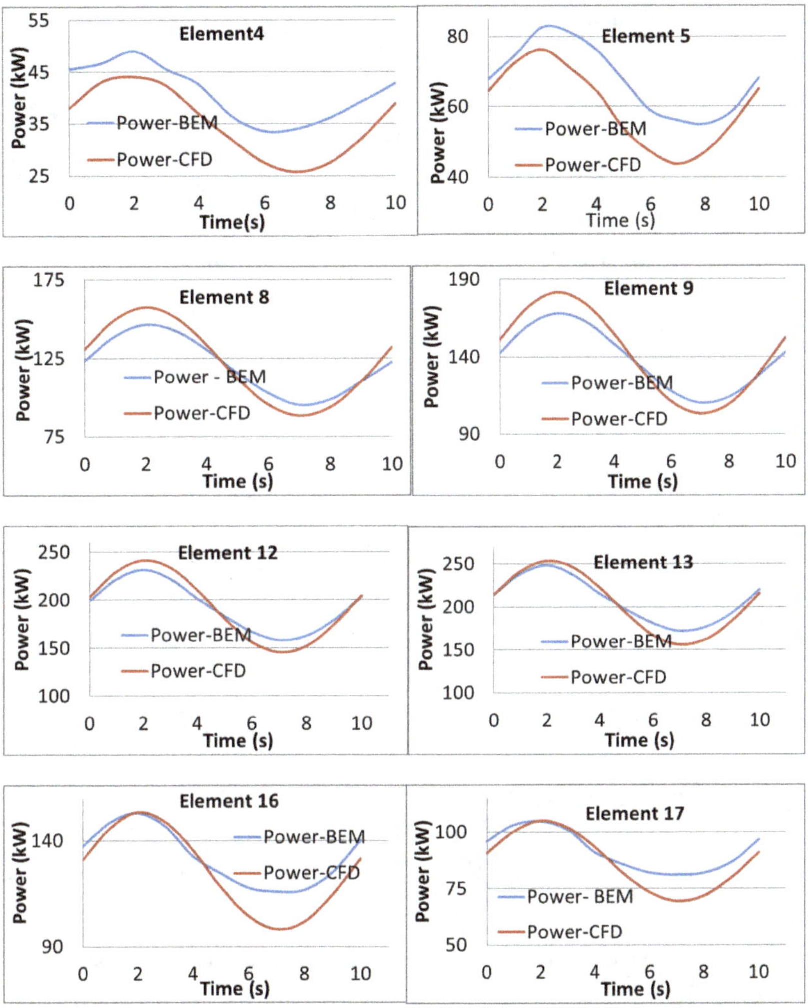

Figure 8.29 Comparison of elemental power for each element over a complete pitching motion cycle.

As three blades are at 120° to each other in the rotor, blade #1 data is used to create the blade #2 and blade #3 data by 120° phase shifting of each other as shown in Figure 8.34 for the AeroDyn BEM power. Figure 8.35 shows the comparison between the total rotor power obtained with

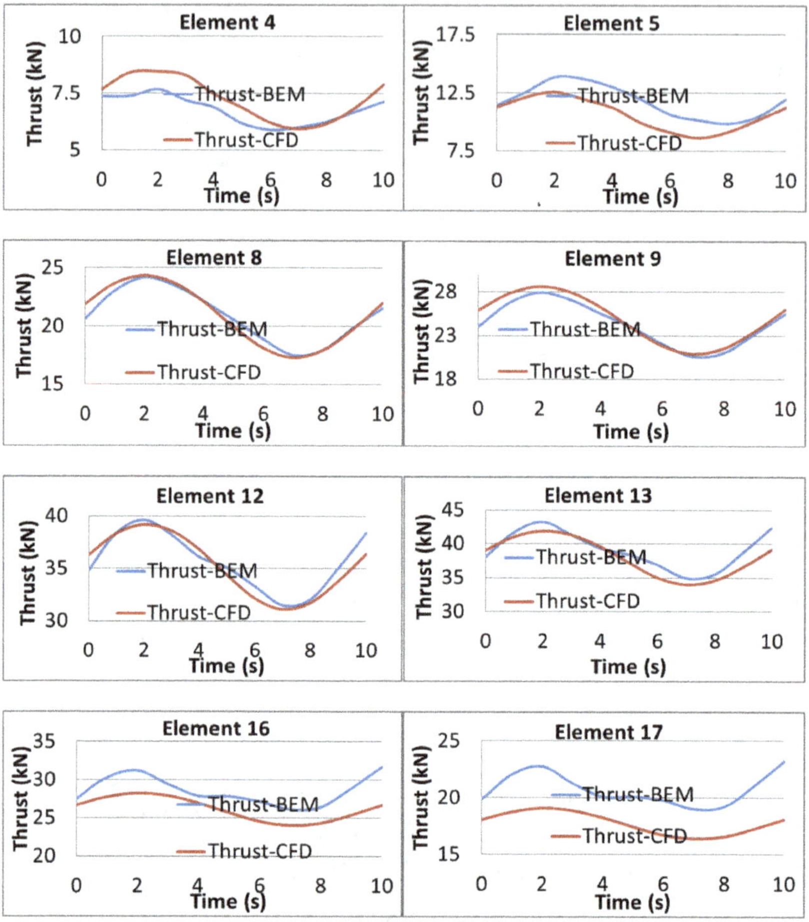

Figure 8.30 Comparison of elemental thrust for each element over a complete pitching motion cycle.

CFD and AeroDyn BEM. AeroDyn BEM predicts an average power of 1870 kW and CFD 1816kW (with respect to the last 10 seconds). Hence, CFD predicts 2.9% less than AeroDyn BEM for average power as shown in Table 8.7. However, the dynamic power fluctuations are higher in CFD compared to the values predicted by AeroDyn BEM. Further investigations are needed in this regard.

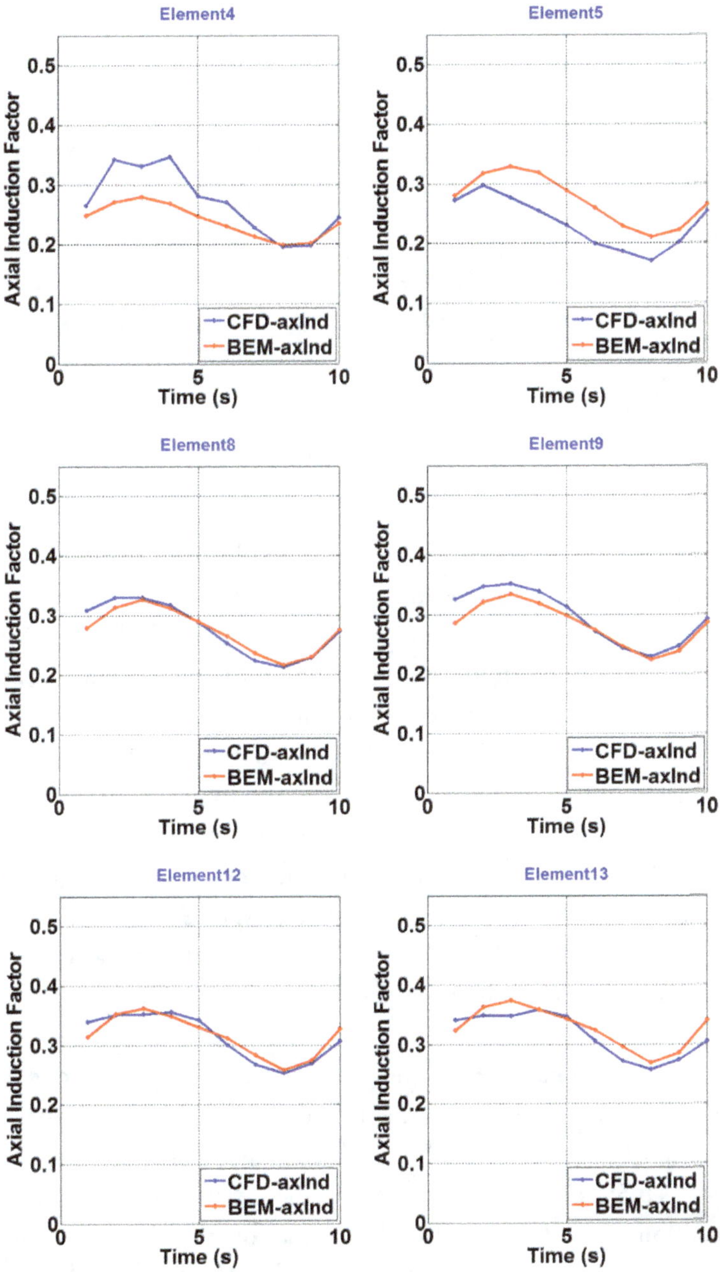

Figure 8.31 Comparison of axial induction factors for one wave period (i.e., 10 s).

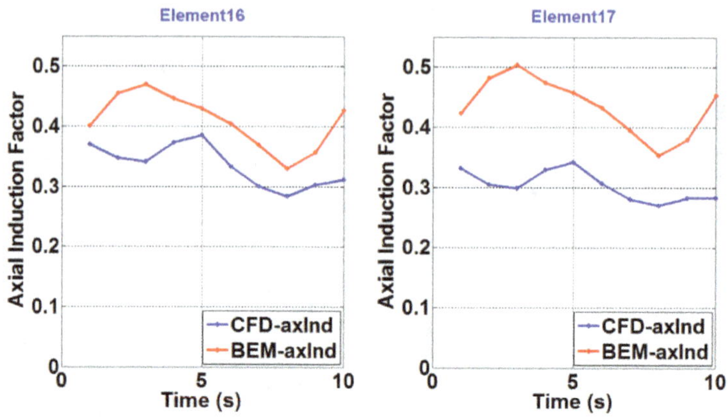

Figure 8.31 (*Continued*)

Profound knowledge on aerofoil design and its limitations are crucial for wind turbine rotor design. As the FOWT rotor experiences rotor plane oscillations due to platform pitching, the severity level of dynamic stall and rotor-wake interaction would depend on the pitch angle (sea state) and wind speed. As the pitch angle is small for OC3 phase IV case, it might be difficult to witness the rotor-wake interaction and dynamic stall as well. One of the ways to visualize the dynamic stall accurately along the length of the blade is to plot the Cp curves at various sections along the blade. In this work, 17 sections have been selected (first three elements are for cylindrical sections) as defined by AeroDyn BEM software. With reference to Figure 8.36, the aerofoil sections that are created at the end of each element are as shown in Figure 8.37.

With reference to Figure 8.38, the Cp curves are obtained for one blade at the various time instances in the transient simulation. Here, the CFD data is chosen from 50 s to 60 s which corresponds to 0–10 s as shown in Figures 8.39 to 8.45. To comprehend the variation of Cp on pressure and suction sides of the aerofoil during platform pitching motion, the Cp values are plotted against the fraction of chord length for each element.

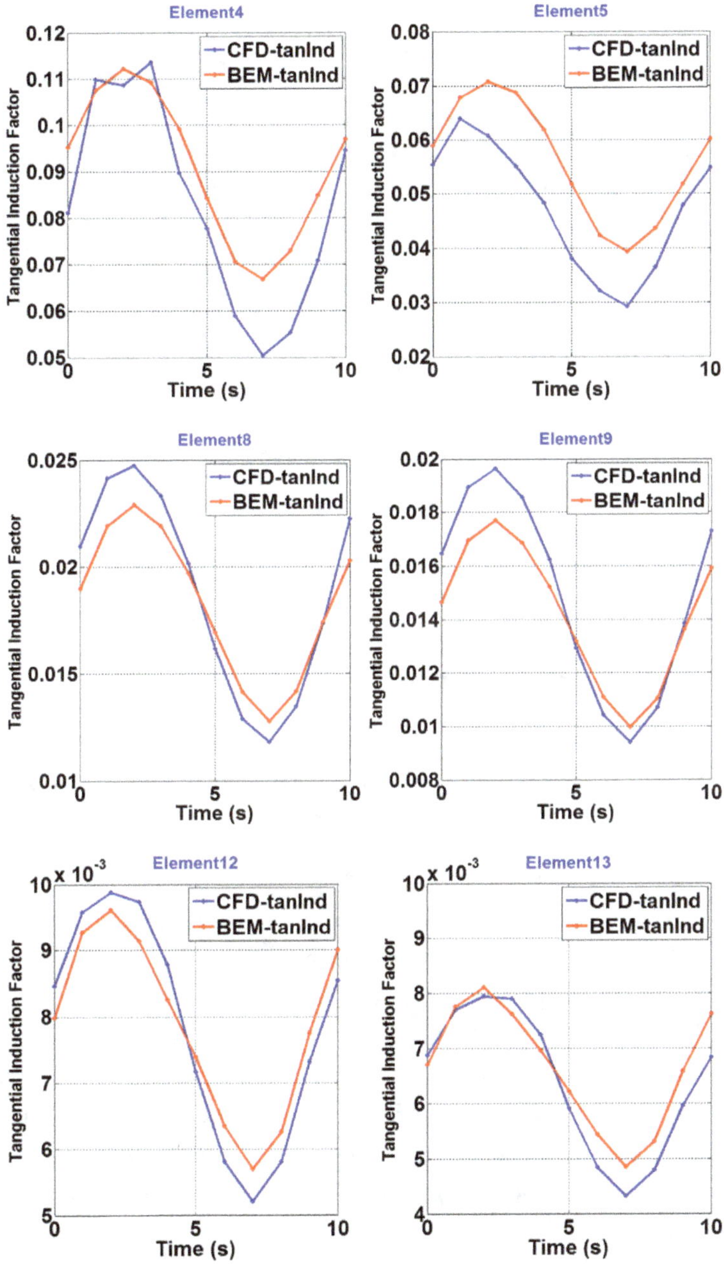

Figure 8.32 Comparison of tangential induction factors of each element.

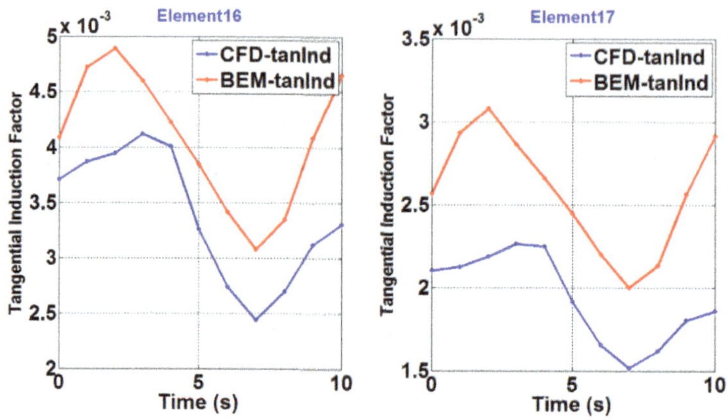

Figure 8.32 (*Continued*)

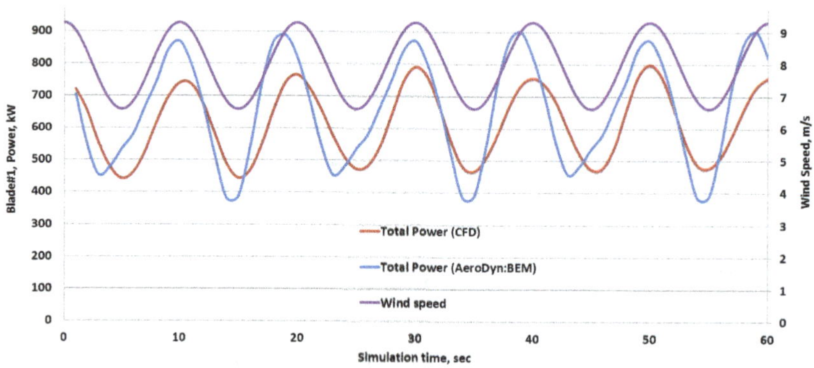

Figure 8.33 Power at blade #1: comparison against the relative wind speed experienced by blade #1 due to platform pitching motion.

Flow separation extends up to 20 m on the inboard section of the blade and it can be correlated to the abrupt change in the slope on the suction side after the leading edge in the Cp curve. From the Cp curve plots, it is apparent that the OC3 phase IV case 5.1 does not undergo dynamic stall condition.

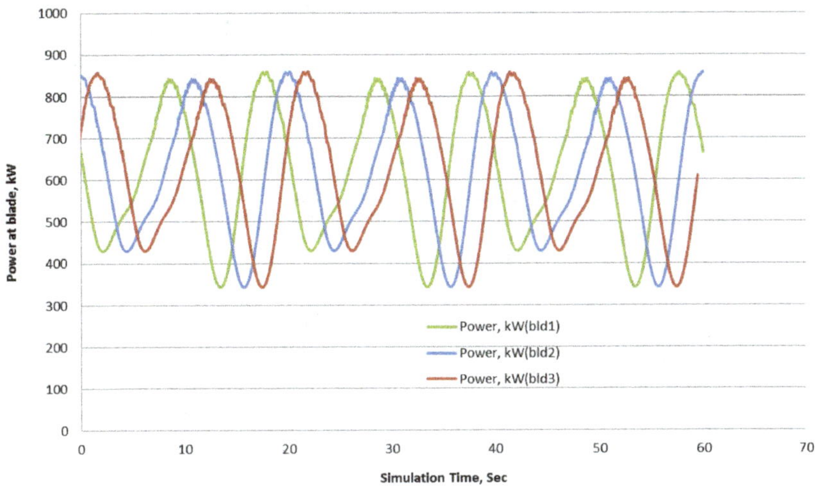

Figure 8.34 Power at blades #1, #2 and #3: experienced by the rotor due to platform pitching motion.

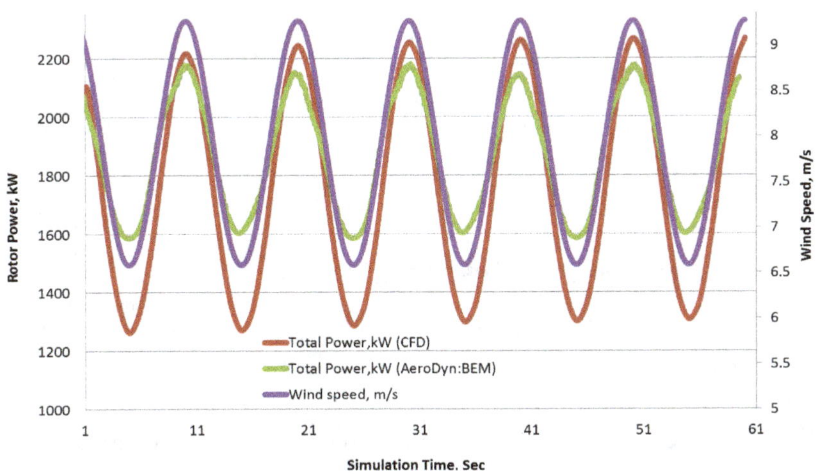

Figure 8.35 Total power at the rotor: comparison against the relative wind speed experienced by the rotor due to platform pitching motion.

Table 8.7 Comparison between AeroDyn BEM and CFD.

	Mean Power (kW)	% Difference with respect to AeroDyn BEM	Standard Deviation (kW)
AeroDyn BEM	1870	0.00	198.32
CFD	1816	2.89	339.20

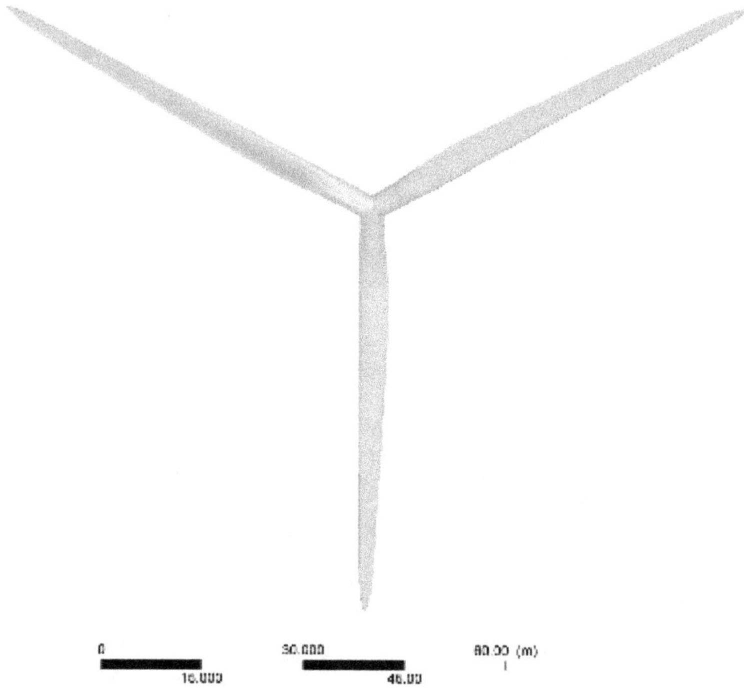

Figure 8.36 3D rotor CFD model.

The sample velocity and pressure contour plots at two seconds are plotted in Figure 8.46 to provide an overview of flow pattern over the entire blade. The flow separation near the blade roots can be observed from these contours.

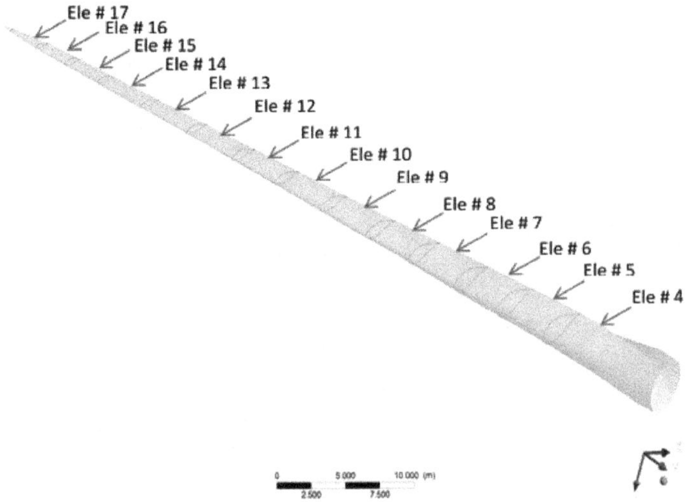

Figure 8.37 2D Aerofoil elemental section details.

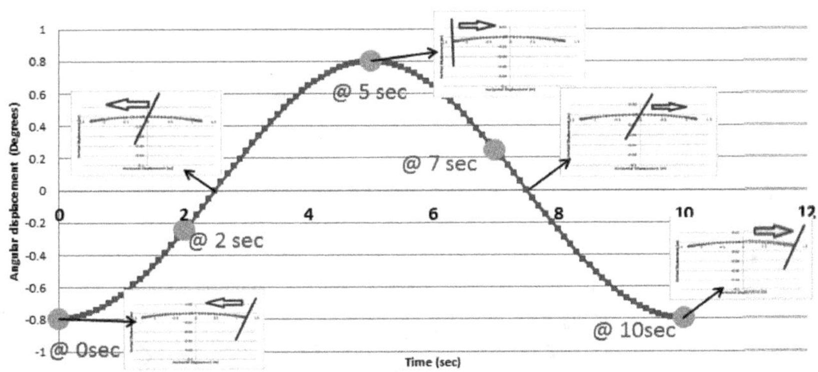

Figure 8.38 Transient time step selection for obtaining Cp curve.

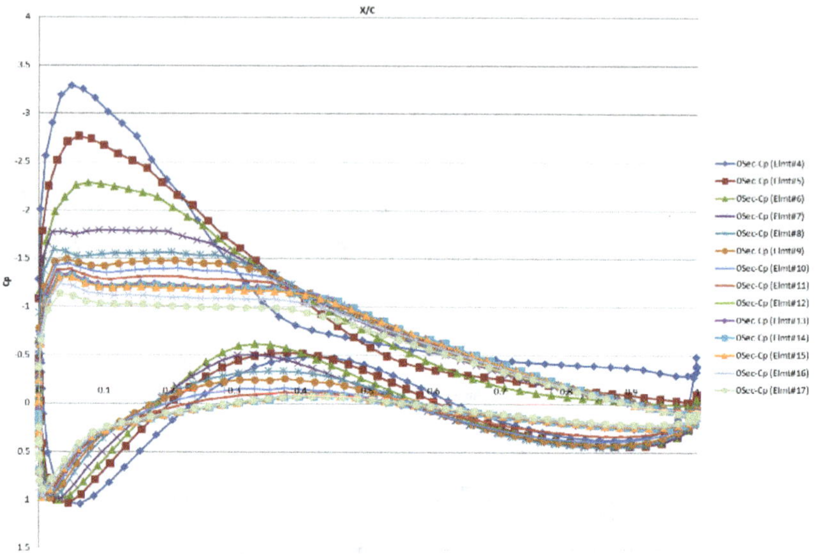

Figure 8.39 CFD — Cp curves for all Aero profiles of blade #1 at 0 s: CFD.

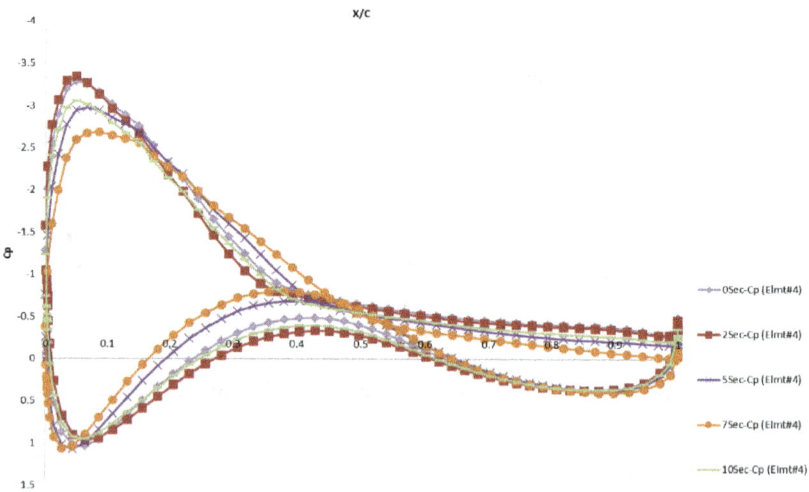

Figure 8.40 CFD — Element #4 (DU-40) Cp curves for five different times in one cycle of pitching motion.

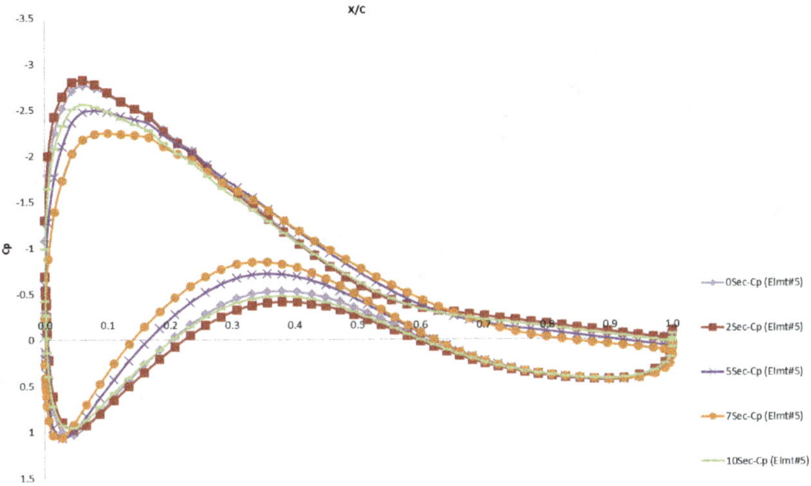

Figure 8.41 CFD — Element #5 (DU-35) Cp curves for five different times in one cycle of pitching motion.

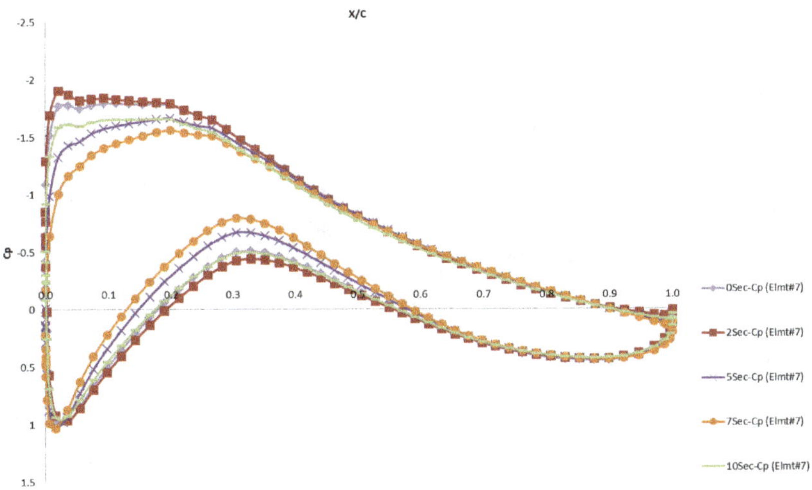

Figure 8.42 CFD — Element #7 (DU-30) Cp curves for five different times in one cycle of pitching motion.

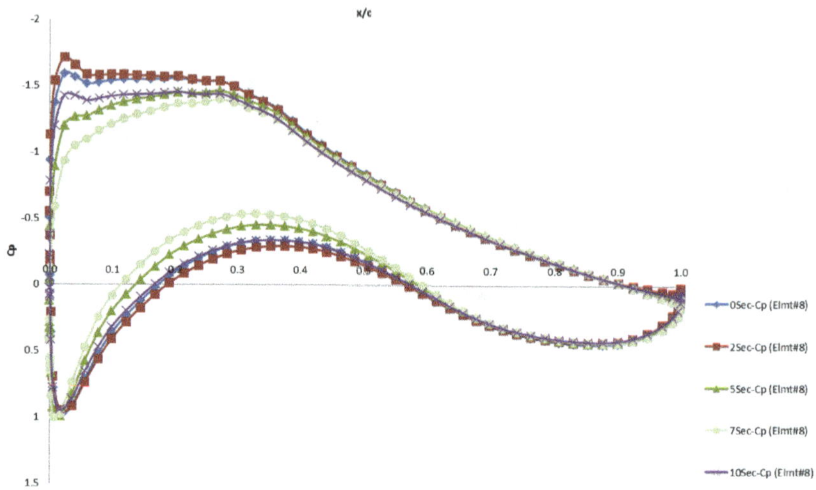

Figure 8.43 CFD — Element #8 (DU-25) Cp curves for five different times in one cycle of pitching motion.

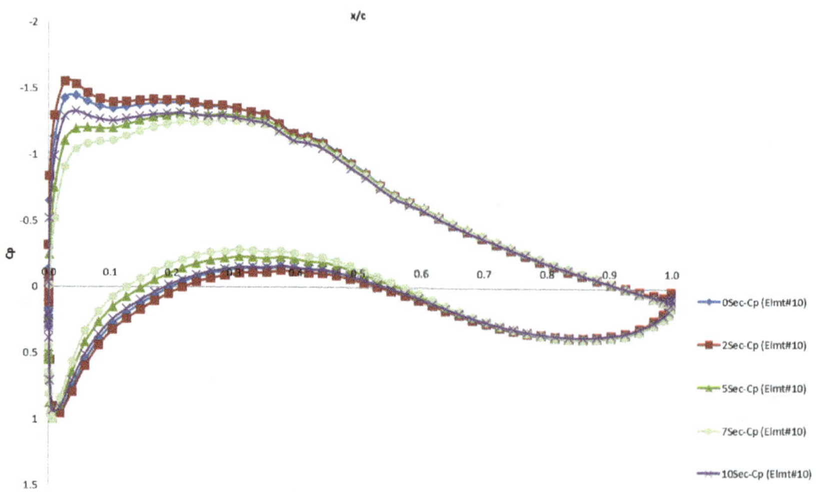

Figure 8.44 CFD — Element #10 (DU-21) Cp curves for five different times in one cycle of pitching motion.

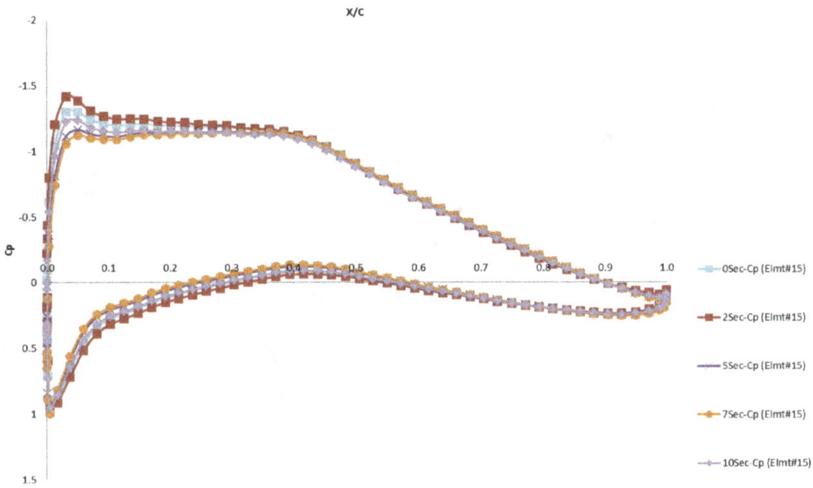

Figure 8.45 CFD — Element #15 (DU-64) Cp curves for five different times in one cycle of pitching motion.

8.4 OC3 Phase IV Case 5.1 — with Theoretical Sea State for Turbulent State Operating Condition Assessment

This section focuses on the dynamic effects of the platform pitching motion on the rotor aerodynamics for OC3 phase IV case 5.1 with modified or theoretical wave height. High-fidelity CFD model was employed along with FAST tool by assuming the wind turbine as a rigid body.

The hydrodynamic effects leading to the pitching motion of the turbine platform are obtained from FAST. These pitching motions are coupled with the rotating blades to study transient flow behaviors using CFD. The results are compared with the standard BEM-based methods. The results show that the increased wave height induces very high velocity and acceleration of the platform motion and thereby on the rotor plane,

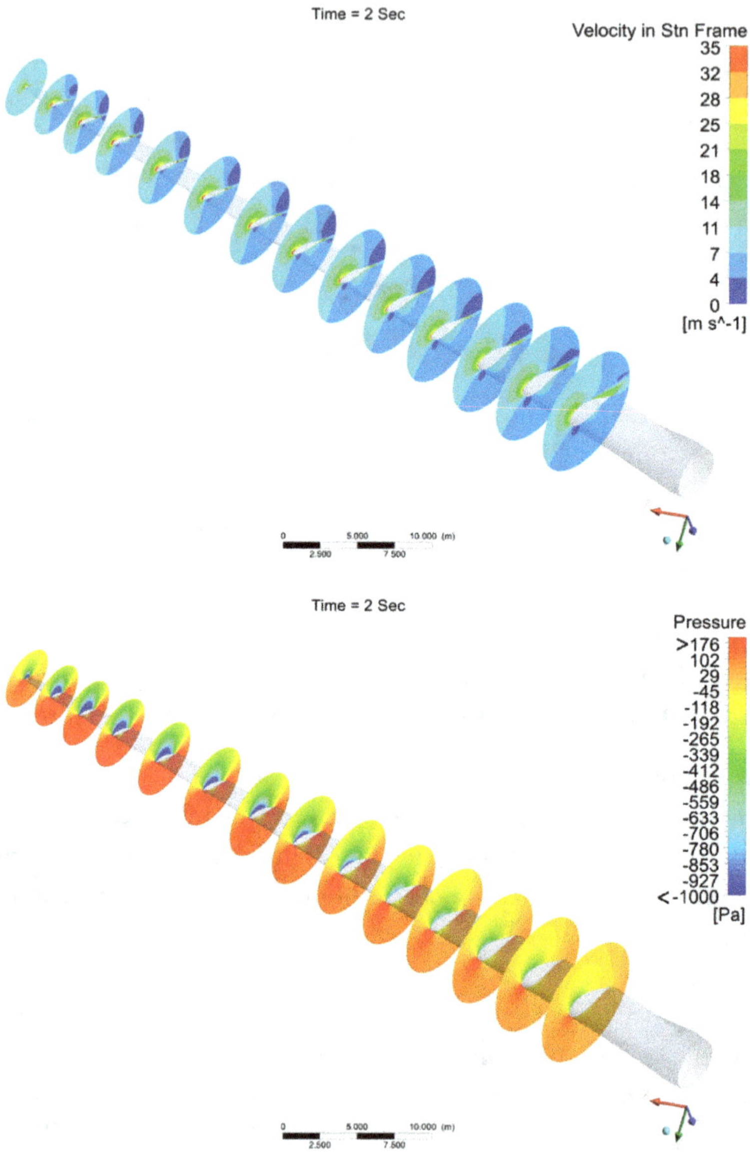

Figure 8.46 Velocity (top) and pressure (bottom) contour profiles at each element of aerofoil sections at 2 s.

which confirms that the turbine is operating both in windmill and turbulent states under such conditions. BEM validity with Glauert correction and appropriate tip loss model is to be further assessed for their application on FOWT performance and design predictions.

8.4.1 *Introduction*

FOWT is a complicated system as it has several components with complex interactions. Total loading of the FOWT is a resultant of all interactions. In order to derive accurate numerical models, all aspects of the wind turbine must be included in the same model.

Numerical performance prediction of FOWT is critical and challenging to accurately assess the aerodynamic loads. Most of the commercial and open-source wind turbine softwares employ BEM methods to determine the loads. The BEM method is well proven as it has been outlined in several publications[6] and successfully applied for fixed-bottom wind turbine during its design cycle.[7] As has been discussed,[7] the BEM method can be extended for FOWT as long as the assumptions are valid for a range of wave and wind conditions to predict FOWT rotor performance.

The effect of rotor plane pitching motion on the aerodynamic performance of FOWT due to the regular 6 m wave height and a CFD-based induction factor development methodology have been developed by Sivalingam *et al.*[8] for the OC3 phase IV case 5.1. It was shown by Sivalingam *et al.*[8] that 6 m, 10 s sea state allowed the turbine to operate in the windmill state as long as the axial induction factor of complete rotor did not exceed 0.4. The thrust loading of the rotor is increased purposely by varying the rotor plane velocity to force the turbine to operate in the turbulent state or beyond. The forced transition from windmill state to turbulent state is to validate the accuracy of BEM by comparing with fine details of the flow from CFD results. The transition can be achieved by increasing the wave height from 6 m to 12 m and maintaining the 10 s wave period theoretically for the OC3 phase IV case 5.1 as shown in Figure 8.47. Increase in the wave height will intensify the pitch rate of the spar buoy floating platform leading to an increased angular velocity and acceleration of the rotor plane. This continuously changes the relative wind speed of the rotor and thereby affects the various resultant forces. When the turbine is working in different operating conditions for a particular pitching motion, it is important to quantify the error associated in predicting the FOWT performance by the BEM method which is basically derived for fixed-bottom turbine. As BEM is mostly used in wind turbine design certification processes where normally more than 6000 simulation cases (CFD and FEA tools cannot be employed for all simulations) are to be performed, the purpose of this research scope is vital. Moreover, the prediction of fine variation of rotor power/load due to dynamic pitching of the floating platform is also equally important for an accurate fatigue life assessment. As the study is focused on the applied forces on the rotor to check the validity of the BEM model for FOWT applications, no blade control strategy is considered.

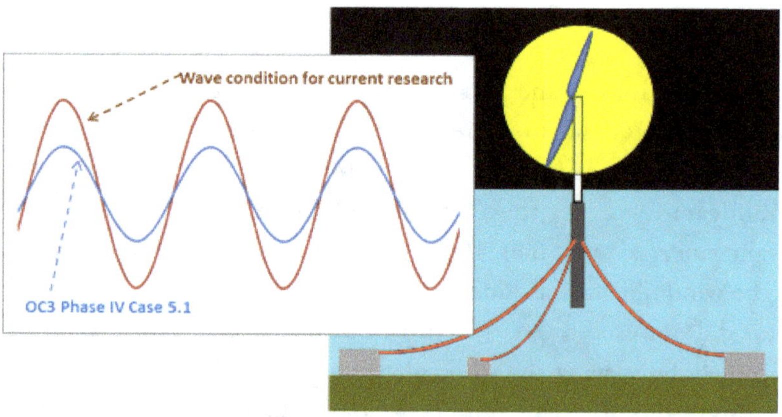

Figure 8.47 Comparison between OC3 phase IV case 1 wave condition and the regular wave condition adopted for this work.

8.4.2 *Methodology and Approach*

Previous studies reveal that pitching motion of the spar buoy floating platform for OC3 phase IV case 5.1 conditions does not significantly influence the wind turbine operating state with specified wave characteristics.[8] The induction factor has a direct correlation with the wind turbine operating state. The preferred windmill state will be retained as long as the induction factor does not exceed 0.5 even with the platform pitching motion. The relation between axial induction factor and coefficient of thrust with the corresponding wind turbine operating state based on the momentum theory is shown in Figure 8.48. Some sector of the wind turbine rotor is expected to operate in the turbulent wake state, when the axial induction factor tips over 0.5 due to increase in wave height to 12 m. The accuracy of the traditional BEM can be validated when the FOWT is operating in the turbulent wake state, though this is impractical in real world scenarios.

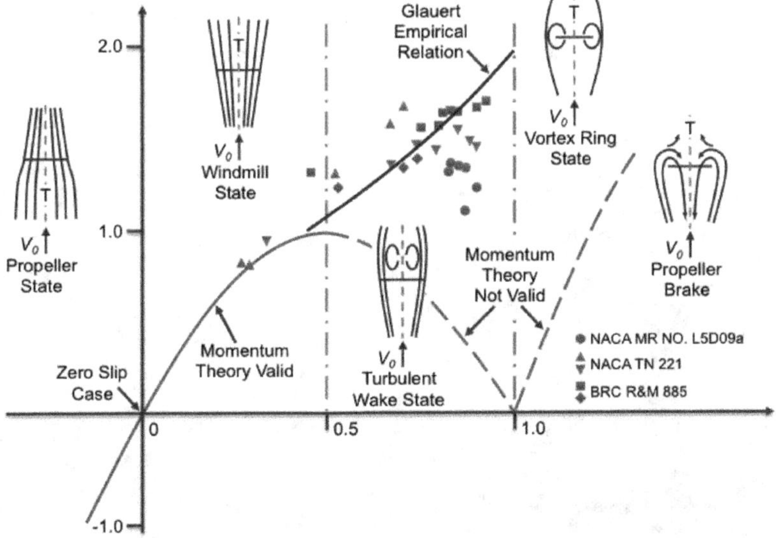

Figure 8.48 Axial induction factor vs thrust/thrust coefficient.[2]

The default operating state of a wind turbine is windmill state when there is a steady flow field. Since FOWT is not constrained in any of the six DoFs, the continuous pitching motion at the rotor plane will result in the transient flow field causing the wind turbine to oscillate between windmill and propeller states. Wave-induced floating platform pitching motion causes a rapid drop in effective wind speed at the rotor plane resulting in higher tip-speed ratios. The BEM theory is not able to model the constant transitions between the operating states of the wind turbine accurately. The problem is further compounded by the platform surge and high pitch rate. Thus, the application of the BEM method for modeling unsteady aerodynamics for FOWT is uncertain.

The hypothetical FOWT[6] shown in Figure 8.49 (from A to D) illustrates how this transition occurs between different states of the wind turbine during platform pitching motion and the wind speed. Initially, the turbine will operate in the windmill state (A) when the platform pitching motion is minimal, extracting energy from the flow field. Wind turbine rotor may interact with its own wake during rotor plane pitching along the flow direction, which may result in the formation of a turbulent region (B), predominantly near tip regions. At some point, illustrated in the

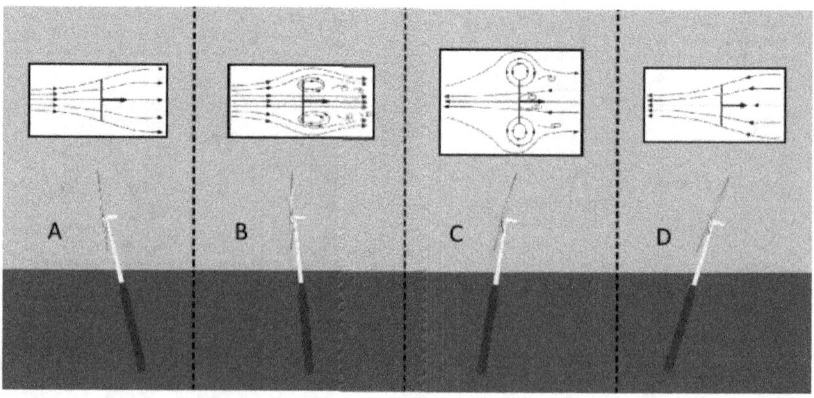

Figure 8.49 Extreme platform pitching motion and the resulting flow field.[6]

region (C), a toroidal recirculation flow path normal to the rotor disk may develop, named as vortex ring state (VRS). In the VRS state, thrust and torque are driven by the rate of energy dissipation into the vortex ring and the momentum balance equations may break down as with scenario B (when axial induction factor exceeds 0.4). At extreme pitching motion, the relative wake may reverse and the wind turbine may impart energy into the flow field and behave like a propeller. In addition to the periodic changes in aerodynamic loads during pitching motion, it may also violate the momentum slip-stream assumption.[7] The accuracy of wind turbine design tools such as FAST is questionable, as FAST uses AeroDyn which in turn is based on momentum balance equations for its BEM-based calculations by Suzuki.[10] All industrial codes developed for wind turbine design are only applicable for fixed-bottom wind turbine for static rotor plane.

The induction factors are incorporated in the BEM equations to deduce the aerodynamic loads on the rotor. By accurately predicting the aerodynamic load, the induction factor can be obtained from the BEM equations. CFD simulations can determine the aerodynamic loads more precisely than BEM theory for the prescribed conditions. As the axial induction factor is expected to go beyond 0.4, CFD-based induction factor derivation methodology[8] cannot be applied as such. Axial induction factors above 0.4 have to incorporate Glauert correction and accurate tip loss model. The outcome of this study is to compare power and thrust values obtained from traditional BEM-based and CFD-based axial induction factors in the turbulent wake state.

The CFD simulation was set up based on chosen case scenarios from an existing analytical FAST tool for one-to-one comparison. The CFD simulations were carried out with known steady state case at the uniform wind speed of 8 m/s with available ANSYS CFD code and the rotor power was compared with the value given by the design curve to gain confidence on the results.[8] The simulation setup includes the mesh parameters, and boundary condition initialization as per the calculations were carried out for the chosen turbine motion scenario of 12 m wave height and 10 s wave period.

FAST solver was modified in this study to extend its applicability to FOWTs. BEM theory is opted for this study again, as it is flexible in accommodating various new models such as hub and tip loss models and Glauert correction term. The two tip loss models (Prandtl and Georgia Tech) embedded in the AeroDyn are not accurate compared to CFD results for the chosen scenario.

As in OC3, 5MW NREL wind turbine was chosen for this study as required data for comparison are readily available. To compare and benchmark the results, OC3 phase IV cases (with minor modifications and increased wave height) were considered.[9] A 3D blade model was created with chord and twist details obtained from NREL 5MW wind turbine using lofting option, and as said before, the rotor is a rigid body rotating in a fixed rotor plane. The input for the pitching motion characteristics is achieved through CCL, a command prompt for CFX tool. The simplified approach enables us to focus on the aerodynamics of the blades, as the objective of the current work is to characterize the induced velocities of FOWT rotor and to compare against BEM results to quantify the accuracy. This will enhance the quality of FAST tool in the rotor design process as CFD tools are laborious and computationally intensive. But CFD helps to visualize and understand the basic flow phenomena, which FAST is not capable of (such as turbulence region development, vortex ring state and propeller state as in Figure 8.49).

8.4.3 *FAST Simulation Scenario for Computational Fluid Dynamics Simulation*

The basis for the current study is OC3 phase IV FAST model. The model was refined to cover the requirements of current simulations as listed below in the FAST model settings. This model includes mooring line and aero-hydrodynamic models to simulate time series of the turbine and platform responses to environmental and operating conditions.

In phase IV of OC3, case 5.1 injects regular sea wave motions (modified with 12 m wave height instead of 6 m) and steady wind excitation. The tower is initialized at its static position, and wind and waves are introduced with parameters shown in Table 8.8.

Wind and wave data from OC3 phase IV case 5.1 were used to define the sea state in terms of wave height, H, and wave period, T, enabling time series simulations as described in Table 8.8 with each simulation lasting 1860 seconds. The outputs generated by the initial 1800 seconds of each simulation were omitted in the analysis to avoid initialization errors until the flow was stabilized.

The initial run was performed to extract the motion of the platform for the above specified OC3 case. Figure 8.50 represents the roll, pitch and yaw motions of the platform for the wave conditions with 12 m wave height.

Table 8.8 FAST simulation scenario.

Uniform Wind Speed (m/s)	Sea State	Rigid Body Assumption	Rotor Speed (rpm)
8	12 m wave height and 10 s wave period	Yes	9.16

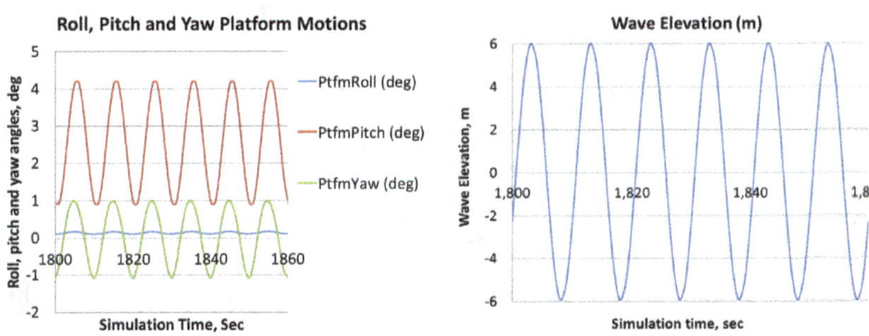

Figure 8.50 Platform motions and wave elevation profile for Table 8.8 case.

As seen in Figure 8.50, pitching motion is only considered for the study as it is dominant when compared to yaw and roll and hence the latter is omitted. For the chosen scenario the mean rotor tilt was found to be 2.54°. As the study focuses on platform pitching motions on the rotor plane aerodynamics, the platform rotations were transformed into the rotor plane motions at the hub height.

8.4.4 *Coupled Dynamic Mesh Motion in Computational Fluid Dynamics*

The transient simulations were carried out at a wind speed of 8 m/s. The rotor pitching motion starts with the mean rotor tilt of 7.54°. As the platform and tower are not modeled in this study, the pure angular motion at the platform is converted into the translational and axis tilting motion at the rotor. Figures 8.51 and 8.52 show the angular motion cycle at the platform (provided by FAST) and the corresponding rotor motion. It can

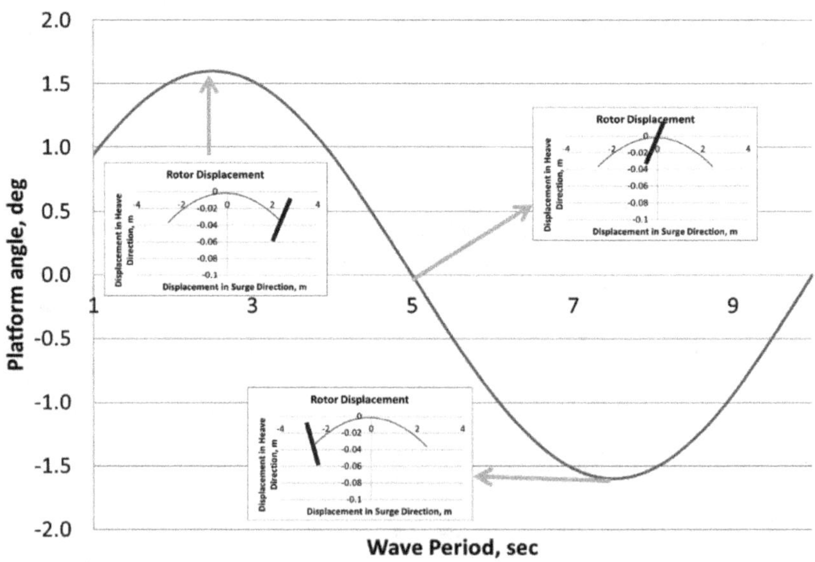

Figure 8.51 Platform pitching motion: angular displacement for 12 m wave height condition with rotor position.

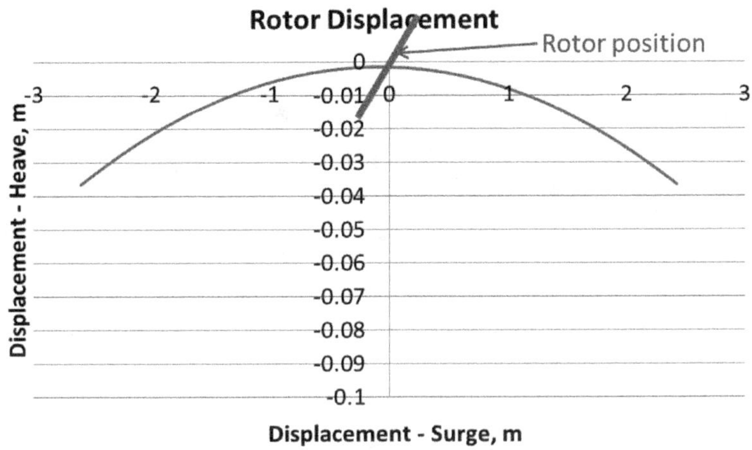

Figure 8.52 Translational displacements at the rotor for 12 m wave height condition.

be seen that, for the chosen pitching motion case, the vertical displacement is negligible and hence the horizontal displacement alone is considered in the current study. The horizontal amplitude is found to be 2.6 m.

The wind speed in the vicinity of the rotor fluctuates due to the pitching motion of the platform. In order to calculate the induction factors, the wind speed at any point on a pitching rotor is obtained using Eq. (8.14).

8.4.5 *Transient Motion Pitching Results*

The transient pitching motion study provides an insight into transient wake effects on every time step for one pitching cycle. Elemental thrust, power and induction factors are obtained for all elements at 8 m/s wind speed with 12 m wave height for 10 equally spaced time steps in one cycle (from 41–50 s of a 70 s simulation). Results are obtained for both CFD and BEM (Prandtl tip loss and modified Prandtl tip loss model by Georgia Tech (GT)) to compare on the same substrate. The comparison of elemental power and thrust values obtained with CFD simulations and BEM-based equations are given in Figure 8.53 for all different elements.

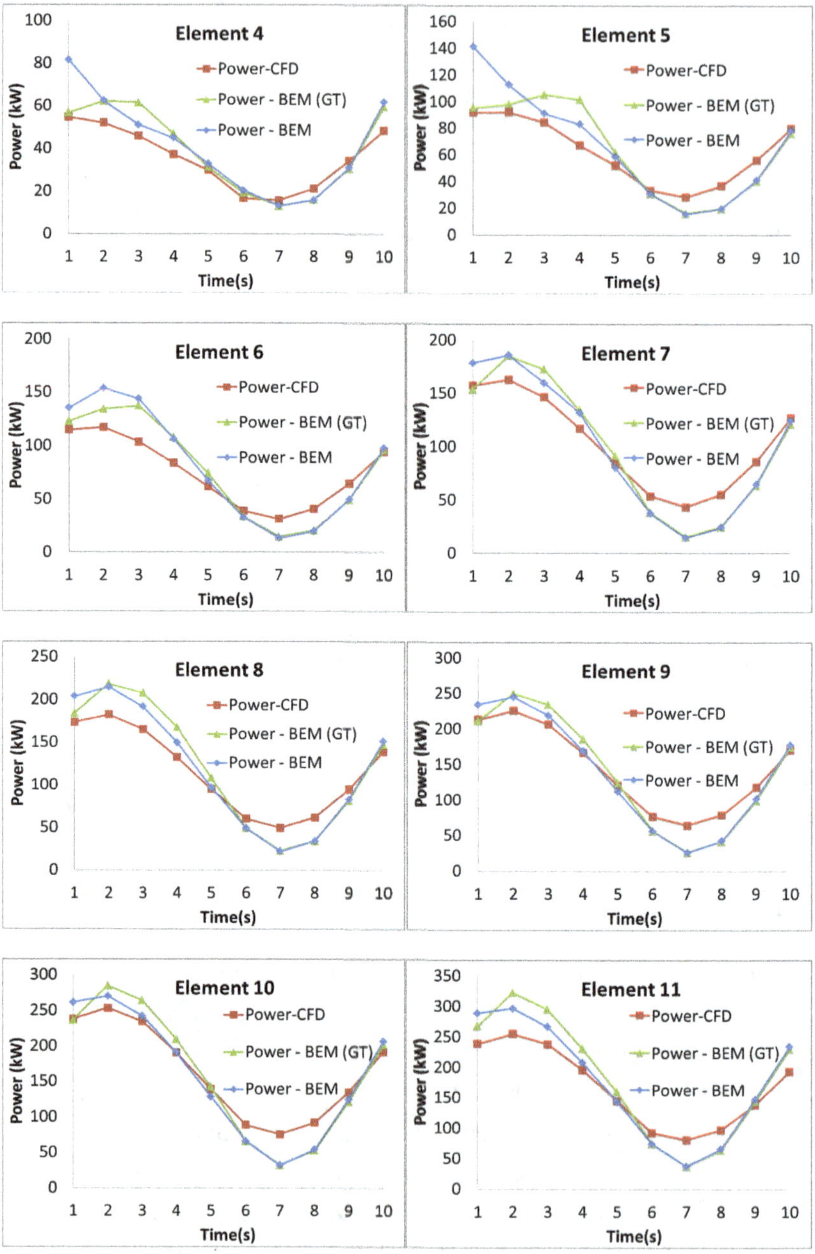

Figure 8.53 Comparison of elemental power over a complete pitching motion cycle at $U\infty = 8$ m/s (Elements 4–17).

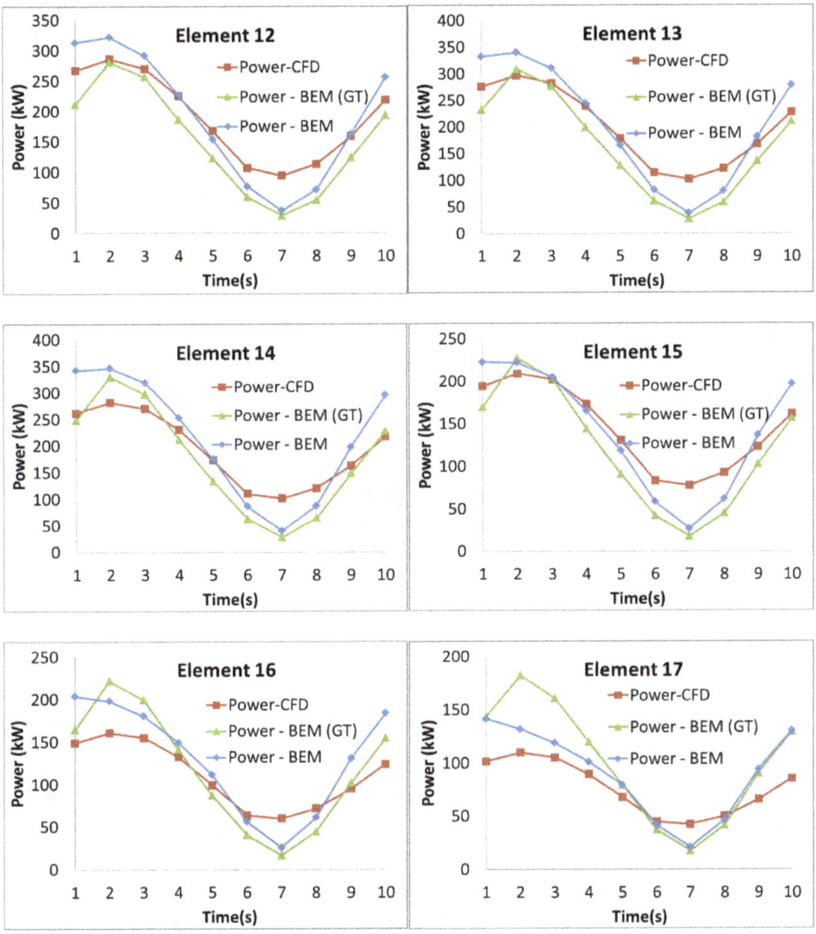

Figure 8.53 (*Continued*)

It is observed from the above plots that the elemental power values calculated with the BEM equations and CFD results agree qualitatively well with each other for most sections of the blade except near the tip. Though the trend is similar, the values are largely different. The variation in the power prediction is higher, close to the tip and this is mainly due to the effect of tip loss models used in BEM.[8] CFD-predicted values are more accurate than BEM for high rate of platform pitching motions. As seen

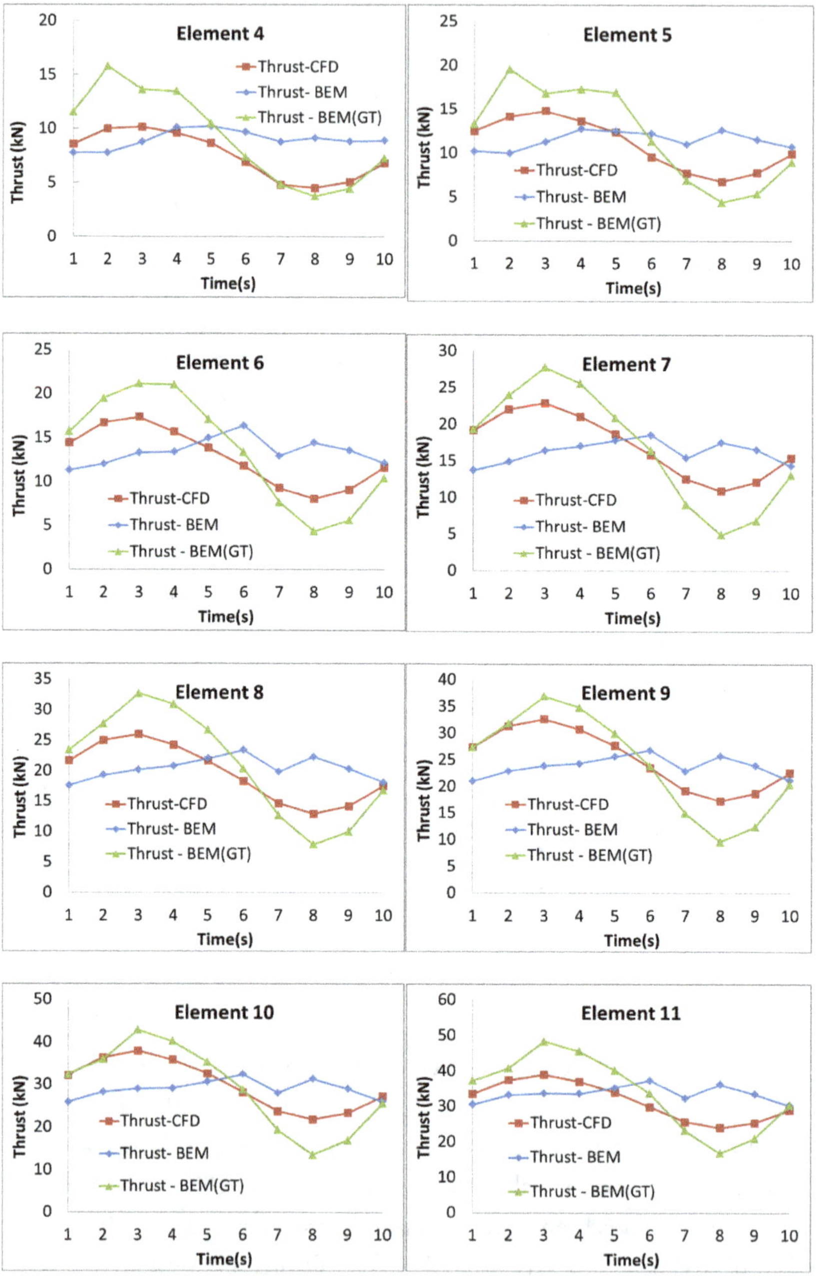

Figure 8.54 Comparison of elemental thrust over a complete pitching motion cycle at $U\infty = 8$ m/s (Elements 4–17).

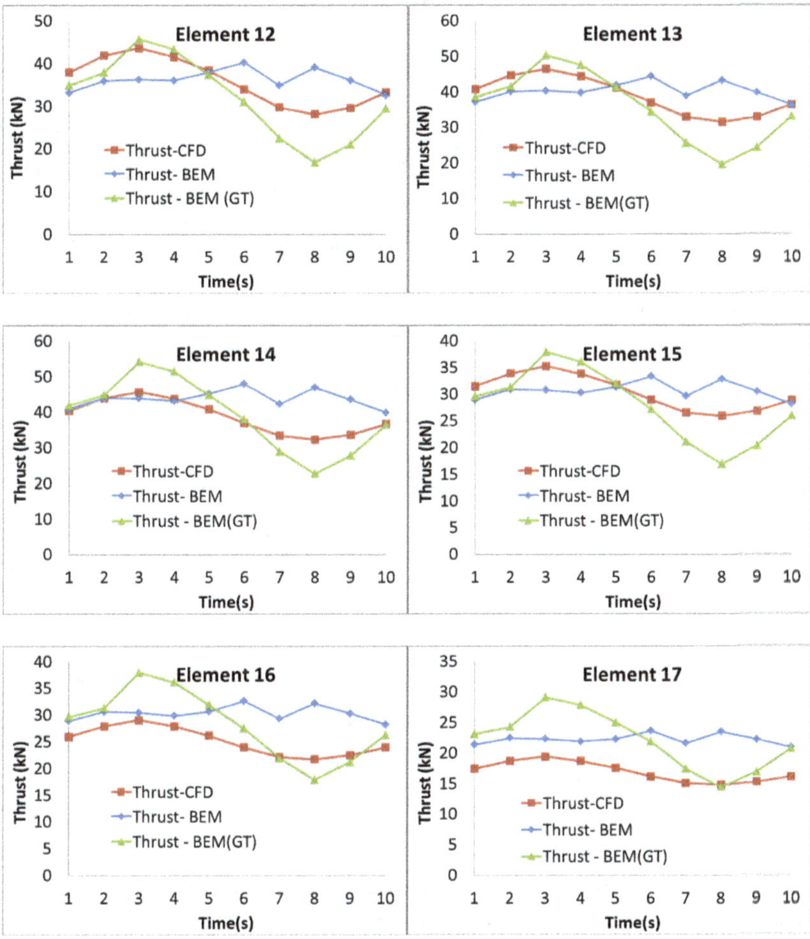

Figure 8.54 (*Continued*)

in Figures 8.48 and 8.49, validity of BEM method is debatable when the axial induction factor exceeds 0.5. Hence, BEM-based axial and tangential induction factors obtained from FAST simulations are compared between elements for one wave period (Figures 8.55–8.57).

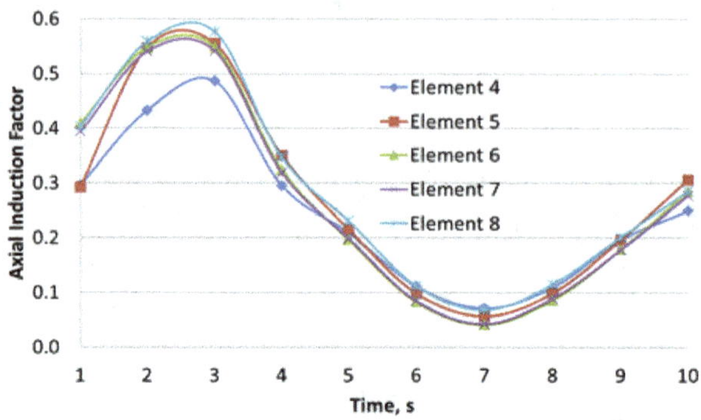

Figure 8.55 Comparison of axial induction factor for the elements from 4–8 over a complete pitching motion cycle at $U\infty$ = 8 m/s.

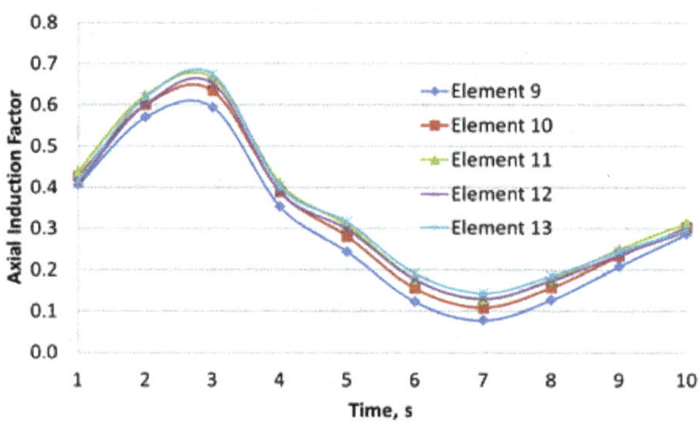

Figure 8.56 Comparison of axial induction factor for the elements from 9–13 over a complete pitching motion cycle at $U\infty$ = 8 m/s.

From the plots in Figures 8.55–8.57, it can be concluded that 50% of the rotor sections are at 0.5 induction factor or beyond for 25% of a wave period (approximately). This confirms the assumption that FOWT is operating in the 'turbulent wake' state. This leads us to further assess

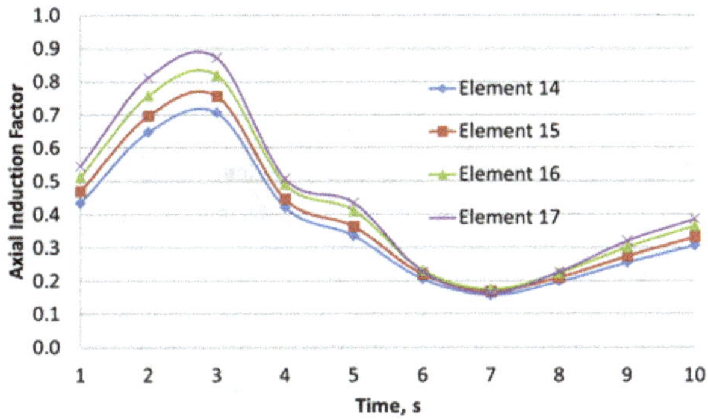

Figure 8.57 Comparison of axial induction factor for the elements from 14–17 over a complete pitching motion cycle at $U_\infty = 8$ m/s.

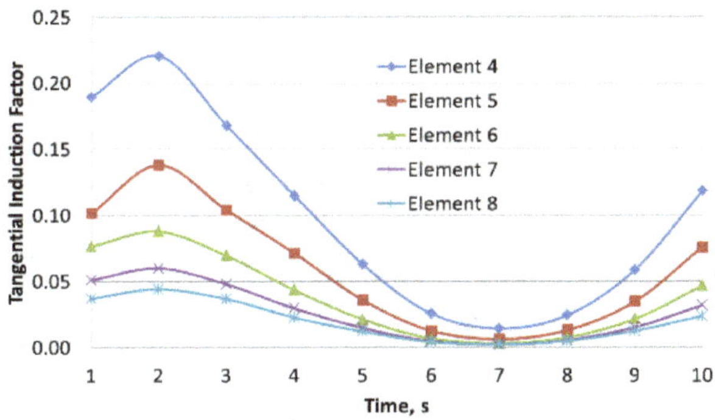

Figure 8.58 Comparison of tangential induction factor for the elements 4–8 over a complete pitching motion cycle at $U_\infty = 8$ m/s.

BEM equations and Glauert correction terms for FOWT applications as these are based on analytical approaches and are built upon with some basic assumptions. The tangential induction factors were also obtained as shown in Figures 8.58–8.60. From the axial and tangential induction factors plot, one can observe a significant variation between each element and also a non-uniform pattern in relation to the wave period.

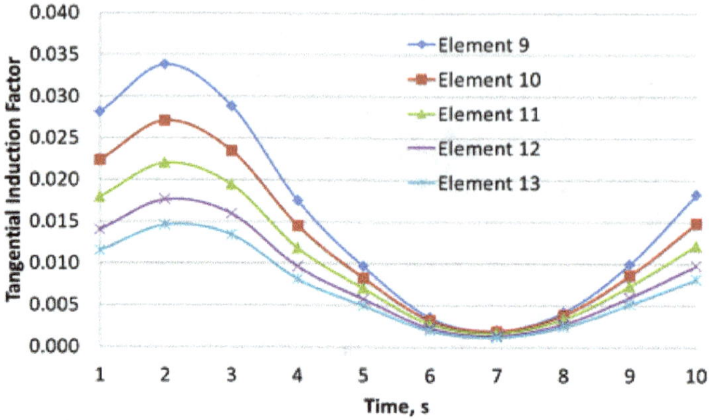

Figure 8.59 Comparison of tangential induction factor for the elements 9–13 over a complete pitching motion cycle at $U\infty = 8$ m/s.

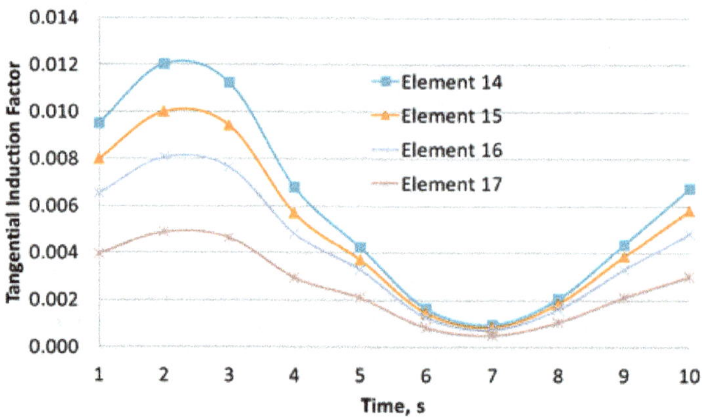

Figure 8.60 Comparison of tangential induction factor for the elements 14–17 over a complete pitching motion cycle at $U\infty = 8$ m/s.

The substantial variations are due to blade-wake interactions during the platform pitching. In order to ensure the validity of BEM, all the empirical equations are to be further studied and compared against CFD and experimental simulations.

8.4.6 *Comparison of Rotor Power in High Wave Elevation Pitching*

FOWTs are simultaneously subjected to aerodynamic, hydrodynamic and mooring line forces which are dynamic in nature inducing the platform motion in six DoFs. These motions potentially amplify the unsteady aerodynamic effects (e.g., blade-vortex interaction, dynamic stall, skewed inflow), which cannot be accurately simulated by current design methods based on the blade element momentum theory with common corrections as shown in Figures 8.48 and 8.49. Blade-wake/vortex interaction effects

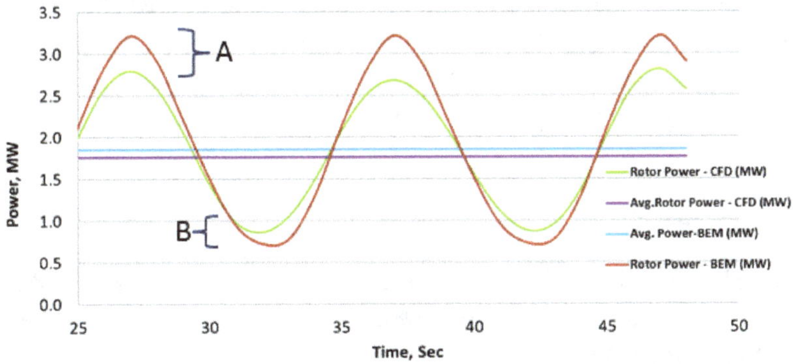

(a)

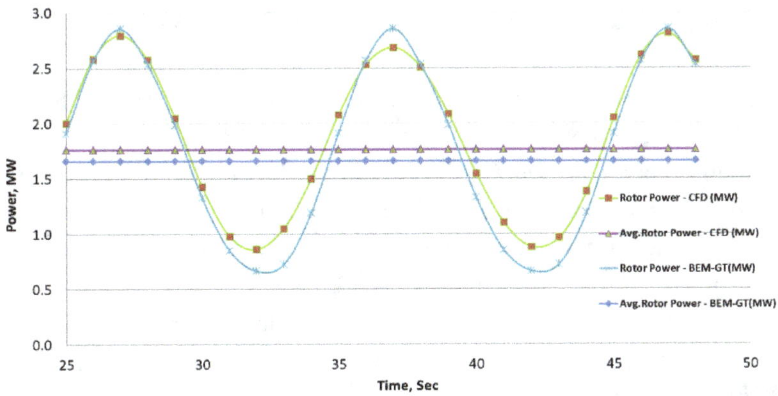

(b)

Figure 8.61 Total power at the rotor: comparison against the relative wind speed experienced by the rotor due to platform pitching motion (a) with BEM and (b) BEM with modified Georgia Tech tip loss model.

are observed due to significant rotor plane displacement rate for the chosen operating conditions for the current turbine. BEM- and CFD-based rotor power predictions and their fluctuations are compared by choosing 25–50 s simulation results in one cycle. As seen from Figure 8.61, BEM-predicted power is higher compared to CFD-based results, when the rotor pitches from the leeward to windward direction. This is marked as regions A and B, respectively. When the rotor pitches towards the wind, the difference between peak power prediction by BEM and CFD is 0.5 MW due to strong interaction between blade and its own wake, whereas 0.1 MW difference is minimum power prediction when the rotor pitching aligns with the wind direction. BEM (AeroDyn)-based rotor power/load fluctuation is high, which will have a direct impact on the blade design, as it will affect the fatigue loading. The average rotor power is 1.761 MW and 1.853 MW for CFD and BEM, respectively.

References

1. Jonkman, J.; Butterfield, S.; Musial, W.; Scott, G. Definition of a 5-MW reference wind turbine for offshore system development. *Technical Report* **2009**, NREL/TP-500-38060.
2. Sebastian, T. *The Aerodynamics and Near Wake of An Offshore Floating Horizontal Axis Wind Turbine* (PhD Thesis). University of Massachusetts Amherst **2012**.
3. Moriarty, P. J.; Hansen, A. C. *AeroDyn Theory Manual.* National Renewable Energy Laboratory **2005**, NREL/TP-500-36881.
4. Matha, D.; Schlipf, M.; Cordle, A.; Pereira, R.; Jonkman, J. Challenges in simulation of aerodynamics, hydrodynamics, and mooring-line dynamics of floating offshore wind turbines. *International Offshore and Polar Engineering Conference.* Hawaii **2011**, NREL/CP-5000-50544.
5. Chow, R.; van Dam, C. P. Verification of computational simulations of the NREL 5 MW rotor with a focus on inboard flow separation. *Wind Energy* **2012**, *15*, 967–981.
6. Sebastian, T.; Lackner, M. Offshore floating wind turbines — an aerodynamic perspective. *49th AIAA Aerospace Sciences Meeting including the New Horizons Forum and Aerospace Exposition.* Florida **2011**, AIAA 2011–720.

7. Peters, D. A.; Chen, S. Y. Momentum theory, dynamic inflow, and the vortex-ring state. *Journal of the American Helicopter Society* **1982**, *27*, 18–24.
8. Sivalingam, K.; Bahuguni, A.; Gulman-Strand, J.; Davies, P.; Nguyen, V. T. Effects Of platform pitching motion on Floating Offshore Wind Turbine (FOWT) rotor. *Offshore Technology Conference.* Texas **2015**, paper number OTC-25962-MS.
9. Jonkman, J.; Musial W. Offshore code comparison collaboration within IEA Wind Task 23: Phase IV results regarding floating wind turbine modeling. *Technical Report* **2010**, NREL/CP-500-47534.
10. Suzuki, A. *Application of Dynamic Inflow Theory to Wind Turbine Rotors* (PhD Thesis). University of Utah **2000**.

Chapter 9

Numerical Validation of Floating Offshore Wind Turbine Scaled Rotor for Surge Motion

9.1 Introduction

Numerical simulations play an important role in every stage of the wind turbine product life cycle development from conceptual design to operation, maintenance and decommissioning to optimize capital and operating expenditures. In addition to their significant contribution in predicting the wind loads on the turbine, numerical simulations also help to develop radically new designs. During the design review, compliance and certification process, thousands of simulations are normally performed. Hence, typical engineering analytic codes like NREL FAST are often used to assess the performance behaviors of the turbine.

As the assumptions in BEM are not applicable[1] for floating offshore wind turbines (FOWTs) where the pitching and surging motions[2] of the platform leads to the turbine rotor going through different wake states, various non-linear effects like rotor-wake interactions, wake-wake interactions, wake meandering, etc., are not accounted for in BEM theory. Experiments on scaled rotor provide an insight on the aerodynamic loading of the rotor and validation of the non-linear models. As platform motions of FOWTs have greater influence on the unsteady aerodynamic response, a scaled model of the rotor was designed to study the effect.[3] The platform motions significantly affect the power and thrust characteristics of the rotor almost to the same order as the effect of tip-speed ratio (TSR) and the blade angle. A scaled-down model of the NREL 5MW rotor was designed and tested in the wave tank (immersed in the water) to match the Re, as in field operations. The experiments are conducted with the focus on surge motion[4] and its effects on aerodynamic performances. The BEM-based AeroDyn model was modified (LR AeroDyn) to predict the hydrodynamic load onto the scaled rotor. LR has enhanced the BEM code (LR-uBEM) to simulate the unsteady aerodynamic behavior of the rotors. The LR-uBEM model implements the dynamic inflow model based on the elemental aerodynamic loads prediction of individual blades. To gauge the accuracy of engineering numerical codes LR AeroDyn and LR-uBEM, high-fidelity CFD model was developed and validated against the experimental simulation.

9.2 Scaled Rotor for Unsteady Aerodynamic Experiments

To predict the global loads in well-ordered experimental conditions by scaling down the wind turbine rotor is challenging. This can be attributed to the complexities involved in accomplishing the three scaling laws, which have to be followed to design a scaled rotor that matches the performance of full-scale reference rotor.[5–8] It is highly challenging to match the performance of the scaled rotor for the coefficients such as coefficient of power (C_p), coefficient of torque (C_q), and the coefficient of thrust (C_t) to a full-scale reference rotor due to the incompatibility of the three primary scaling criteria: the maintenance of geometric, kinematic, and dynamic similarities between model and full scale. Geometric and kinematic similarity are readily achieved by the application of constant geometric scale factor and the maintenance of the TSR, respectively. However, in doing so the dynamic similarity criteria, defined principally as the maintenance of the Re for the flow over the blade at model and full scale, is typically not met. In order to address this, a number of researchers have proposed a global performance matching approach for the design of model scale wind turbine rotors where the primary objective is not the study of the rotor in normal operating conditions. Examples of such are the MARIN Stock Wind Turbine for the evaluation of the performance of floating wind turbine foundations scaled down under wind and wave actions and the evaluation of advanced control strategies under unsteady operating conditions.[5–8] In both cases, the design objective was to match the coefficients of the model scale to their respective full scales for a range of operating TSRs. This is achieved by changing the aerofoil profile in the scaled-down model along the length of the blade so as to increase the lift coefficient, increasing the chord distribution along the length of the blade by a constant factor, and altering the twist distribution to maximize the lift-to-drag ratio for the

TSRs of primary interest. The criteria for geometric similarity is abandoned, kinematic similarity is maintained, and the deficit in the dynamic similitude is corrected for.

As is the case for the two rotor designs discussed, the objective of the experiments in this case is to evaluate the rotor performance subjected to external influence relative to the normal operating state, specifically, the quantification of the time-varying coefficient of thrust resulting from the periodic motion of a rotor at a number of frequencies and amplitudes typical of floating wind turbine systems. Hence a similar approach to the scaling criteria has been taken. However, in this case, in addition to the global performance matching model-scale rotor design objective, additional objectives have been defined in an attempt to maintain the critical local performance measures. The model-scale rotor requirements are defined as follows:

1. The coefficient of thrust of the model rotor must be similar to the full-scale reference for a range of TSRs.
2. The chord must be scaled by the same geometric scale factor as the diameter of the rotor.
3. The twist distribution along the non-dimensional length of the model blade must be the same as the full-scale reference.
4. The axial induction factor along the non-dimensional length of the model blade must be the same as the full-scale reference for a range of TSRs.

The full-scale reference rotor used for this study is the NREL 5MW reference wind turbine,[9] which has been studied extensively and well reported in literature. Figures 9.1 and 9.2 show the torque and thrust coefficient of the 5MW reference turbine, respectively. The axial and tangential induction factors of TSR 7 for the rotor blade of 63 m length is given in Figure 9.3.

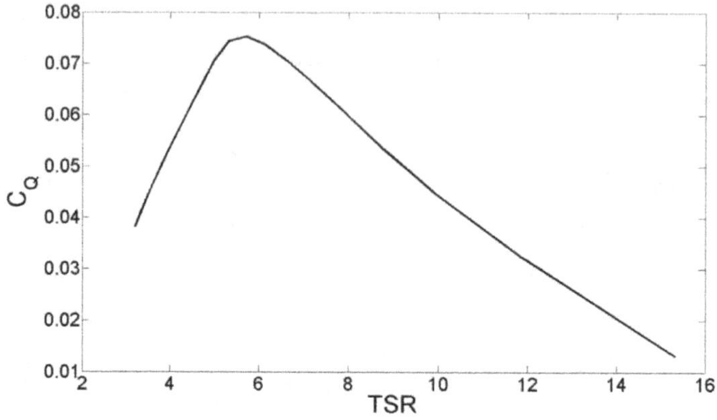

Figure 9.1 NREL 5MW Baseline Wind Turbine rotor C_Q curve.

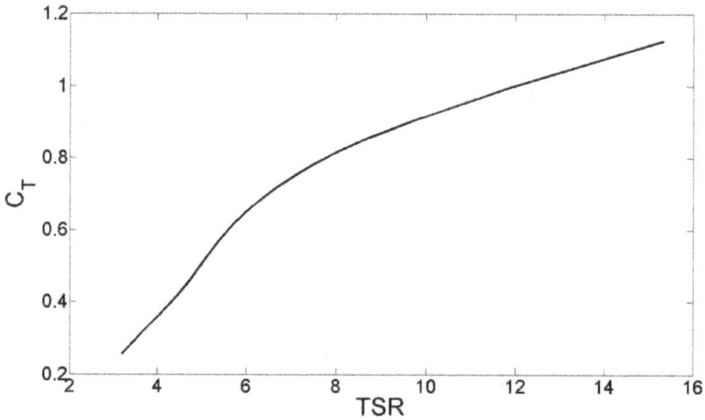

Figure 9.2 NREL 5MW Baseline Wind Turbine rotor C_T curve.

As the C_Q and C_T curves are non-dimensional, they can be used as a reference for scaled rotor comparison purposes. Figures 9.1 and 9.2 show that these respective curves for the NREL rotor will have identical C_Q and C_T curves if the model-scale rotor has a similar steady state performance as the reference. The design of the blade incorporates multiple aerofoil profiles. NREL defines locations along the blade at which each

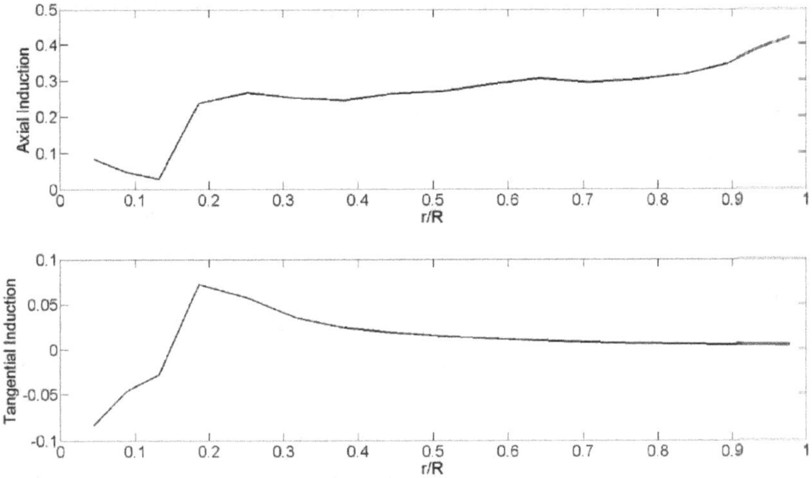

Figure 9.3 NREL 5MW Baseline Wind Turbine rotor axial (top) and tangential (bottom) induction factors, TSR = 7.

profile is located (Table 9.1). Cylinder1 and Cylinder2 comprise the structural region of the blade, and the diameter of the cylinder at a specific point is a function of the chord length at that position. The only difference between the two is their respective drag coefficient: Cylinder1 has a drag coefficient of 0.5, Cylinder2 0.35, as reported by Kooijman *et al.*[10] The remaining section of the blade is constructed of various aerofoil profiles as listed in the table. The non-dimensional geometry of each of these is provided as part of the DOWEC project.

This was also the basis for the design of the MARIN Stock Wind Turbine. The diameter of the scaled model of the rotor has been chosen to be 1.0 m in order to minimize experimental blockage effects, resulting in a geometric scale factor of 126.0. As aerodynamic coefficients largely influence the rotor performance, a new methodology has been adopted to match the scaled airfoil Re to the full-scale airfoil Re. Significant effort has been dedicated to optimizing the airfoil profile in order to match the lift curves even in low Re,[3] by retaining the twist of 5MW NREL rotor

Table 9.1 Scaled rotor airfoil details.

Element/ Node	Element Center Radius (m)	Twist (°)	Element Center Chord (m)	*Cross-sectional Profile
1	0.023	13.308	0.028	Cylinder
2	0.044	13.308	0.031	Cylinder
3	0.066	13.308	0.033	Cylinder
4	0.093	13.308	0.036	DU40 (SMA3540)
5	0.126	11.48	0.037	DU35 (SMA3540)
6	0.158	10.162	0.035	DU35 (SMA3540)
7	0.191	9.011	0.034	DU30 (SMA2130)
8	0.223	7.795	0.032	DU25 (SMA2130)
9	0.256	6.544	0.03	DU25 (SMA2130)
10	0.288	5.361	0.028	DU21 (SMA2130)
11	0.321	4.188	0.026	DU21 (SMA2130)
12	0.354	3.125	0.024	NACA64 (SMA64)
13	0.386	2.319	0.022	NACA64 (SMA64)
14	0.419	1.526	0.02	NACA64 (SMA64)
15	0.446	0.863	0.018	NACA64 (SMA64)
16	0.467	0.37	0.017	NACA64 (SMA64)
17	0.489	0.106	0.011	NACA64 (SMA64)

*The name in the bracket is scaled rotor airfoil with corresponding airfoil of 5MW NREL rotor airfoils.

blade. As the lift coefficient heavily influences the non-dimensional rotor thrust and the axial induction factor, the primary objective is to achieve the required lift coefficients of airfoil sections and in turn complying with the design requirements 1 and 4. Considering the blade length of the scaled rotor, three significant airfoil profiles are employed in comparison to the six profiles in the full-scale model. Table 9.1 indicates the full-scale rotor airfoils and the corresponding airfoils of the scaled rotor. The shape of SMA (Strathclyde Model rotor Airfoil) series airfoils are shown in comparison to the full-scale model airfoils in Figure 9.4 by Sivalingam *et al.*[12] The methodology followed to accomplish the airfoil

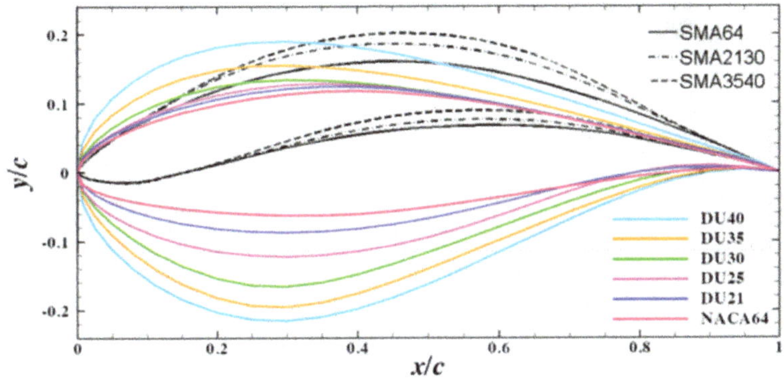

Figure 9.4 Scaled rotor model airfoil shapes in comparison with Sivalingam *et al.*[12]

profiles and further analysis on the scaled rotor model performances are described in more detail in Martin *et al.*[3] and in Martin and Day.[11]

The XFOIL-generated lift coefficients of the SMA airfoils at model Re were compared to the wind tunnel results of the reference airfoils at full-scale Re. Good correlation of lift coefficient exists at lower angles of attack (AoA), while an early onset of flow separation at higher AoA at the lower model Re indicates a low lift coefficient. Due to the significant difference in Re, the SMA airfoils generate higher drag for most of the AoA. The comparisons between the SMA airfoils and the DU40, DU25 and NACA64 airfoils are shown in Figures 9.5 to 9.7.

Lift coefficients for the SMA airfoils are computed using XFOIL for the model Re, while for the full-scale airfoils, the generated lift coefficients from XFOIL are verified through wind tunnel tests. The XFOIL results compare well with the experimental curves at lower AoA, but at higher AoA XFOIL overpredicts the coefficients. This can be attributed to the earlier onset of flow separation in the experimentation. A 3D rendering of the fully assembled rotor is shown in Figure 9.8. The blades are fixed to the hub at a fixed pitch angle of 0° using a flange machined in the root end of the blades; as a result the accuracy of the pitch angle is defined

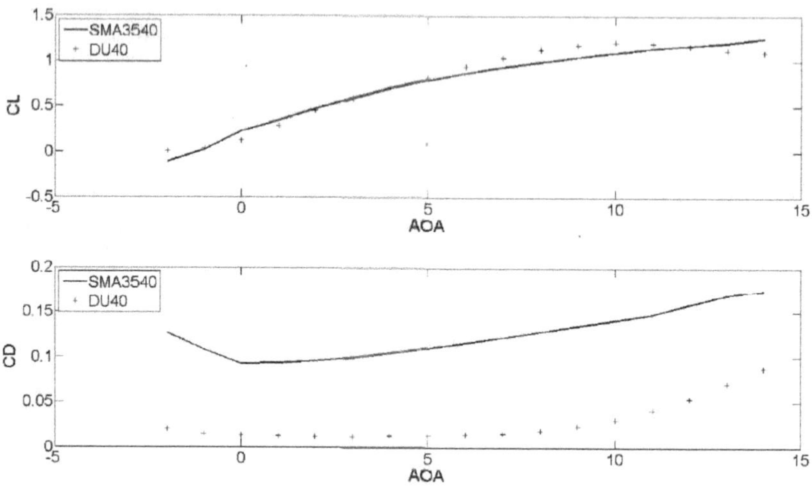

Figure 9.5 Comparison of coefficients of lift (top) and drag (bottom) between SMA3540 and DU40.

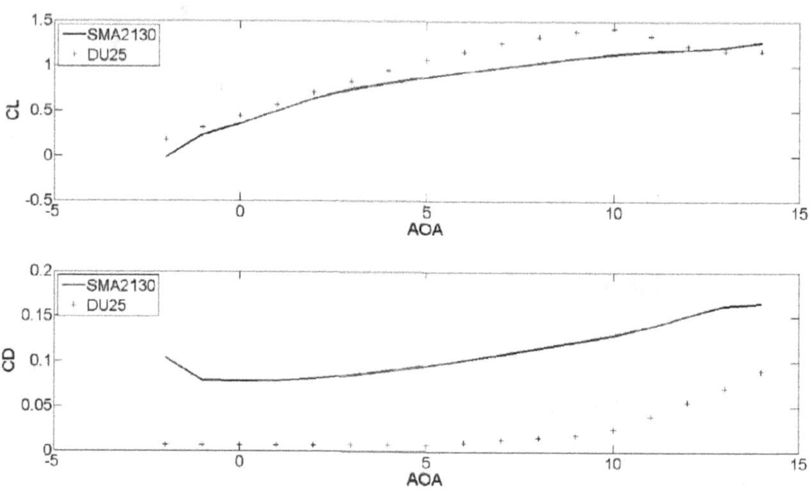

Figure 9.6 Comparison of coefficients of lift (top) and drag (bottom) between SMA2130 and DU25.

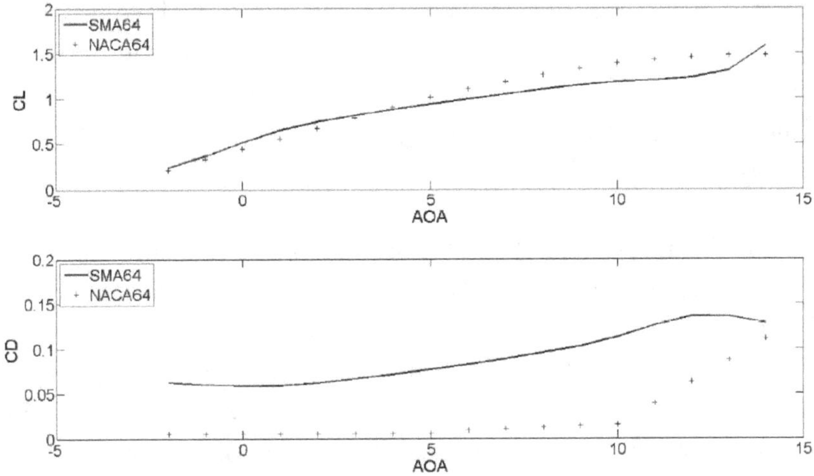

Figure 9.7 Comparison of coefficients of lift (top) and drag (bottom) between SMA64 and NACA64.

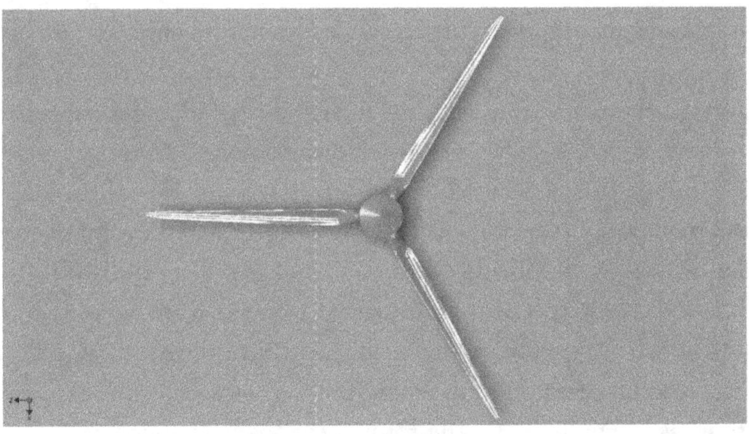

Figure 9.8 3D model of the scaled rotor.

by the machining tolerance of the stainless steel components and not that of the measurement of the pitch angle itself, as would be the case if a locking bearing-type fixture was used. With this approach, the model-scale hub is non-dimensionally larger than the full-scale reference to create sufficient space for blade-to-hub and hub-to-shaft connection. However, the increased diameter of the hub does not necessitate the blade geometry to be altered in any section other than the cylindrical section detailed in Table 9.1 and thus will have no impact on the aerodynamic performance of the blades.

During the design stage, the global model rotor performance was compared to the full-scale reference for non-dimensional thrust and torque. The thrust force on the rotor is primarily influenced by the aerofoil lift and hence the match between the model and reference for this property is favorable. The increased aerofoil drag at the experimental Re results in a reduced torque as compared to the full-scale rotor. This was an expected result of implementing this scaling methodology. The local blade performance was assessed for the model-scale rotor by comparing the axial and tangential induction factors and the AoA along the non-dimensional length of the blade for different TSRs by Martin *et al.*[3] The scaled rotor with its surge motion dual carriage is as shown in Figures 9.9 and 9.10.

In order to achieve Re similarity the model was tested under water at the University of Strathclyde's Kelvin Hydrodynamics Laboratory towing tank (76 m length, 2.5 m depth and 4.6 m width). The cross-sectional area of the tow tank was sufficiently large compared to the rotor diameter under investigation assuring minimum blockage. The prescribed steady and periodic flow speeds were imparted into the rotor by pulling the rotor through the stationary water using a dual carriage system (primary carriage was for constant motion and secondary carriage for surge cyclic motion) as in Figure 9.10. The rotor speed was controlled using a DC motor and PID controller directly coupled to the rotor through a shaft fixed to a stiff bedplate using three low-friction bearings. The torque and thrust forces are measured in-line with the shaft using a bi-directional force transducer at a frequency of 137 Hz. Scaled rotor was designed by University of Strathclyde researchers and experimental scenarios are conducted for the requirements of numerical validation purposes.

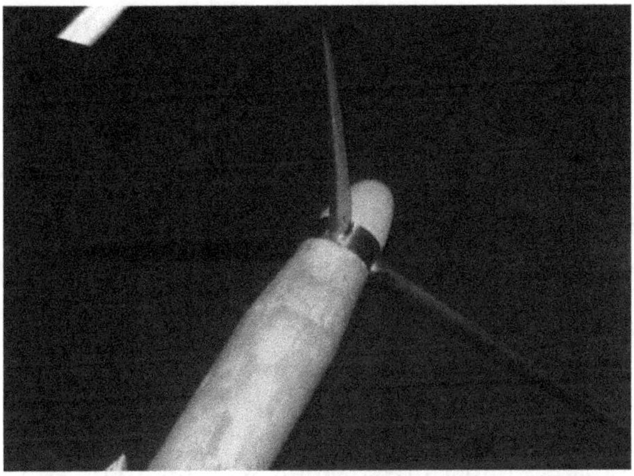

(a)

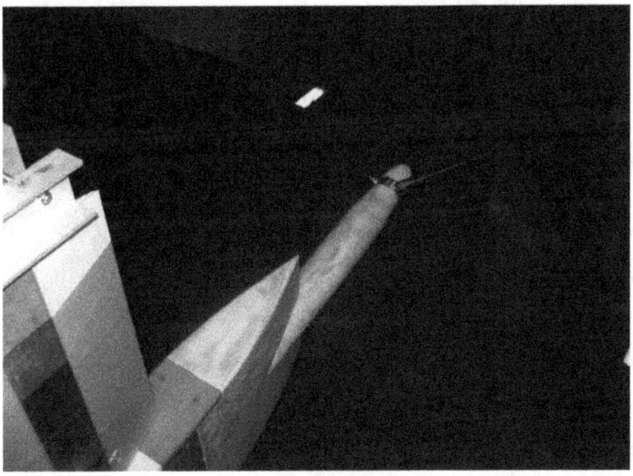

(b)

Figure 9.9 Physical scaled rotor with a rig setup (for surge motion unsteady effects) on the carriage of the towing tank. (a) Scaled FOWT rotor. (b) Scaled FOWT rotor with carriage.

9.2.1 *Experimental Design of Surge Motion*

The surge motion of the rotor does not represent a specific floating wind turbine system, but rather a generalized matrix of operating points, from which conclusions can be drawn regarding the rotor thrust loading.

Figure 9.10 The tow tank at the Kelvin Hydrodynamics Laboratory and carriage systems.

The three variables under investigation were the rotor TSR, the surge amplitude, and the surge frequency. Similar to the static analysis of the rotor, the tow speed was constant and the rotor speed was varied to alter the TSR. The operating range of values for these parameters were specified such that they mimic the full-scale rotor operating conditions limited to the experimental set-up and surge carrier motion.

Three prior studies have been conducted with the objective of studying the aerodynamic performance with the surge motion. The initial values for the parameters were set according to these studies for the current unsteady experimentation of the model-scale rotor. Liu *et al.*[4] studied the NREL 5MW Baseline Wind Turbine at three different wind speeds, surge velocities, and surge frequencies numerically using a BEM-based aerodynamic model with a dynamic inflow correction. Also, de Vaal *et al.*[13] compared the aerodynamic performance of the NREL 5MW turbine subjected to a surging motion using four different numerical approaches: a quasi-static BEM model, a BEM model with an integrated Pit-Peters dynamic inflow correction, a BEM model with an integrated Stig Oye dynamic inflow correction, and a CFD-based actuator disc model. Similarly, Micallef and Sant[14] studied the NREL 5MW rotor in surge using a quasi-static BEM model, a generalized dynamic wake model, and a CFD-based actuator disk model. The valuable conclusion taken from these studies is that the wake-induced effects of the rotor thrust as a result of the rotor surging into the flow depend upon the TSR, surge frequency, and surge displacement amplitude.

The experimental surge operational points considered are drawn primarily from the work of Liu *et al.*[4] and de Vaal *et al.*[13] The surge frequencies studied by Liu *et al.*[4] and de Vaal *et al.*[13] were 0.1, 0.2, and 0.3 rad/s, and between 0.127 and 1.0 rad/s, respectively; the corresponding surge displacement amplitudes were 3.0, 6.0, and 9.0 m, and between 2.0 and 16.0 m. de Vaal considers only the rotor operating at its rated wind speed ($V = 11.2$ m/s), while Liu *et al.*[4] and Micallef and Sant[14] both consider this same rotor operation in addition to off-design points.

The investigated rotor operational states, defined by Liu *et al.*[4] and Micallef and Sant[14] in terms of inflow wind speed and TSR were 8.0, 11.2 and 16.0 m/s, and 4.0, 7.0 and 11.0, respectively. The TSR of the NREL 5MW Baseline Wind Turbine at its rated wind speed of 11.2 m/s, as defined by Jonkman *et al.*,[9] is 7.0.

The model-scale surge motion displacement and frequency were derived from the studies discussed by Liu *et al.*[4] and de Vaal *et al.*[13] by applying a suitable scaling methodology. This was achieved by following the scaling relationships defined by Jain *et al.* for the experimental analysis of the combined effect of aerodynamic and hydrodynamic force components on a model floating wind turbine rotor and foundation in a wind and wave basin. The model-scale surge frequency and displacement were scaled following the relationships defined in Eqs. (9.1) and (9.2), respectively.

$$\omega(\text{surge, model}) = \omega(\text{surge, full}) \times \sqrt{\lambda} \tag{9.1}$$

$$A(\text{surge, model}) = \frac{A(\text{surge, full})}{\sqrt{\lambda}} \tag{9.2}$$

where A is amplitude, ω is the angular frequency and λ is the geometric scaling factor, which is 126.0 for the present set of experiments.

A comprehensive static and unsteady test matrix was performed. The dimensionless parameters C_t and C_q that determine the amount of momentum and energy extracted from the water were computed for static cases. The coefficient of thrust is defined by Eq. (9.3):

$$C_t = \frac{T}{0.5\rho AV^2} = 4a(1-a) \tag{9.3}$$

where T is the rotor thrust force, ρ is the fluid density, A is the rotor plane area and V is the fluid velocity. The coefficient of torque is defined by Eq. (9.4):

$$C_q = \frac{Q}{0.5\pi \rho R^3 V^2} \tag{9.4}$$

where Q is the torque and R is the rotor radius. The torque is measured when the scaled rotor is at constant rotational speed, constant tow speed of primary carriage for the static test as in Table 9.2. Time history of thrust and torque are captured in unsteady surge motion test scenario cases (as in Table 9.3).

During the unsteady test scenario, secondary carriage oscillates while primary carriage is at constant speed. For different rotor rotational speeds, experiments were carried out and results plotted in Figures 9.11 and 9.12.

Table 9.2 Static state test condition.

Tow Speed (m/s)	TSR	Rotational Speed (rad/s)
1.000	3.000	6
0.931	4.500	9
1.000	7.000	14
0.822	8.519	14
0.735	9.522	14

Table 9.3 Unsteady experimental test case scenarios chosen for numerical validation at the rotor rotational speed of 14 rad/s.

TSR	Tow Speed, m/s	Surge Frequency (f), rad/s	Amplitude (A), M	Scenario Name for Reference
8.5	0.824	1.12 (0.18 Hz)	0.0238	SFA1
8.5	0.824	1.12 (0.18 Hz)	0.1190	SFA2
8.5	0.824	3.37 (0.54 Hz)	0.0238	SFA3
8.5	0.824	3.37 (0.54 Hz)	0.1190	SFA4
8.5	0.824	5.61 (0.89 Hz)	0.0238	SFA5
8.5	0.824	5.61 (0.89 Hz)	0.0952	SFA6
7.0	1.000	1.12 (0.18 Hz)	0.0238	SFA7
7.0	1.000	1.12 (0.18 Hz)	0.1190	SFA8
7.0	1.000	3.37 (0.54 Hz)	0.0238	SFA9
7.0	1.000	3.37 (0.54 Hz)	0.1190	SFA10
7.0	1.000	5.61 (0.89 Hz)	0.0238	SFA11
7.0	1.000	5.61 (0.89 Hz)	0.0952	SFA12
6.0	1.167	1.12 (0.18 Hz)	0.0238	SFA13
6.0	1.167	1.12 (0.18 Hz)	0.1190	SFA14
6.0	1.167	3.37 (0.54 Hz)	0.0238	SFA15
6.0	1.167	3.37 (0.54 Hz)	0.1190	SFA16
6.0	1.167	5.61 (0.89 Hz)	0.0238	SFA17
6.0	1.167	5.61 (0.89 Hz)	0.0714	SFA18

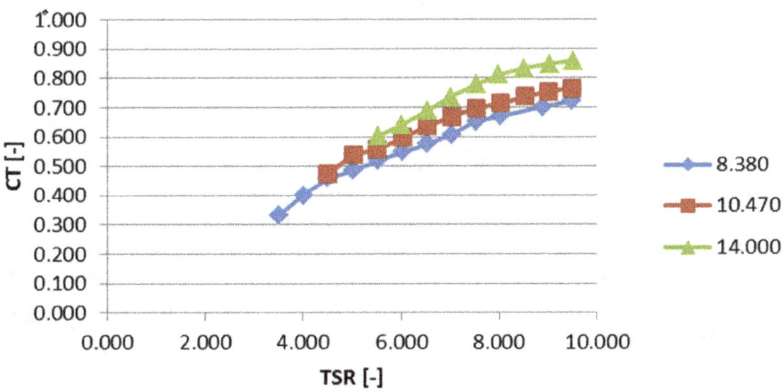

Figure 9.11 The tow tank static test for various rotor rotational speeds — Ct curves.

9.3 Numerical Methodology

Due to wind shear between ground or sea level to tip of the rotor blade, gust, turbulence, and flow field around the rotor and near wake regions are highly complex even in land-based installations. In FOWT, the complexity is further compounded by the additional motions at the rotor plane induced by the floating platform subjected to the three translational (heave, sway, and surge) and three rotational (yaw, pitch, and roll) motions as shown in Figure 9.13. Among these six motions, surge and pitch motions are responsible for pushing the rotor to interact with its own wake. This interaction significantly modifies the wind turbine's operating state and in turn varies the axial induction velocity field and unsteadiness. As the BEM theory is constructed based on the axial induction factor range, it has to be validated to gain sufficient confidence before applying to FOWT. In the current study, scaled rotor surge motion experimental results were exploited for validating the numerical codes.

CFD, LR-AeroDyn and LR-uBEM have been chosen for the validation study. CFD simulations are carried out with reasonable accuracy to gain insight on surge motion's flow field and wake interactions. As AeroDyn is widely employed in the industry for its accuracy, Marine and Hydrokinetic code is modified to incorporate the surge motion scenarios

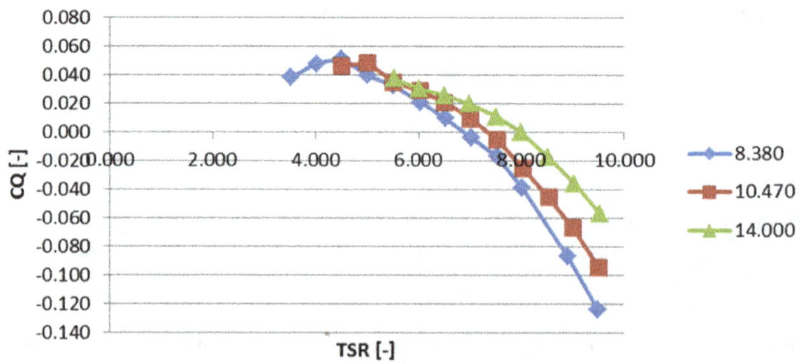

Figure 9.12 The tow tank static test for various rotor rotational speeds — C_q curves.

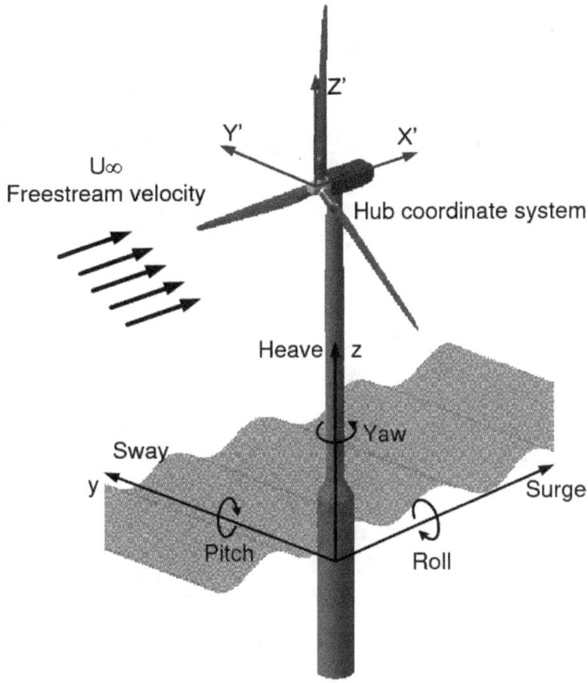

Figure 9.13 Floating platform motions (three linear and three rotational). Reproduced with permission from Tran, T., Kim, D., Song, J. Computational fluid dynamic analysis of a floating offshore wind turbine experiencing platform pitching motion. *Energies* **2014**, 7(8), 5011–5026.

(LR-AeroDyn). LR's research on unsteady aerodynamics of FOWT resulted in the development of LR-uBEM code accounting for dynamic inflow model with other unique features. The following sections discuss in detail these methodologies.

9.3.1 *Computational Fluid Dynamics Model*

The tow tank is defined in the CFD numerical environment as in Figure 9.14 to solve the unsteadiness on the rotor due to surge motion. The primary or main carriage motion (constant) of the rotor is represented as uniform inlet water boundary condition and the secondary carriage surge motion is represented as MRF (Multiple Reference Frame) and sliding mesh motion on the rotor domain.

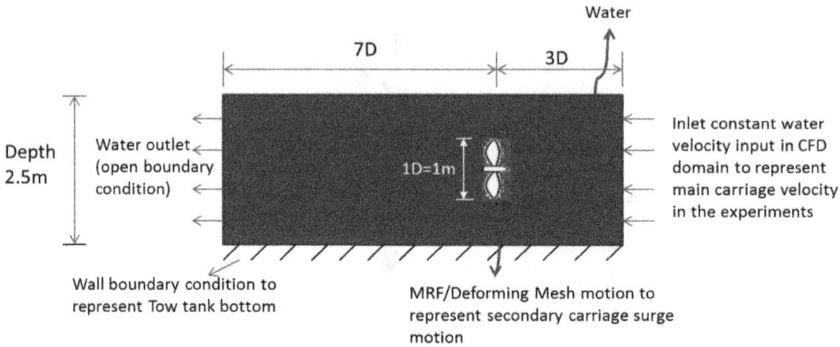

Figure 9.14 Schematic representation of tow tank in CFD domain.

Figure 9.15 Scaled rotor 3D model generated in Ansys ICEM CFD environment.

As shown in Figure 9.15, the scaled rigid 3D rotor blade geometry was generated through commercial meshing software Ansys ICEM CFD by sweeping through all the 17 aero profiles of the scaled model. The hub is modeled to the dimensions used in the tow tank to eliminate any inaccuracies introduced due to the flow variation near the blade root.

A multi-block meshing strategy has been employed to create hexahedral mesh in the outer domain. 'O-grid'-type mesh is chosen for the blade domain. A hexahedral structured mesh around the scaled rotor of one blade and the surrounding water of turbine rotor was generated and for other blades mesh was generated using periodicity option. As shown in Figures 9.16 to 9.18, the 3D scaled rotor model and three kinds of mesh-clustering methods (log-scale change in near wall and linear towards perpendicular direction) around the rotor were used in the mesh sensitivity study as the thrust force is more sensitive to the method of mesh clustering.

The entire computational domain is shown in Figure 9.18. The mesh was sufficiently discretized to capture the near wake and far wake fields reasonably. The scaled rotor was refined to yield a y^+ of ~1 near the tip and less than 2 for the rest of the domain as shown in Figure 9.19.

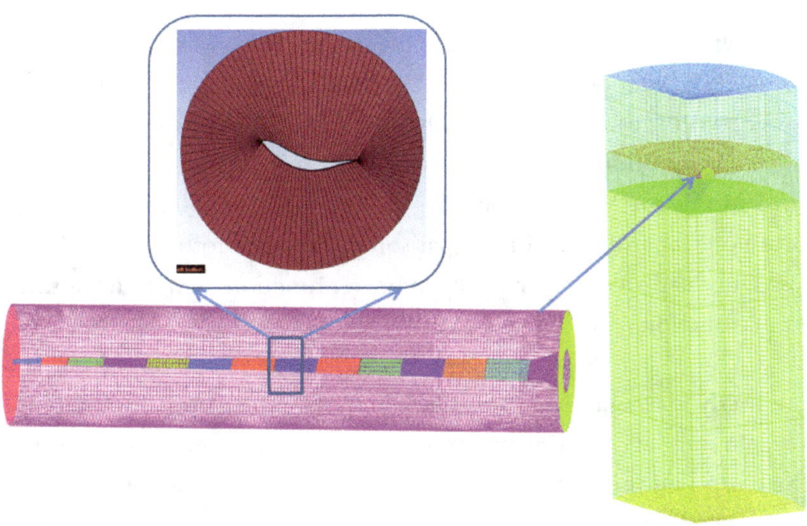

Figure 9.16 Scaled rotor mesh model and 1/3rd model of rotating domain with single rotor.

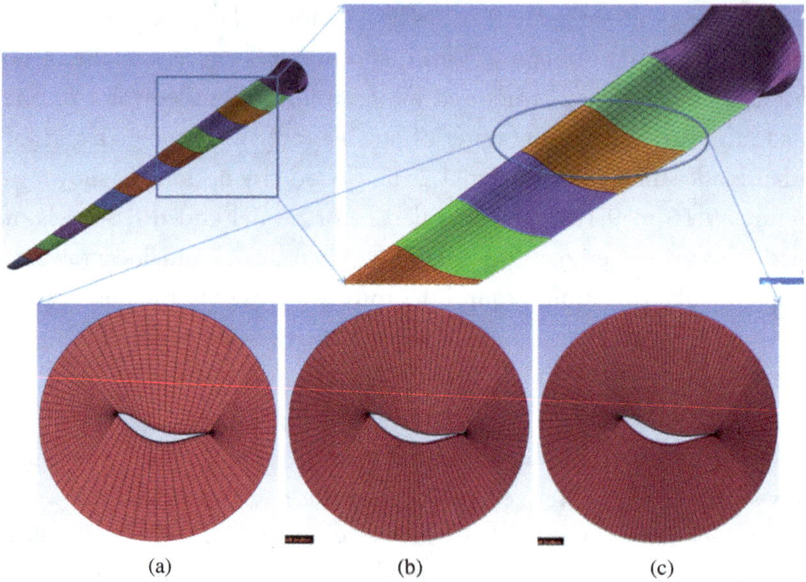

Figure 9.17 Mesh sensitivity study: (a) coarse with 0.59 million nodes, (b) medium with 1.65 million nodes, (c) fine with 2.38 million nodes.

As shown in Figures 9.20 to 9.22, the deviation between medium and fine mesh CFD model is about 1.2%. Hence, medium-size CFD model was chosen for the validation study with static and unsteady test experimental cases. Moreover, the computational time is also less when compared to fine mesh CFD model for industrial-level applications. Thrust force alone is considered for comparison as the scaled rotor was developed based on the thrust force-related parameter function used in the optimization function.

The mesh mathematical model was imported into Ansys CFX CFD solver to simulate experimental scenario cases of the rigid 3D scaled rotor. The *k*-ω SST (Shear Stress Transport) turbulence scheme was employed without transition modeling. A high-order advection scheme and first-order numerical method were chosen for turbulence modeling. The *k*-ω SST

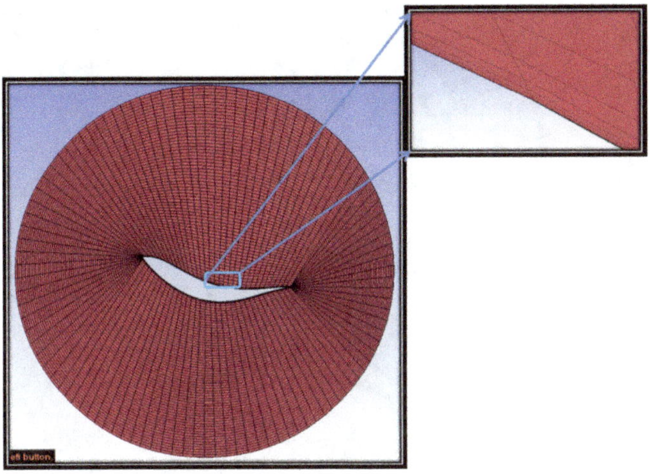

(a)

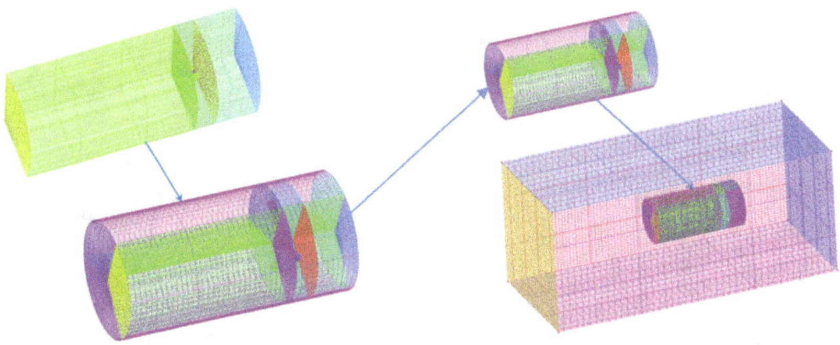

(b)

Figure 9.18 CFD domain. (a) 1/3rd rotor model is swept for 360° with near wake domain. (b) The complete near wake region mesh model inside the global model.

model employs the *k-ω* for the near surface treatment and the *k-ε* model in the free-shear layers. In this methodology a blending function is adopted to bridge these two models. The *k-ω* SST model accounts for the transportation of the turbulent shear stress and is capable of accurately predicting the onset and magnitude of flow separation under adverse pressure gradients in the rotor blade, near wake and far wake regions.

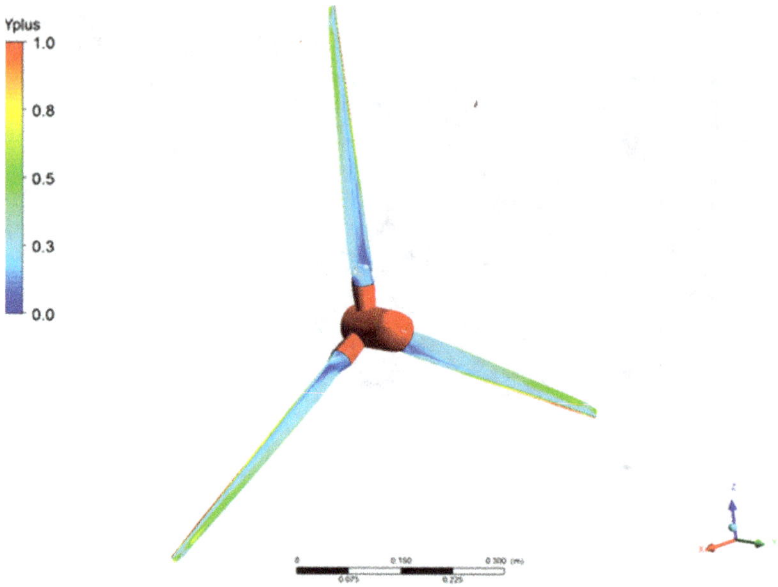

Figure 9.19 Scaled rotor — y^+ contour.

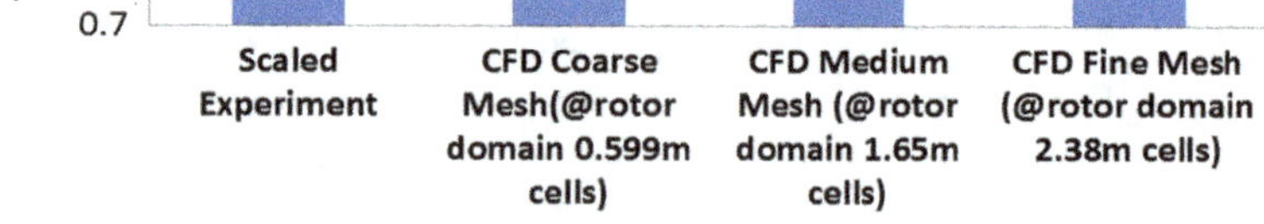

Figure 9.20 Thrust coefficient of various mesh sizes.

For the aforementioned reasons, the two-equation *k-ω* SST model is widely exploited in the wind industry for its accurate predictions of flow over blunt body.

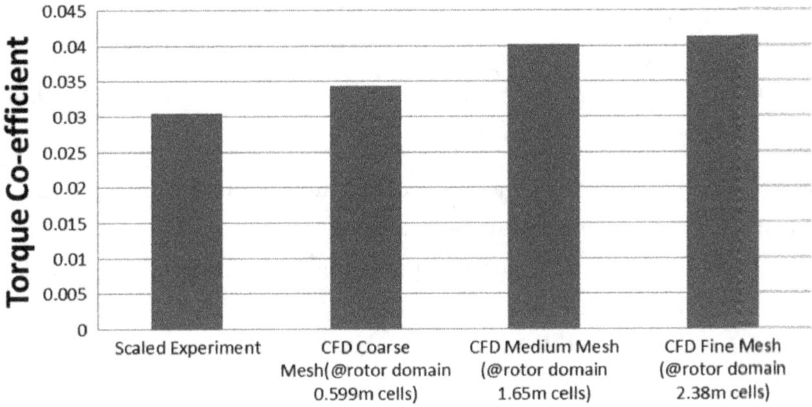

Figure 9.21 Torque coefficient of various mesh sizes.

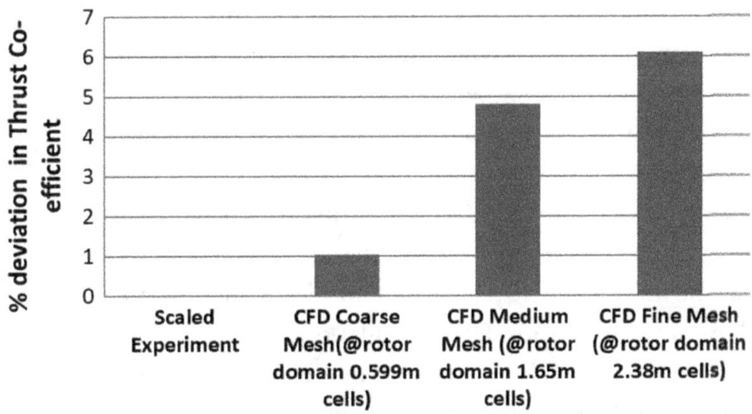

Figure 9.22 Percentage deviation in thrust coefficient for various mesh sizes.

The *k-ω* SST model has a similar form to the standard *k-ω* model:

$$\frac{\partial}{\partial x}(\rho\kappa)+\frac{\partial}{\partial x_i}(\rho\kappa u_i)=\frac{\partial}{\partial x_j}\left(\Gamma_\kappa\frac{\partial\kappa}{\partial x_j}\right)+\widetilde{G}_\kappa-Y_\kappa+S_\kappa \tag{9.5}$$

and

$$\frac{\partial}{\partial x}(\rho\omega)+\frac{\partial}{\partial x_i}(\rho\omega u_i)=\frac{\partial}{\partial x_j}\left(\Gamma_\omega\frac{\partial\omega}{\partial x_j}\right)+\widetilde{G}_\omega-Y_\omega+D_\omega+DS_\omega \quad (9.6)$$

where $\widetilde{G}_\kappa$ represents the generation of turbulence kinetic energy due to mean velocity gradients, while $\widetilde{G}_\omega$ represents the generation of ω. Γ_k and Γ_ω represent the effective diffusivity of κ and ω, respectively. Y_κ and Y_ω represent the dissipation of κ and ω due to turbulence, respectively. D_ω represents the cross-diffusion term. S_κ and S_ω are user-defined source terms.

The CFD simulations were carried out with uniform water velocity at the inlet boundary for the static case scenarios (without surge motion) as shown in Table 9.2 before carrying out the unsteady test case simulations. For mesh sensitivity study, the rotor thrust coefficient was compared against the chosen case of 14 rad/s with 7 TSR and 1 m/s water speed, which is representative of 1 m/s tow speed of the carriage mounted with scaled physical rotor as in Table 9.2. The overall medium mesh model domain contains 7.32 million nodes. The chosen mesh model yields y^+ of less than 1 near the tip and less than 2 everywhere else as shown in Figure 9.19. MRF and sliding mesh simulation methodology was used in the CFD simulations to validate the thrust and torque predictions of the scaled rotor model. The CFD-predicted values were then validated against the measured values of the experiments as in Table 9.2.

The static runs were carried out to compute the C_t and C_q in relation to TSR as benchmark values to compare the rotor subjected to static test case conditions. In order to operate the rotor at a wide range of TSR, i.e., from 3 to 9.5, the two-speed is varied, though a constant TSR is aimed for static scaling. The singular speed of 14 rad/s of the scaled rotor models compounds to NREL 5MW rated RPM of 12.10, while other model rotor speeds are linearly proportional to the NREL rotor RPM.

Thus, for a model rotational speed of Ω_m in rad/s, the full-scale rotational speed Ω_f in rpm is given by Eq. (9.7):

$$\Omega_f = \frac{\Omega_m}{14\left(\frac{\text{rad}}{\text{s}}\right)} \times 12.1 \text{ rpm} \tag{9.7}$$

The full-scale wind speed, V_f, is found from the TSR of the experiment:

$$V_f = \frac{63 \neq \Omega_f}{30 \text{ TSR}} \tag{9.8}$$

To validate the static environment, the full-scale rotor RPM of 12.10 and 11.4 m/s rated wind speed are modeled as 14.0 rad/s and 1 m/s tow speed for scaled model rotor. Hence, the coefficient of thrust C_t and the coefficient of torque C_q in relation to the TSR are computed analytically (designed values), numerically (CFD) and experimentally. The results are plotted as shown in Figure 9.23.

The thrust coefficients from all the three methods were comparable and the deviation found to be within 10%, while the torque coefficient displays a notable difference in analytical prediction (design values) from the well-correlated experimental and CFD results. XFOIL predictions of lift coefficient C_l and drag coefficient C_d would have contributed significantly to the deviation. It was experimentally verified that the XFOIL over-predicts the drag in the low Re range as the flow over the airfoil is extremely complex and non-linear to compute the drag by panel method. The over-prediction is exacerbated by the extrapolation for higher AoA from the initial set of lift and drag coefficients computed at lower AoA. A possible way of curtailing the error is to experimentally obtain the initial set of C_l and C_d from the wind tunnel test and extrapolate by Viterna or Montgomerie method. Considerable effort has been dedicated to refining the mesh for improved near-wall treatment to compute the boundary layer separation and to achieve accurate rotor drag. Since optimization of the airfoil for the operating Re of the scaled

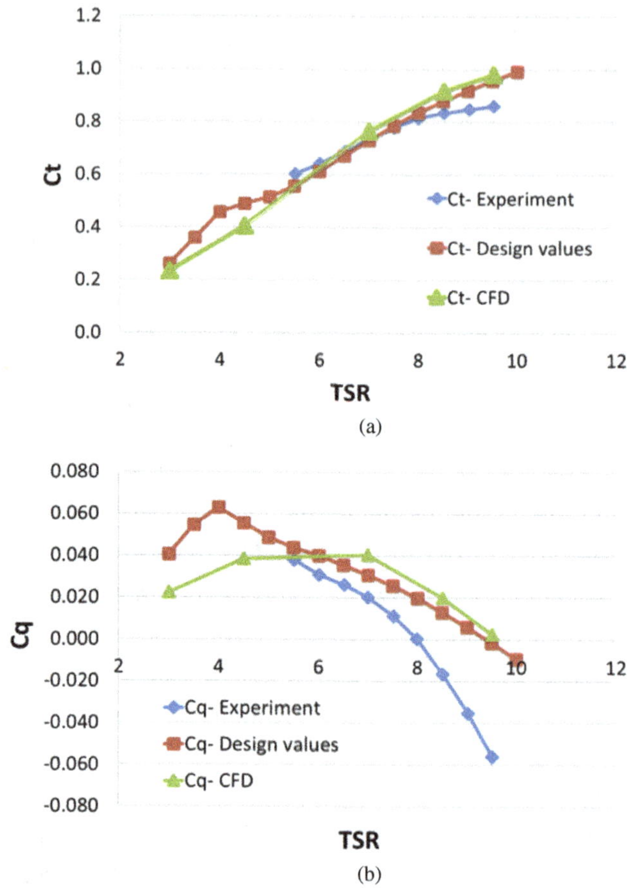

Figure 9.23 Static state validation curves for CFD. (a) Comparison of thrust coefficient. (b) Comparison of torque coefficient.

rotor is not in the scope of the current study, the effect of low Re is anticipated. The discrepancies due to drag are inevitable, especially for the scaled prototype due to premature flow separation on airfoils induced by low Re because of smaller chord length.

The surge motion methodology was developed in Ansys-CFX software. Medium-size mesh model with 7.32 million nodes was adopted for the unsteady simulations. The unsteady runs include sinusoidal surging motions of the rotor from low to extreme frequencies (low, medium, and high) and amplitudes (lower and higher) to demonstrate the effect on the rotor unsteady aerodynamics as experimentally simulated. Simulations were carried out with MRF and deforming mesh set-up in the CFD environment, where constant water velocity from the inlet (with a turbulence intensity of 5%) of the computational domain represented constant primary carriage motion. The deforming mesh was set up such that the mesh in the blade and near-wake regions was given a high rigidity, so that deformation is limited to the outer region with bigger cells, and negative volumes would not occur. A high-resolution scheme was used for advection and turbulence numerics, while a second-order backward Euler scheme was used for time stepping. Pressure-velocity coupling was solved using a fourth-order Rhie-Chow interpolation. The surge motion is defined by Eq. (9.9):

$$-A + A \cos 2\pi f \tag{9.9}$$

where A is the amplitude of motion and f is the frequency.

The entire blade and near-wake regions were set to oscillate as defined by Eq. (9.9). The equation assumes that the mesh starts at the forward-most position, since there are more cells downstream of the rotor, thus it would facilitate smoother deformation. As per Table 9.3, simulations are carried out to compare with the experimental values.

For unsteady test cases, the sinusoidal surging motions for the model rotor are designed as per the studies by Liu *et al.*[4] and de Vaal *et al.*[13] As the surge motion is predominant in FOWT applications,[14] detailed surge motion experiments were carried out with the scaled rotor. The scenarios were chosen with utmost care to validate the effects of unsteadiness by the various numerical simulations. Mean thrust and torque values were considered for validation purposes initially. For each of the three different TSRs, three different surge frequencies and two amplitudes (minimum and maximum) were chosen to study the unsteady behavior. For the scenarios given in Table 9.2, the steady state MRF simulations are first carried out by equating the water inlet in CFD to rotational speed and towing speed. The results of the steady state calculations were used to initialize the solutions for complete transient case involving the surge motion of the scaled rotor unsteady test case scenario. The domains were structured in such a way that both the blade rotations and rotor surging motions could be handled smoothly by the solver. The rotor domain had a sliding mesh interface for blade rotations and the mesh motion (deforming mesh) applied for surging motion had an extremely high stiffness in this domain which was relaxed gradually towards the outer domain to preserve the fine boundary layer mesh. This high stiffness ensures that the mesh in the rotor domain has no relative nodal displacement, as the mesh on the blade had the first node in the order of microns to yield a $y^+ \sim 1$.

9.3.2 LR-AeroDyn Model for Unsteady Experimental Scenario

Though the usual practice is to include the dynamic stall while studying unsteady aerodynamics, for a FOWT, the unsteadiness is high even for a fully attached flow due to the rotor interacting with its own wake. Theories based on the 2D assumptions for predicting the unsteady aerodynamics that were developed in the 1930s assume that the flow is incompressible.[15–17] This can be due to small relative velocity between blade elements. But the study by Leishman[18] argues that incompressible flow assumption is not valid under specific conditions such as blade passing through tower shadow or discrete vorticity of blade tip vortices. Above conditions have to be accounted for in CFD simulation at the expense of higher computational resources.

The incompressible assumption holds good for most scenarios. The theories formulated by Theodorsen[15] are based on incompressible flow, which provide analytical solutions for predicting unsteady effects on airfoils. At low frequencies unsteady effects will be incited in fully attached flow leading to phase variations at higher amplitude compared to quasi-steady loads.[18]

However, Theodorsen's theories are applicable only to fully attached flow. To account for the full spectrum of unsteady aerodynamics, FAST was used to simulate OC3-Hywind[19] turbine, which is the original 2.3MW Hywind wind turbine scaled to 5MW. Being an aero-elastic structural code, FAST simulation includes flexible structures (tower and blades), a controller specifically designed for floating wind turbines, and coupled aerodynamic and hydrodynamic effects.

Initially, each DoF was simulated separately by setting acceleration to zero and keeping the initial prescribed displacement constant for each disabled DoF. This allowed for comparison of response of the system in each DoF for quasi-steady and unsteady aerodynamic assumptions. Two different value sets were used — quasi-steady and fully unsteady. The similarities and differences between both aerodynamic modeling techniques are summarized in Table 9.4.

A simplified flowchart of AeroDyn's working principle is given in Figure 9.24.

Table 9.4 Similarities and differences between the quasi-steady and unsteady codes.

EQUIL & Steady	BEDDOES & DYNN
Differences	
Look-up tables in quasi-steady	Beddoes-Leishman model: dynamic airfoil
No dynamic inflow (no time lag)	Dynamic inflow time lag
Prandtl's model (tip/hub)/GTECH model for tip loss	Highly unsteady
Glaurt's model for heavily loaded rotors	At low wind speeds it switches to BEM (does not account for wake rotation, *a*')
Similarities	
Based on flat disc assumptions (no deflection)	

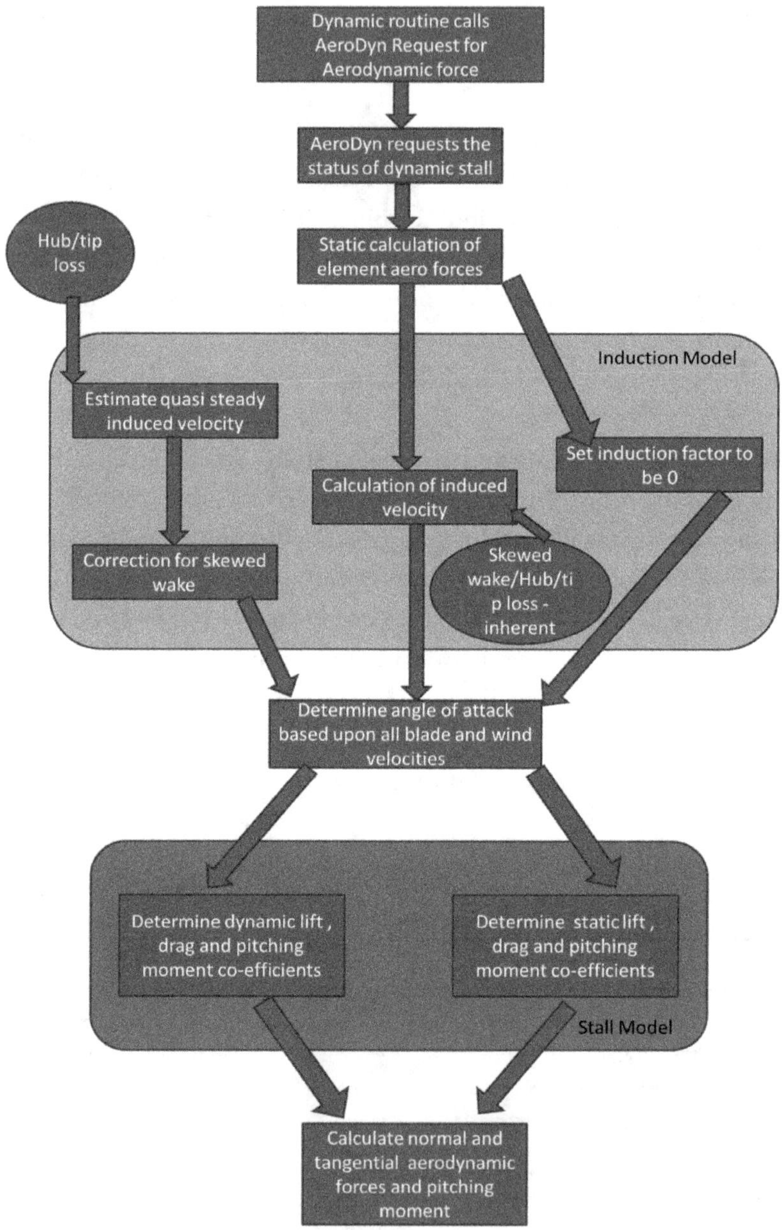

Figure 9.24 Working principle of AeroDyn in simplified flowchart form.

AeroDyn (V15.04) is an open source (NREL) time-domain wind turbine aerodynamics module that has been tailored for Marine Hydrokinetic Turbines (MHT).[20] MHT AeroDyn module is customized for scaled rotor performance prediction of unsteady experimental test case runs as shown in Table 9.3 as the model was refined to cover the whole scenario.

9.3.2.1 *FAST Model Settings*

AeroDyn calculates aerodynamic loads on both the blades and tower. Aerodynamic calculations within AeroDyn are based on the principles of actuator lines, where the 3D flow around a body is approximated by local 2D flow at cross-sections, and the distributed pressure and shear stresses are approximated by lift forces, drag forces, and pitching moments lumped at a node in a 2D cross-section. Analysis nodes are distributed along the length of each blade and tower, the 2D forces and moment at each node are computed as distributed loads per unit length, and the total 3D aerodynamic loads are found by integrating the 2D distributed loads along the length. The actuator line approximations restrict the validity of the model to slender structures and 3D behavior is either neglected, captured through corrections inherent in the model (e.g., tip-loss, hub-loss, or skewed-wake corrections), or captured in the input data (e.g., rotational augmentation corrections applied to airfoil data).

AeroDyn assumes the turbine geometry consists of a one-, two-, or three-bladed rotor atop a single tower. For blades, the 2D cross-sections where the aerodynamic analysis takes place may follow the out-of-plane curvature, but in-plane sweep is assumed to be accomplished by shearing, rather than rotation of the 2D cross-section. Aerodynamic imbalances are possible through the use of geometrical differences between each blade.

AeroDyn consists of four submodels: (1) rotor wake/induction, (2) blade airfoil aerodynamics, (3) tower influence on the fluid local to the blade nodes, and (4) tower drag. Nacelle, hub, and tail-vane fluid influence and loading, aeroacoustics, and wake and array effects between multiple turbines in a wind plant are not yet available in AeroDyn v15 and newer.

The MHK-based model (LR-BEM) customized for the scaled rotor calculates the influence of the wake via induction factors based on the quasi-steady BEM theory, which requires an iterative Brent's method for non-linear equations. In the quasi-steady or equilibrium wake (EQUIL) model, the induction reacts instantaneously to loading changes. The induction calculation, and resulting inflow velocities and angles, are based on flow local to the analysis node of each blade, which is in turn based on the relative velocity between the fluid and structure (including the effects of local inflow skew, shear, and turbulence). This method has no time lag which means no dynamic inflow. Glauert's empirical correction (with Buhl's modification) is used in the linear momentum balance at high axial induction factors. In the BEM solution, Prandtl tip-loss and Prandtl hub-loss models are applied.

AoA calculations in AeroDyn assume that the axial induction factor a only applies to the freestream wind velocity V_α. To calculate the total relative wind speed V_{rel} normal to the rotor disc, each blade element's structural velocity is added to the freestream wind velocity at the rotor disc,

$$V_{rel} = V_\alpha(1 - a) + V_{rotor} \tag{9.10}$$

where V_{rotor} is the structural velocity of the blade element normal to the disc (measured positive when pointing upwind).

The assumption in AeroDyn is that the structural velocities are the product of structural vibrations at high frequencies, which would result in relatively small induced velocities. However, when considering FOWT, low-frequency motions, particularly in surge, pitch and yaw degrees of freedom which induce change in the freestream velocity, could lead to high induced velocities. In low-frequency motion, a better way of expressing the relative wind speed normal to the rotor disc would be:

$$V_{rel} = (V_{\alpha} + V_{rotor})(1 - a) \tag{9.11}$$

The primary LR AeroDyn input file defines modeling options, environmental conditions (except freestream flow), airfoils, tower nodal discretization and properties, as well as output file specifications. Airfoil data properties are read from dedicated input files (one for each airfoil) and include coefficients of lift force, drag force, optional pitching moment, minimum pressure versus AoA as well as UA model parameters. (Minimum pressure coefficients versus AoA are also included in the airfoil input files in case a cavitation check is requested.) Blade nodal discretization, geometry, twist, chord, and airfoil identifier are likewise read from separate input files (one for each blade).

Different combinations of frequency of secondary carriage and amplitude with corresponding rotor rotational speeds are used. For each case in Table 9.3, the complete scenario was programmed in MATLAB code to generate the LR-AeroDyn input files and solved through MATLAB command for six surge motion cycles. The last two cycles are considered for comparison and validation.

9.3.3 *LR-uBEM Model for Unsteady Experimental Scenario*

The LR-uBEM code was developed to address the issue of uncertainty in predicting the aerodynamics of FOWT. The development of the uBEM code included several features vital to an accurate assessment of FOWT aerodynamics. These include:

1. An improved tip-loss model which accounts for changes in TSR and changes in loss distributions in different wake states.

2. An unsteady airfoil model including a modified Beddoes-Leishman model which models the unsteady circulation using the suction surface shape-dependent time constants.
3. A combined unsteady airfoil and stall delay model, accounting for changes in stall delay due to changes in wind and rotational velocities in real time.
4. An extension of the Glauert correction by Buhl to the propeller and propeller-brake wake states.
5. A dynamic wake model to account for the time lag between aerodynamic forces on the rotor and wake velocities.

In addition, uBEM predicts the forces on each blade section separately, taking into account local velocity changes at any position on each blade. The development of the code was based on CFD simulations of the NREL 5MW virtual wind turbine, used for code comparison studies in OC3, OC4, and OC5. The CFD simulations were carried out for a static and a surging wind turbine under various surge conditions, with a fixed rotor RPM. The code was developed and is modularized to allow for changes to the various corrective models implemented.

The uBEM used in the present study is based on the method previously presented.[21,22] The relative velocity vector is given by:

$$V_{rel} = V_0 + W + \begin{pmatrix} 0 \\ -\Omega r \cos(\theta_c) \\ 0 \end{pmatrix} \tag{9.12}$$

where V_0 is the blade-specific wind speed at the blade element including relative speed due to rotor motion, Ω is the rotor rotational speed in rad/s, and r is the local blade segment radius. The flow angle, ϕ, is the angle between the relative velocity and the normal of the rotor plane.

$$\phi = arctan\left(\frac{V_{rel,y}}{V_{rel,z}}\right) \tag{9.13}$$

The normal (z) and tangential (y) induced velocities are then given by:

$$W_z = \frac{-\left|V_{rel}\right|^2 C_l \cos(\phi) cB}{8F \left| (V_0 + A) + f_g \begin{pmatrix} 0 \\ 0 \\ W_z \end{pmatrix} \right| \pi r} \tag{9.14}$$

$$W_y = \frac{-\left|V_{rel}\right|^2 C_l \sin(\phi) cB}{8F \left| (V_0 + A) + f_g \begin{pmatrix} 0 \\ 0 \\ W_z \end{pmatrix} \right| \pi r} \tag{9.15}$$

where B is the number of blades, F is the tip-loss factor, f_g is the Glauert correction for high axial induction factors and C_l is the coefficient of lift computed using the Du & Selig stall delay model[23] and the shape-specific modified Beddoes-Leishman model with the AoA equal to the flow angle less the blade twist.[24] W_x is set to 0. At every time step, for each blade element in each blade, Eqs. (9.12) and (9.13) are iterated until there is no change in the induced velocity vector. When the steady state induced velocity vector is computed, it is corrected for the dynamic wake effect using the dynamic wake model.[21,22] The force coefficients are then recomputed to determine the unsteady forces at each blade element. The LR-uBEM model is customized and simulated for scaled rotor model unsteady simulations as per the experimental scenarios.

LR-uBEM is designed for the unsteady environment of FOWTs and validated against unsteady CFD simulations of the full-scale NREL 5MW rotor. This applies BEM equations in an unsteady environment and the flow velocities are directly calculated, instead of induction factors. Moreover, it directly accounts for changes in flow direction and rotor orientation through transformation of earth-fixed, platform-fixed, rotor-fixed and blade-fixed coordinate systems. Blade elemental forces are calculated independently for each blade and in response to the flow changes on each blade. Global momentum changes are inherently modeled. Table 9.5 shows the difference between conventional BEM (LR-AeroDyn) vs LR-uBEM.

LR developed modification to Prandtl tip-loss and addressed the dependency on TSR which accounts for rotor operation beyond windmill state, and the correction applies to both steady and unsteady BEM. Figure 9.25 shows the comparison between four methods for three different TSRs.

As discussed, LR-uBEM incorporates the Beddoes-Leishman model. The Beddoes-Leishman model includes eight time constants which are optimized over a variety of airfoils:

- Limited accuracy for individual airfoils
- LR-modified Beddoes-Leishman model: adapts to airfoil shape

Table 9.5 Scaled rotor airfoil details.

LR-BEM (AeroDyn)	LR-uBEM
Rotor-averaged velocities	Blade-specific and blade element-specific velocities
Azimuthally averaged forces (may be corrected for yaw)	Blade-specific forces (corrections limited to skewed wake effects)
Local unsteady aerodynamic modeling ineffective due to averaging of velocities	Effective modeling of local unsteady aerodynamics due to local velocities
General unsteady airfoil model	Airfoil shape-specific unsteady airfoil model
Prandtl tip-loss model	Corrected tip-loss model based on TSR and wake state
Pre-applied stall delay model	Real-time stall delay model
General dynamic wake model	Calibrated dynamic wake model
Range of applicability (windmill and turbulent wake states ($0 < a < 1$))	Range of applicability: propeller to propeller brake wake states Theoretical: −infinity to +infinity Tested: $-20 < a < 5$

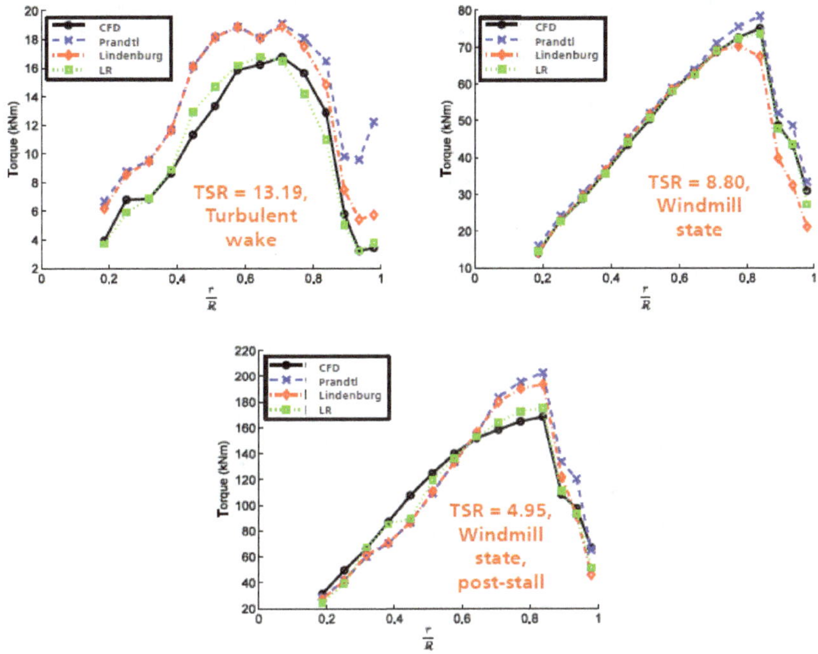

Figure 9.25 LR tip-loss model consistently produces an accurate distribution of shaft torque.[21,22]

- Suction surface parameterized based on a shape equation: two constants (thickness and curvature)
- Improved predictive capability vs experimental data

Figure 9.26 shows the results of Beddoes-Leishman implementation with LR-uBEM.

Combined stall delay and unsteady airfoil model are implemented in LR-uBEM. Rotational speed of rotor and TSR may change during operation due to speed control, motion or changing wind conditions as the level of stall delay constantly changes. Presently, stall delay is applied as a treatment to 2D airfoil data beforehand, which does not account for changing conditions during wind turbine motion. LR combined model

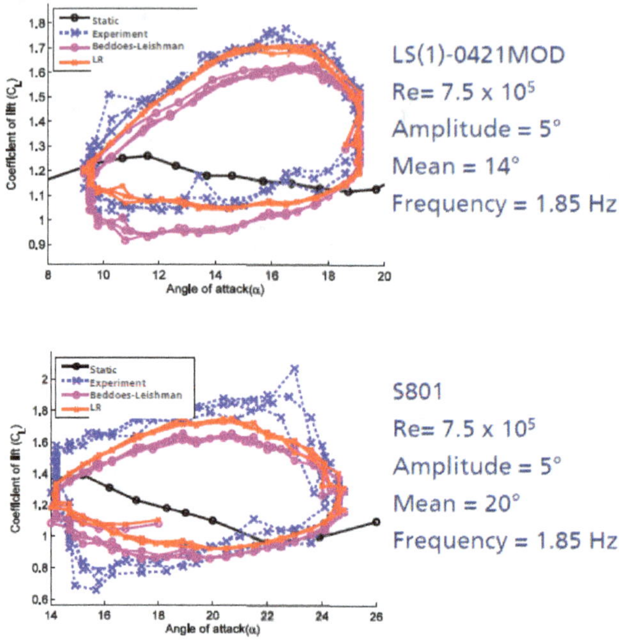

Figure 9.26 Beddoes-Leishman model implementation in LR-uBEM.[21,22]

accounts for real-time changes in TSR and inputs the changing lift and drag curves into LR-modified Beddoes-Leishman model. Figure 9.27 shows the difference between pre-applied stall delay and LR real-time stall delay model.

LR-uBEM model is optimized with the 2-equation Øye model which works in tandem with other unsteady models. Moreover, it is modified to suit the FOWT unsteady environment. The previous model was with limited amount of unsteady modeling of other effects as it was based on limited blade-pitching experiments, not rotor motions. Figure 9.28 shows the difference between the static and dynamic wake models in comparison with CFD.

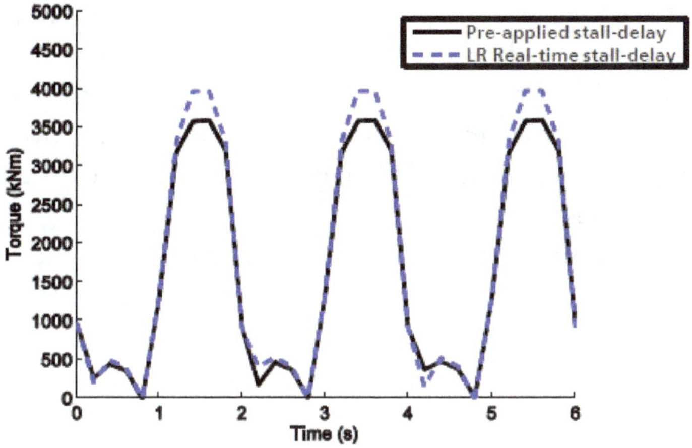

Figure 9.27 NREL 5MW in sinusoidal surge motion: frequency = 0.5 Hz, amplitude = 6 m.[21,22]

Figure 9.28 NREL 5MW in dynamic wake model implementation.[21,22]

A vortex ring (toroidal vortex) is a torus-shaped vortex in which flow spins around an imaginary axis that forms a closed loop. Vortex rings can be formed by a spinning rotor. These have been thoroughly studied by the helicopter industry. A vortex ring formation around the rotor of helicopter is known as vortex ring state (settling with power). This is a very dangerous condition of flow recirculation that can result in significant loss of lift and a potential accident.

A vortex ring state can develop in FOWTs as they switch between turbine (extracting energy) and propeller (inducing energy into the flow) operation states due to the system's motion (particularly in surge, pitch and yaw). When operating in a vortex ring state, the turbine will experience large, periodic changes in loads and, more importantly, it will lead to the breakdown of momentum balance equations, which form the base of the majority of the industry's simulation codes. A detailed description of the breakdown of moment theory in vortex ring state is given by Sebastian.[25]

Using FAST, Sebastian[26] analyzed how much time is spent in vortex ring state under three different operating conditions (below-rated, rated and above-rated) for monopile and three FOWTs (barge, spar buoy and TLP) with the NREL 5MW wind turbine and no dynamic inflow. For the OC3-Hywind spar the time spent in vortex wind state differs significantly between the operating conditions. Below-rated it varied from near zero close to the blade root to more than 50% at the blade tip, at rated from 0.1% at the root to slightly more than 1% at the blade tip, and for above-rated it was less than 0.25% overall for the whole blade.

Wake simulation in FAST (AeroDyn) is modeled using either BEM or GDW. Both of these use momentum theory to compute induced velocities at the rotor, meaning that vortex ring state cannot be accounted for when using FAST.

Sebastian and Lackner[27] identified pitch and yaw DoFs of OC3-Hywind as the most likely to lead to unsteady loads compared to monopile.

The LR-uBEM code is made possible to predict almost all kinds of wake states, even for vortex ring state and propeller brake wake state as it is a concern at high axial induction ($a > 1$). Moreover, the propeller has issues with negative effective wind speeds ($a < 0$). At present, momentum models do not account for these states. But LR-uBEM extends the Glauert correction to these wake states. Figure 9.30 shows the full-scale 5MW

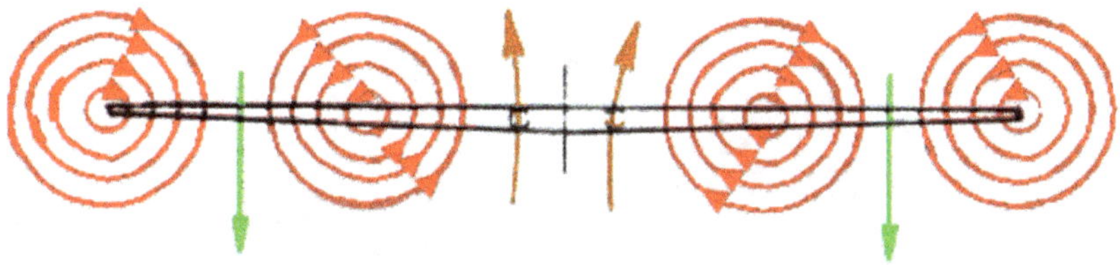

Figure 9.29 Vortex ring state on a helicopter. Reproduced with permission from https://www.copters.com/aero/settling.html.

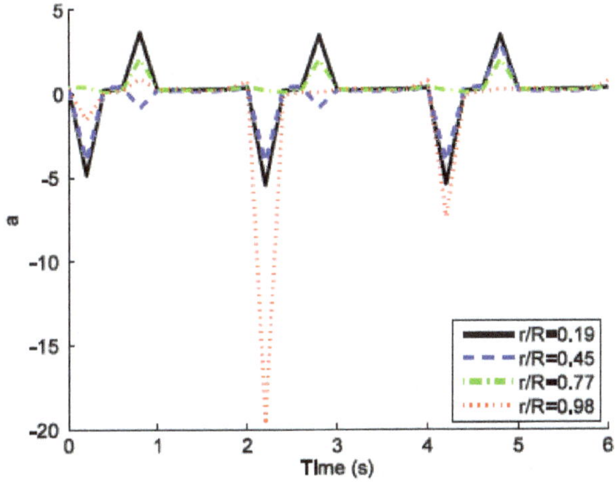

Figure 9.30 5MW NREL turbine rotor blade's axial induction factor at various stations for the sinusoidal surge motion of 0.5 Hz frequency and 6 m amplitude.[21,22]

NREL turbine rotor blade's axial induction factor at various stations for the sinusoidal surge motion of 0.5 Hz frequency and 6 m amplitude.

The experimental conditions and the scaled rotor properties were used as inputs into LR-uBEM, which is executed to compare the thrust and torque for both the static and unsteady cases. The airfoil aerodynamic data at model Re were obtained using XFOIL,[28] and extended to −180 and 180 using the NREL AirfoilPrep Microsoft Excel tool.[29]

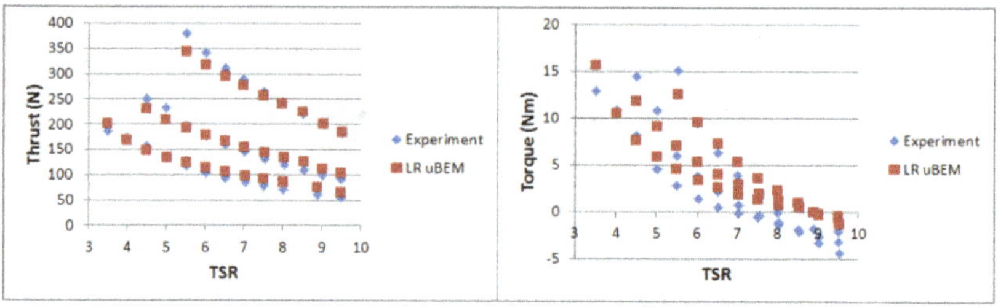

Figure 9.31 Thrust (left) and torque (right) against TSR for experimental and LR-uBEM cases.

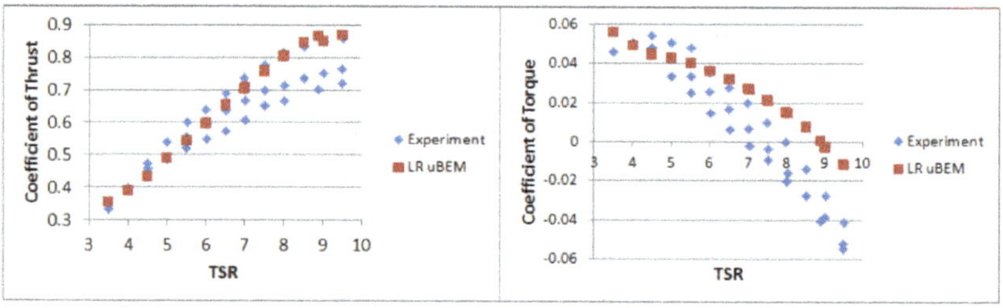

Figure 9.32 Coefficients of thrust (left) and torque (right) against TSR for experimental and LR-uBEM cases.

The static experimental scenarios were simulated in LR-uBEM. The distribution of torque and thrust against TSR predicted by LR-uBEM are compared to the experimental results in Figure 9.31.

The thrust distribution generally matched the experimental results, while some minor differences in torque were observed. This is likely to be due to the usage of XFOIL to generate the lift and drag data for each model airfoil, which is not ideal especially in the experimental Re, where flow separation effects are expected to be significant. It is probable that the drag distribution was not well predicted using XFOIL. The distributions of the experimental coefficients of thrust and torque against TSR are compared to the LR-uBEM predicted results in Figure 9.32.

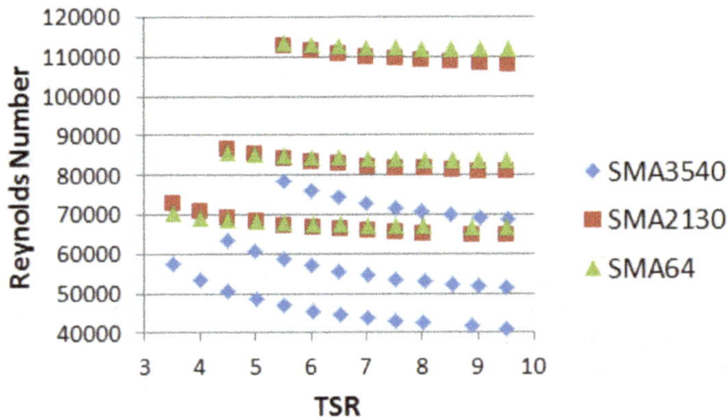

Figure 9.33 Averaged Reynolds numbers for each airfoil used in the scaled model rotor at each static test case.

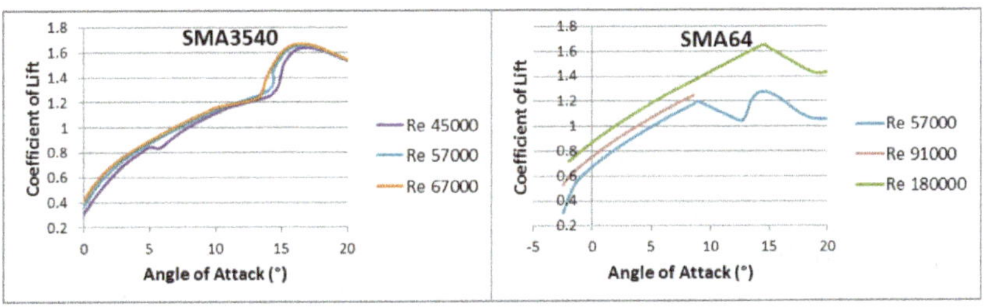

Figure 9.34 Lift coefficient curves for SMA3540 (left) and SMA64 (right) at different Reynolds numbers.

Three distinct C_T and C_Q against TSR curves can be observed from the distribution of experimental thrust and torque coefficients, while the LR-uBEM computed values merge into one curve, which is expected. A possible reason for this is that the three different rotational speeds used in the experiment resulted in three distinct bands of Re, as shown in Figure 9.33.

The changes in airfoil lift coefficients in these ranges of Re are shown for the SMA3540 and SMA64 airfoils in Figure 9.34.

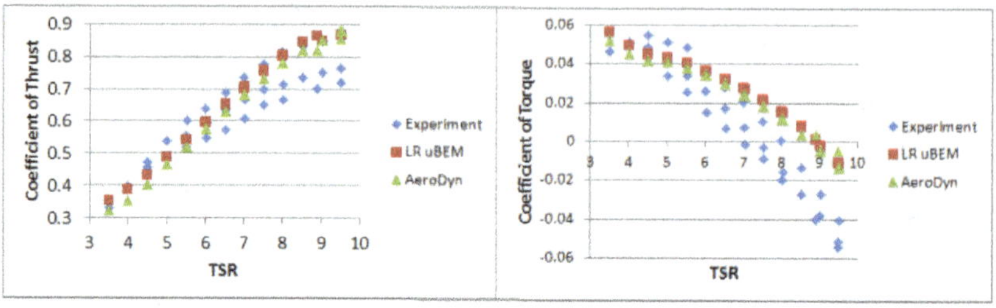

Figure 9.35 Coefficients of thrust (left) and torque (right) against TSR for experimental, LR-uBEM and LR-AeroDyn cases.

Significant differences can be found in the airfoil lift coefficients at different Re, suggesting that multiple tables are required for an assessment of the accuracy of the LR-uBEM code. The same computations were performed in LR-AeroDyn,[30] shown in Figure 9.35.

The similarities between LR-AeroDyn and LR-uBEM suggest that the assumption about the inaccuracy in airfoil cross-sectional coefficient estimation is valid, and an in-depth analysis of the airfoil aerodynamics is further required.

9.4 Results and Discussion of Unsteady State Test Cases

Comprehensive static and unsteady experimental scenarios were performed and results were obtained for numerical evaluation and validation. For the unsteady experimental scenario cases, the results from the last two cycles of six cycles of surge motion simulation were extracted for comparison and further analysis. This is to ensure that all the initial transient factors were eliminated. Results of CFD, LR-AeroDyn and LR-uBEM model computations were compared against the experimental results to evaluate their ability in predicting the aerodynamic forces due to unsteady effects. The variation in time mean thrust values of different numerical simulation methods would determine the capability and limitation to predict the unsteady aerodynamic loads on the scaled rotor.

Although thrust is the only significant parameter for the comparison between the numerical methods in this case (as the scaled rotor is designed against lift curve or thrust of full-scale reference rotor), torque comparisons will further establish their capabilities.

9.4.1 *Hydrodynamic Thrust*

One of the main aerodynamic performance parameters is thrust force which is used for axial induction factor computation in BEM and unsteady-based BEM methodologies. Hence, thrust force was chosen for the model evaluation and validation purposes. Moreover, the axial induction factor was used to predict the type of wind turbine operating wake states as well. Validity of BEM models are pretty much dependent on wind turbine operating wake states and appropriate correction factors involved. A detailed investigation on thrust force prediction by numerical codes and comparison is made against the experimental prediction for 8.5, 7.0 and 6.0 TSR unsteady scenarios. Numerical simulation of the unsteady scenarios with LR-AeroDyn and LR-uBEM are typically executed for a few minutes to complete each scenario, whereas each CFD simulation case requires 34 hours with 36 core parallel processing workstations.

In order to understand the unsteady flow behavior at the below-rated wind speed which is about 16% lower than the rated wind speed (11.4 m/s) of full-scale FOWT, scaled rotor was subjected to experimental and numerical simulations at 8.5 TSR. See Figure 9.36. With a constant angular speed of 14 rad/s for the different prescribed surge amplitude frequency motions the experiments were carried out at a constant primary carriage motion of 0.824 m/s. Owing to the unsteadiness triggered by prescribed surge motion, both thrust and torque fluctuate with time. Lower amplitude surge motion-based numerical models were over-predicting mean thrust force within 5% error against experimental results as shown in Figure 9.36(h). Higher amplitude surge motion-based

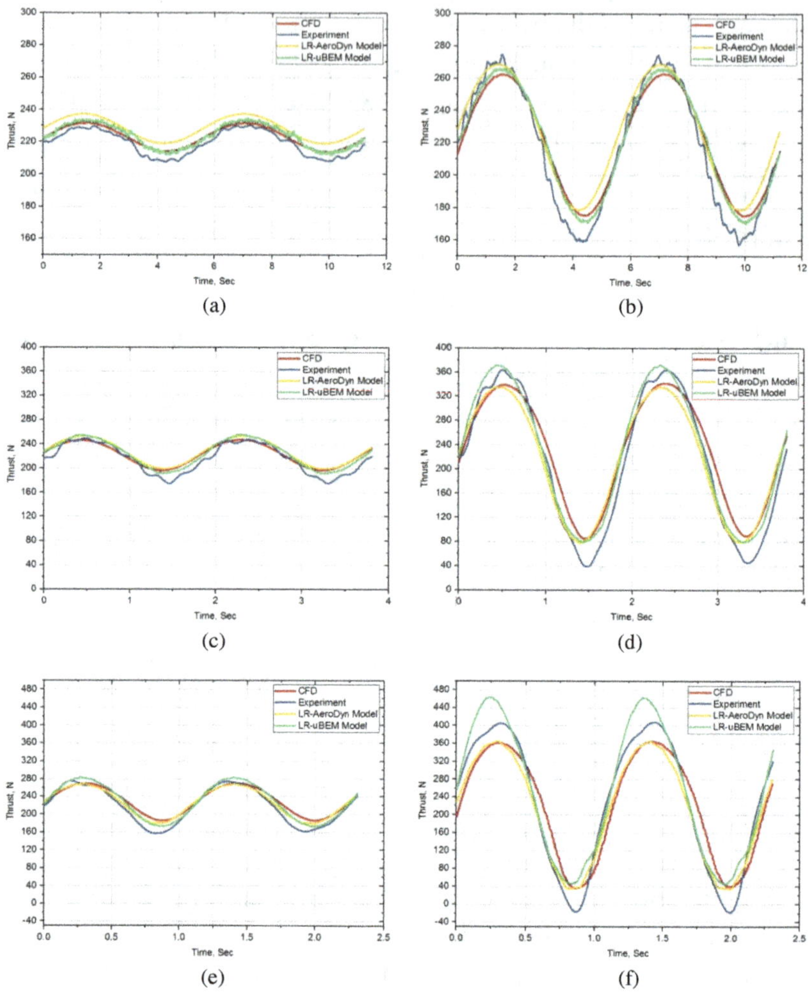

Figure 9.36 TSR 8.5 cases: (a) SFA1 — Low f & lower A; (b) SFA2 — Low f & higher A; (c) SFA3 — Medium f & lower A; (d) SFA4 — Medium f & higher A; (e) SFA5 — High f & lower A; (f) SFA6 — High f & higher A; (g) Mean trust — comparison for all TSR 8.5 cases; (h) Mean trust — percentage error comparison for all TSR 8.5 cases.

numerical models were predicting mean thrust within 10% error except at the high surge frequency motion. Mean thrust slightly increases as surge motion frequency and amplitude increase. As in Figure 9.36(a) to (f), backward and forward motions in surge oscillations case the

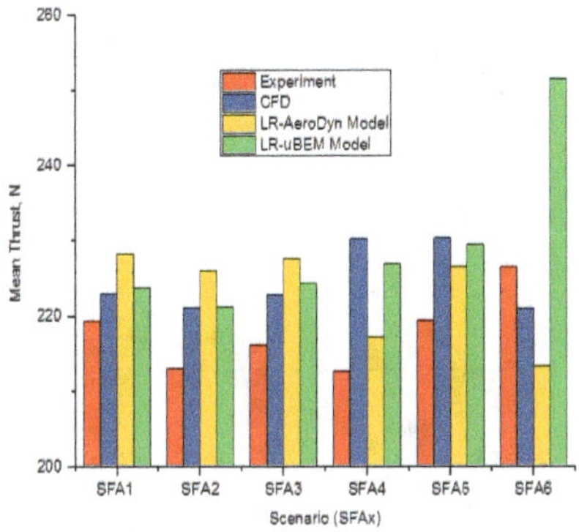

(g)

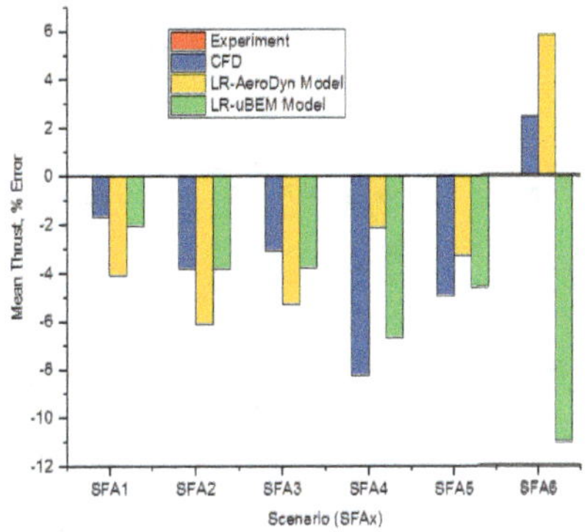

(h)

Figure 9.36 (*Continued*)

maximum thrust to vary from almost 8–100% higher than the mean value while the minimum thrust is about 8–100% lower at TSR 8.5. It is essential to consider the large difference between the extremes for structural stress and related fatigue issues during full-scale design of wind turbine rotors and its pitching control for floating applications. The possible sources of error: (1) Experiments — It is estimated to be around 2–4% error from various sources of uncertainties such as measurement system, random error and systematic error as the model is very similar to Doman *et al.*[31] and yet to be quantified in detail. (2) CFD — Numerical discretization, error due to mesh deformation techniques and time step selection for high-frequency and larger amplitude motions. (3) LR-AeroDyn and LR-uBEM — Load computations are completely based on the design data of scaled 2D airfoil lift and drag data, not on scaled 2D airfoil experimental data and 3D correction errors. A detailed investigation is needed to quantify the uncertainties of these methods.

We carry out an experiment at the prescribed surge amplitude frequency motions of 7.0 TSR, 14 rad/s inside the tow tank water when the primary carriage constant motion is 1 m/s. See Figure 9.37. The same scenario was simulated for all three chosen numerical simulation tools. Lower amplitude surge frequency motion-based numerical models were

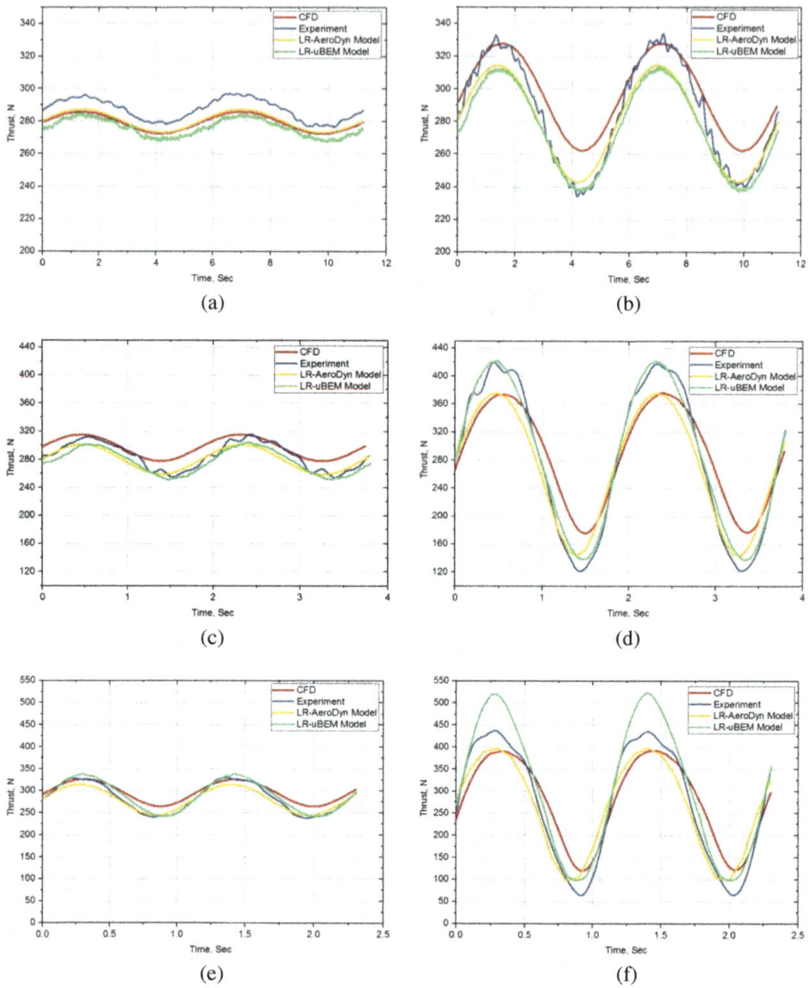

Figure 9.37 TSR 7 cases: (a) SFA7 — Low f & lower A; (b) SFA8 — Low f & higher A; (c) SFA9 — Medium f & lower A; (d) SFA10 — Medium f & higher A; (e) SFA11 — High f & lower A; (f) SFA12 — High f & higher A; (g) Mean trust — comparison for all TSR 7 cases; (h) Mean trust — percentage error comparison for all TSR 7 cases.

predicting mean thrust force within 5% error against experimental results as shown in Figure 9.37(h). Higher amplitude surge frequency motion-based numerical models were predicting mean thrust within 5% error except at the high-frequency surge motion, and close to 10% error at

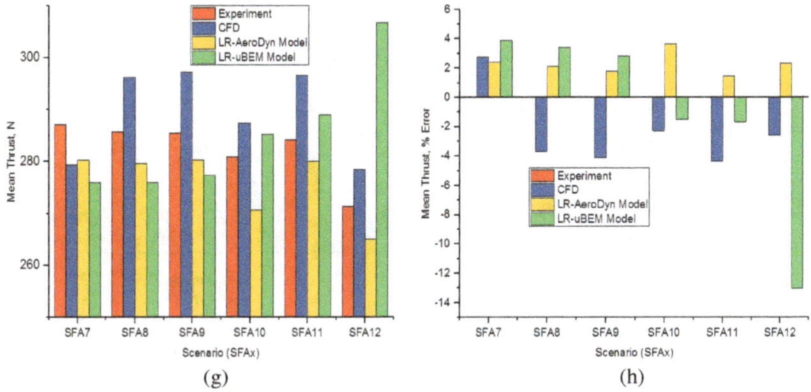

Figure 9.37 (*Continued*)

higher amplitude scenario. In experiments, mean thrust gradually reduces as surge motion frequency and amplitude increase, whereas mean thrust gradually increases in LR-uBEM. Mixed trend is found in both LR-AeroDyn and CFD. As in Figure 9.37(a) to (f), backward and forward motions in surge oscillations cause the maximum thrust to vary between 4–110% higher than the mean value while the minimum thrust is about 4–110% lower at TSR 7.0.

In order to understand the unsteady flow behavior above the rated wind speed (13.2 m/s), which is 16% higher than the rated wind speed of full-scale FOWT, scaled rotor is subjected to experimental and numerical simulations at 6.0 TSR, 14 rad/s for the different prescribed surge amplitude frequency motions when the primary carriage velocity is 1.167 m/s. See Figure 9.38. As shown in Figure 9.38(h), all surge frequency motions were under-predicting the mean thrust force within 8% error against experimental results except at the high-frequency motion for higher amplitude scenario. Mean thrust slightly increases as surge motion frequency and amplitude increases. As in Figure 9.38(a) to (f), backward and forward motions in surge oscillations cause the maximum thrust to vary between 5–60% higher than the mean value while the minimum thrust is about 5–60% lower at TSR 6.0.

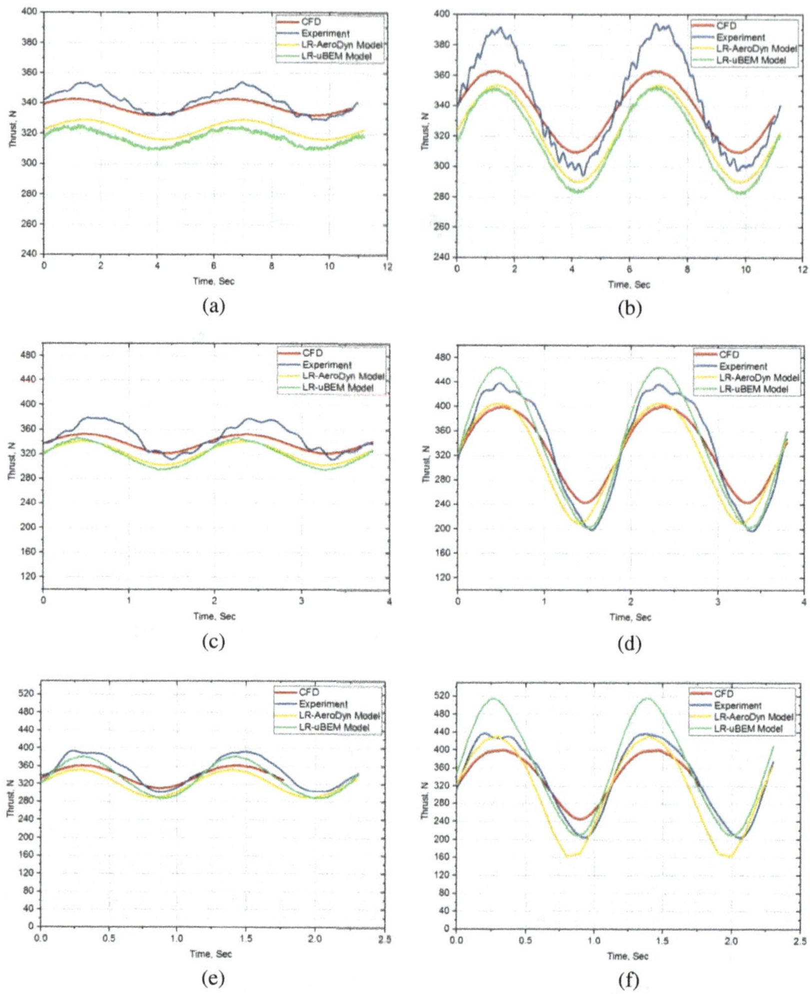

Figure 9.38 TSR 6 cases: (a) SFA13 — Low f & lower A; (b) SFA14 — Low f & higher A; (c) SFA15 — Medium f & lower A; (d) SFA16 — Medium f & higher A; (e) SFA17 — High f & lower A; (f) SFA18 — High f & higher A; (g) Mean trust — comparison for all TSR 6 cases; (h) Mean trust — percentage error comparison for all TSR 6 cases.

The mean thrust variation and the error for all three TSR scenarios were compared together in Figures 9.39 and 9.40. CFD over-predicts the mean thrust in all the cases except in the lower TSR (TSR 6) scenarios.

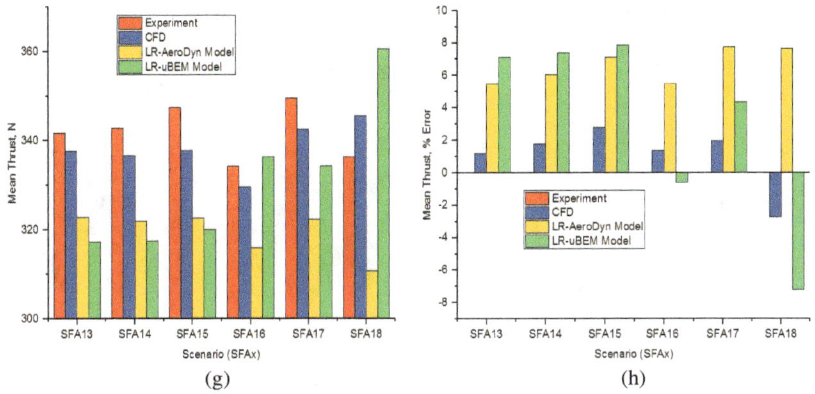

Figure 9.38 (*Continued*)

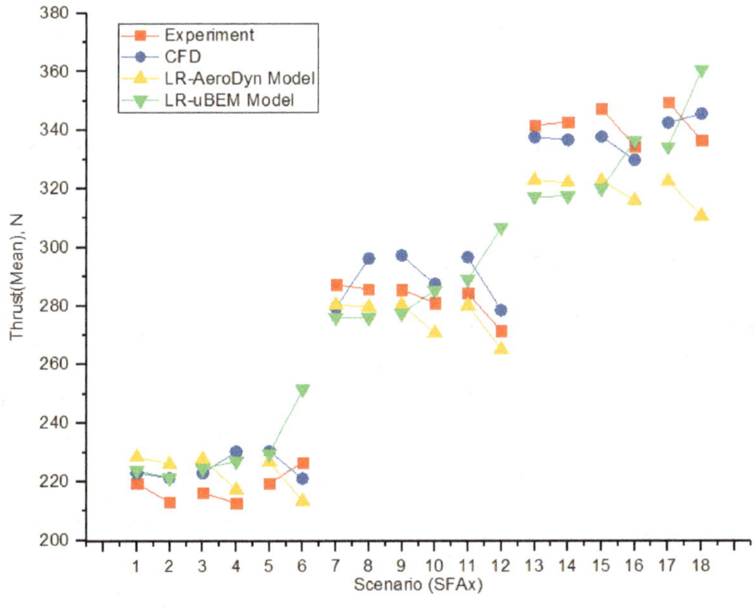

Figure 9.39 Mean thrust comparison for all 18 scenarios.

In order to have a better understanding on the ranges of values in cyclic variation of unsteady aerodynamic effects, Appendix A1 provides box plots with indicated values in 75% and 25% split.

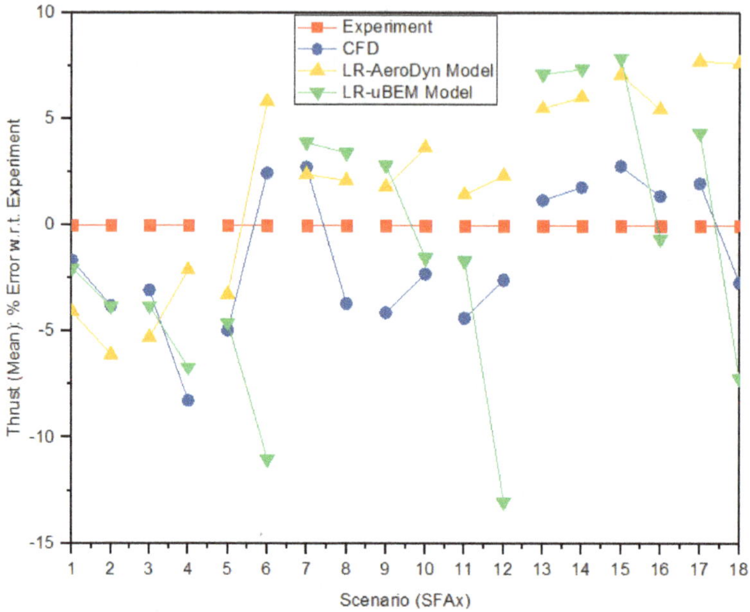

Figure 9.40 Mean thrust comparison: percentage error with respect to experiment.

9.4.2 *Hydrodynamic Torque Comparison*

As power is calculated from the torque values, the torque predictions are compared between the numerical models, though the scaled rotor design is based on the lift curve. In order to understand the unsteady flow behavior at the below-rated wind speed which is about 16% lower than the rated wind speed (11.4 m/s) of full-scale FOWT, scaled rotor was subjected to experimental and numerical simulations at 8.5 TSR. See Figure 9.41. With a constant angular speed of 14 rad/s for the different prescribed surge amplitude frequency motions the experiments were carried out at a constant primary carriage motion of 0.824 m/s. Mean torque value predictions by all numerical models are close to each other, but differ by a maximum of 420% error when compared with experimental measurements. This is a clear indication that the low tow speed (lower wind speed corresponding to full-scale 5MW wind turbine)

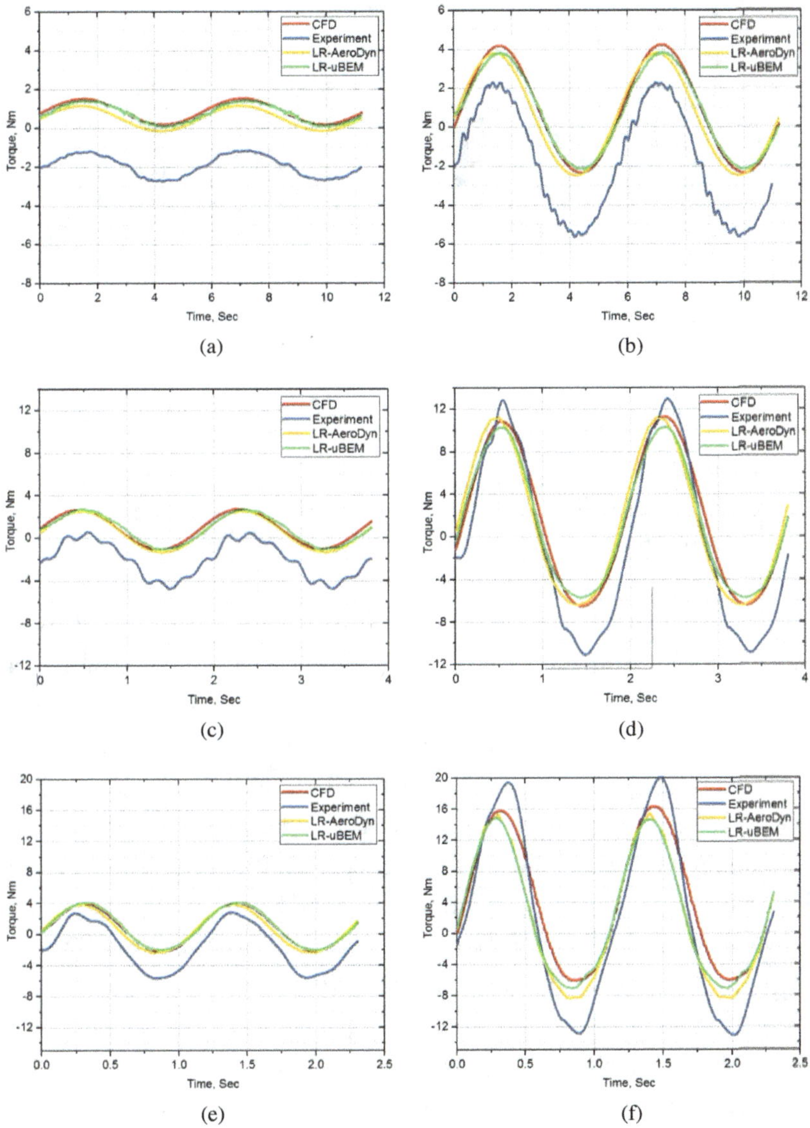

Figure 9.41 TSR 8.5 cases: (a) SFA1 — Low f & lower A; (b) SFA2 — Low f & higher A; (c) SFA3 — Medium f & lower A; (d) SFA4 — Medium f & higher A; (e) SFA5 — High f & lower A; (f) SFA6 — High f & higher A; (g) Mean torque — comparison for all TSR 8.5 cases; (h) Mean torque — percentage error comparison for all TSR 8.5 cases.

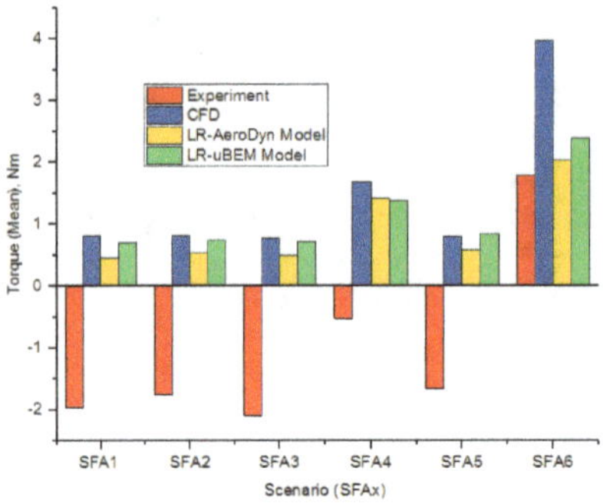

(g)

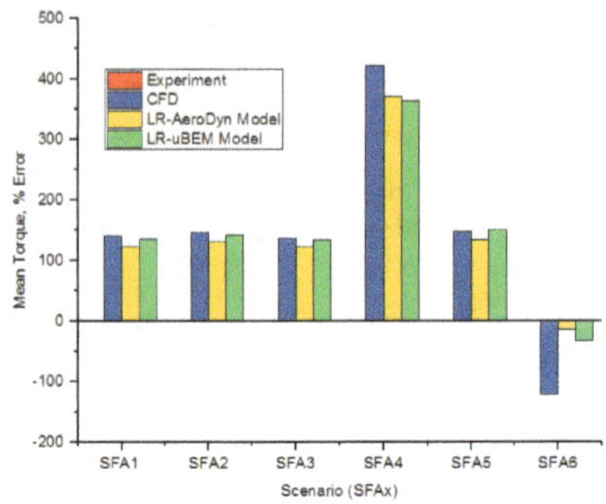

(h)

Figure 9.41 (*Continued*)

introduces considerable experimental error. As in Figure 9.41(a) to (f), backward and forward motions in surge oscillations changes are given.

For the study at TSR 7.0 with wind speed (11.4 m/s) of full-scale FOWT, scaled rotor experimental and numerical model results were compared for the prescribed surge amplitude frequency motions at 7.0 TSR, 14 rad/s rotational speed at primary carriage constant motion of 1 m/s. See Figure 9.42. Mean torque value predictions by all numerical models are close to each other, but differ by a maximum of 45% error compared with experimental measurements. As in Figure 9.42(a) to (f), backward and forward motions in surge oscillations cause the maximum torque to vary between 10–150% higher than the mean value while the minimum torque is about 10–150% lower at TSR 7.0.

At 13.2 m/s, which is 16% higher than the rated wind speed of full-scale FOWT, scaled rotor is subjected to experimental and numerical simulations at 6.0 TSR, 14 rad/s for the different prescribed surge amplitude frequency motions when the primary carriage velocity is 1.167 m/s. See Figure 9.43. As shown in Figure 9.43(h), all surge frequency motions were under-predicting the mean torque within 30% error against experimental results. Mean torque slightly increases as surge motion frequency

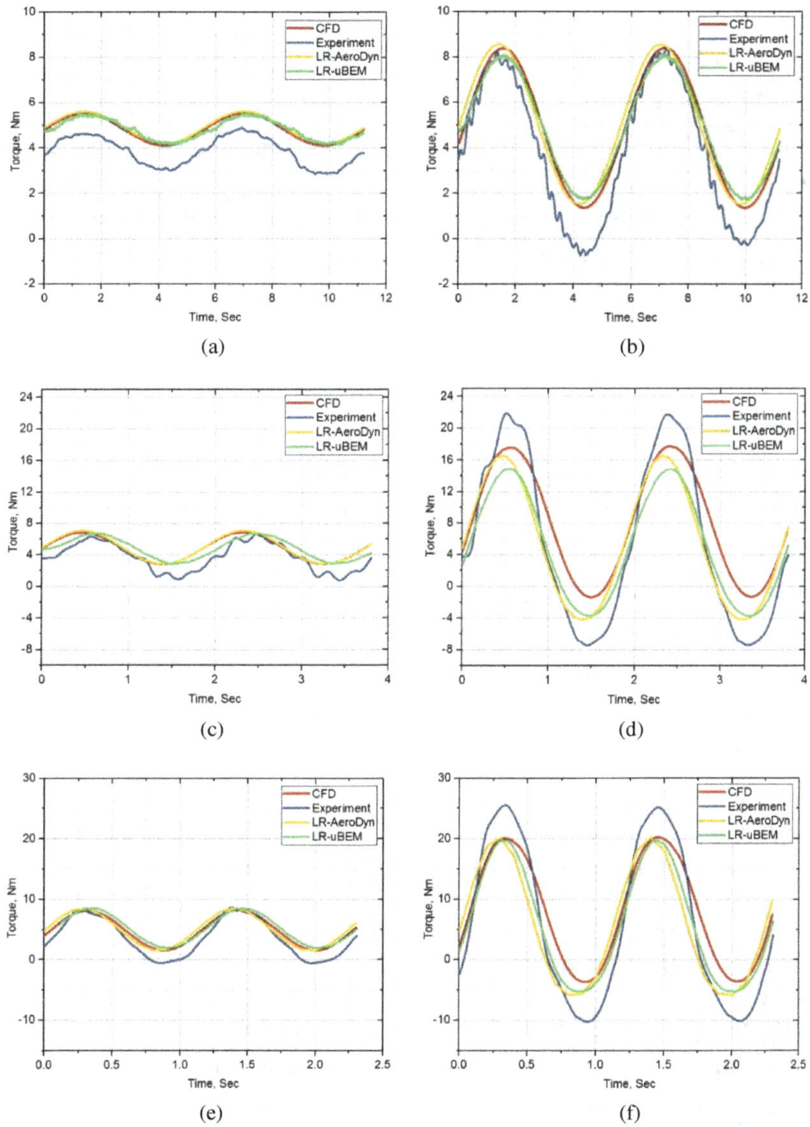

Figure 9.42 TSR 7 cases: (a) SFA7 — Low f & lower A; (b) SFA8 — Low f & higher A; (c) SFA9 — Medium f & lower A; (d) SFA10 — Medium f & higher A; (e) SFA11 — High f & lower A; (f) SFA12 — High f & higher A; (g) Mean torque — comparison for all TSR 7 cases; (h) Mean torque — percentage error comparison for all TSR 7 cases.

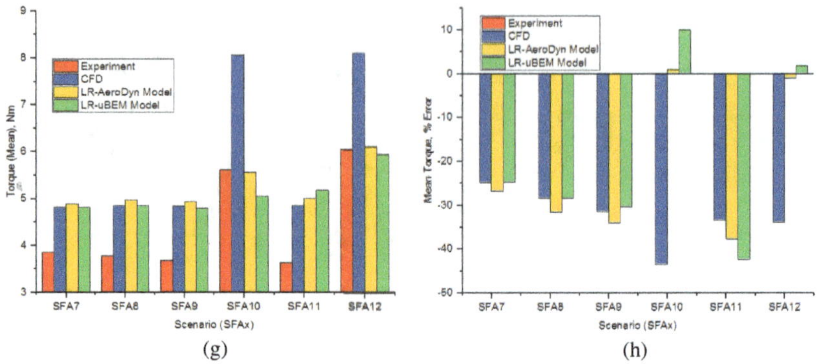

Figure 9.42 (*Continued*)

and amplitude increase. As in Figure 9.43(a) to (f), backward and forward motions in surge oscillations cause the changes in maximum and minimum torque ranges at TSR 6.0.

The mean torque variation and the error for all three TSR scenarios are compared together in Figures 9.44 and 9.45. CFD overestimates the mean torque in almost all the TSR cases.

As shown in Figures 9.46 and 9.47, the cyclic variation prediction by all the numerical codes are close to each other.

9.4.3 *Evaluation of Wind Turbine Operating State*

As shown in Figure 9.48, when the axial induction factor exceeds 0.4, the turbine rotor will no longer be in windmill state and will be operating in turbulent wake state until the induction factor exceeds 1. Two wake states are predominantly developed during normal wind turbine operation based on the drop of the wind speed in the wake. The primary state is the windmill state during medium to high winds, in which the rotor follows the momentum theory with coefficient of thrust as indicated in Eq. (9.3).

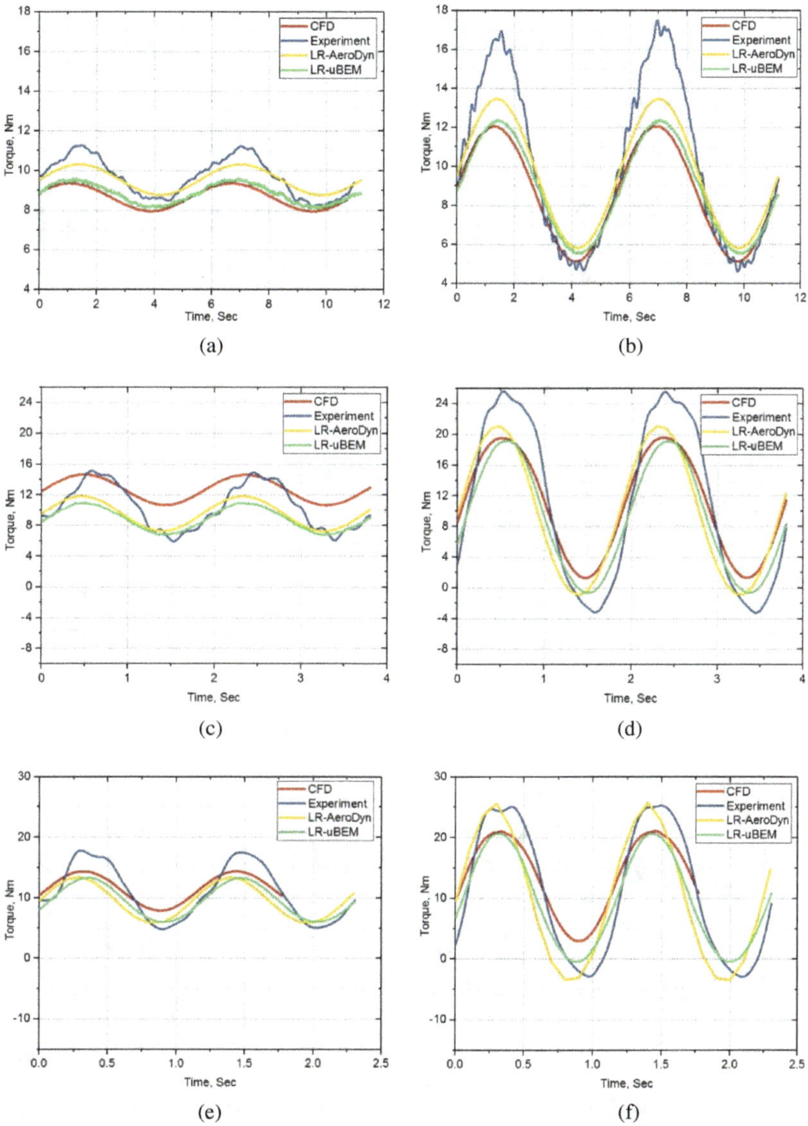

Figure 9.43 TSR 6 cases: (a) SFA13 — Low f & lower A; (b) SFA14 — Low f & higher A; (c) SFA15 — Medium f & lower A; (d) SFA16 — Medium f & higher A; (e) SFA17 — High f & lower A; (f) SFA18 — High f & higher A; (g) Mean torque — comparison for all TSR 6 cases; (h) Mean torque — percentage error comparison for all TSR 6 cases.

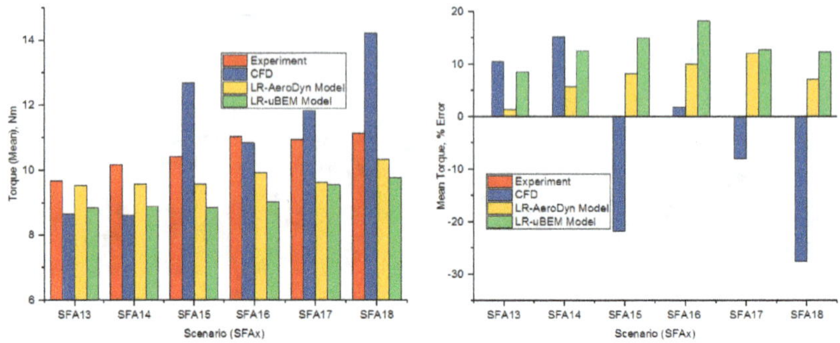

(g) (h)

Figure 9.43 (*Continued*)

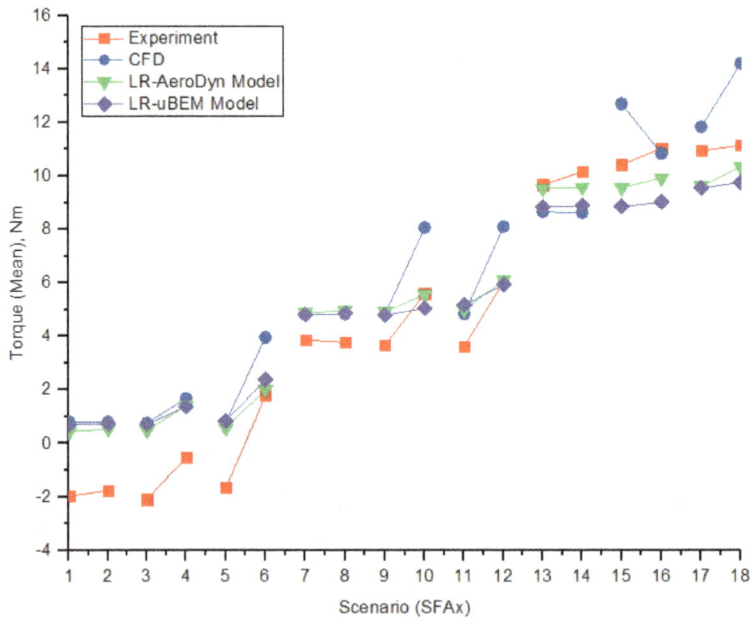

Figure 9.44 Mean torque comparison for all 18 scenarios.

At lower wind speeds, the wind velocity between the freestream wind and the wake differs by a higher magnitude. This leads to the low-energy wake with recirculation leading to the turbulent wake state. The free shear layer between the wake and freestream is unstable, producing eddies that carry

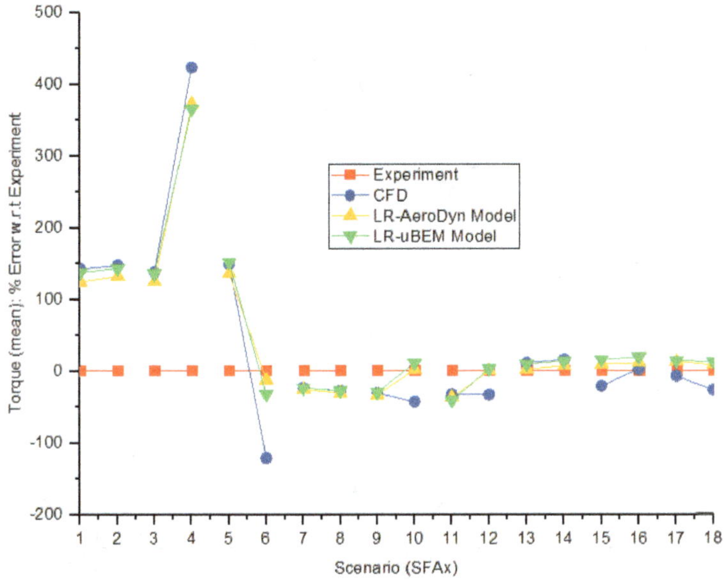

Figure 9.45 Mean torque comparison: percentage error with respect to experiment.

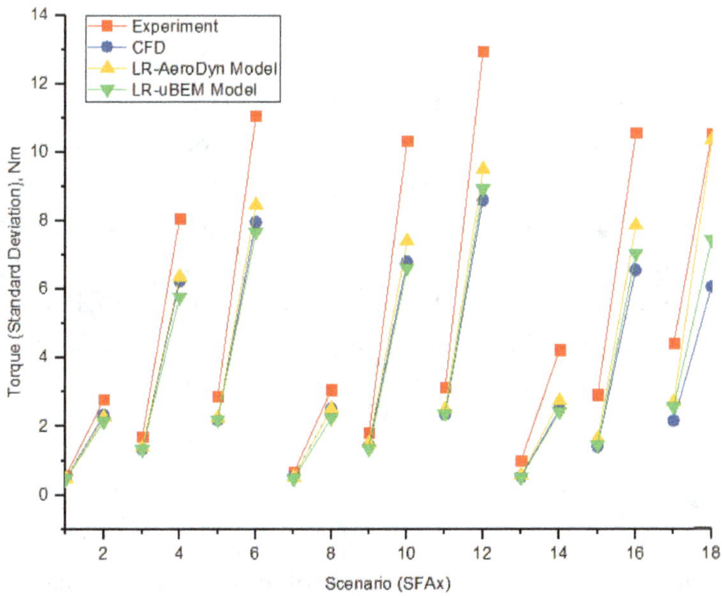

Figure 9.46 Comparison of standard deviation of torque with changes in amplitude in surging cycles.

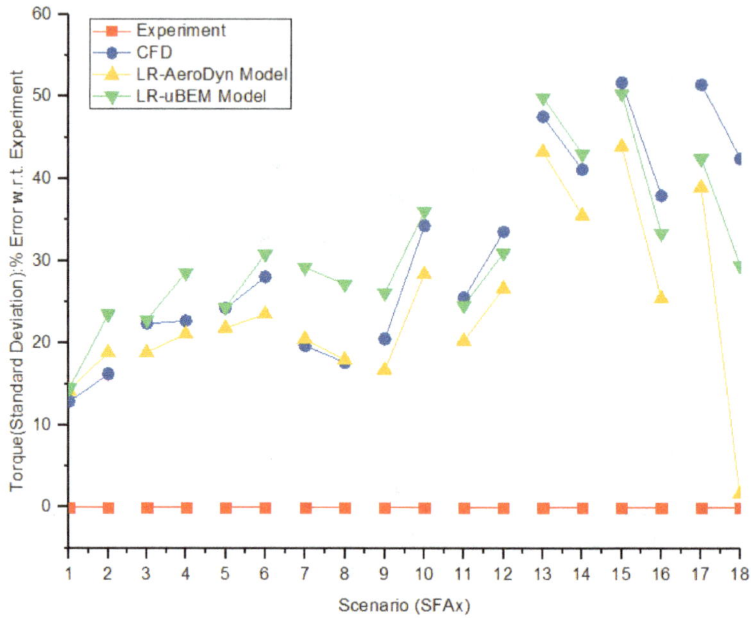

Figure 9.47 Comparison of standard deviation of torque with changes in amplitude in surging cycles: in percentage error.

momentum from the freestream into the wake. Hence the BEM theory following the idealized momentum principle is not able to predict the drop in the thrust force. As rated wind speed differs for each turbine, the occurrence of the turbulent wake state can be represented by using an axial induction factor. Based on the experimental data shown in Figure 9.48, Glauert recommended an equation to model the relationship between C_t and axial induction factor in the turbulent wake state.

$$Ct = 4aF\left(1-\left(\frac{1}{4}\right)(5-3a)a\right),\ a > \frac{1}{3} \tag{9.16}$$

where F is the tip-loss factor.

Even with the inclusion of F in Eq. (9.16), discontinuities nonetheless existed at $a = 0.4$ between the momentum theory and Glauert equation at values of F other than 1. Correction has to be introduced as suggested by Buhl in Eq. (9.17), that can eliminate the discontinuities at $a = 0.4$ and have a value of $C_t = 2$ when $a = 1$, for all values of F.

$$C_t = \left(\frac{8}{9}\right) + (4F - \left(\frac{40}{9}\right)a + \left(\frac{50}{9} - 4F\right)a \times a \tag{9.17}$$

The current study bolsters the fact that the surge cyclic motions are pushing the rotor under investigation to the turbulent wake state of SFA10 scenario as depicted in Figure 9.48. The axial induction factor for the elemental sections, which crosses 0.4, are given in Appendix A2 for all other scenarios. From the scaled rotor blade element number

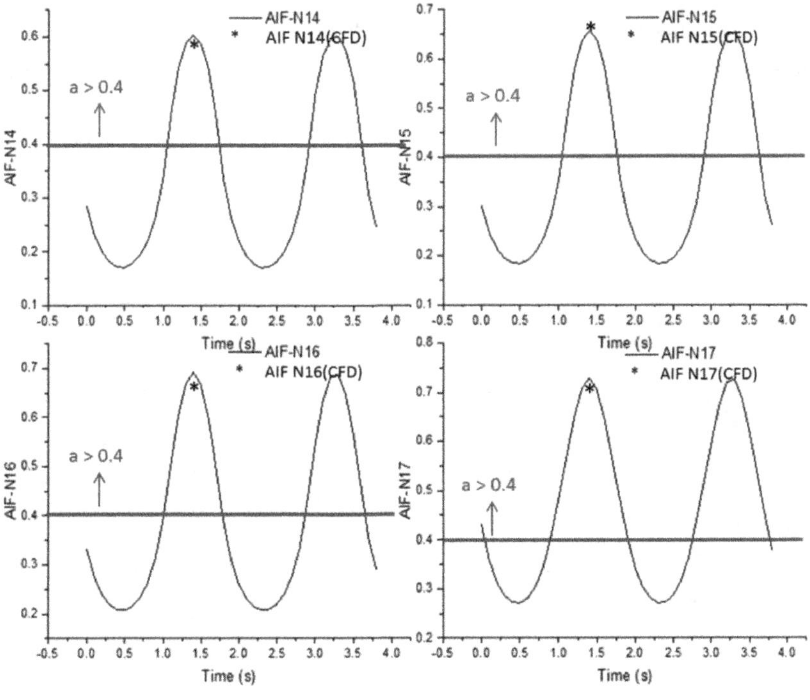

Figure 9.48 Axial induction factor of elements 14–17 of the scaled rotor for the SFA10 scenario with LR-AeroDyn model simulation and corresponding point-wise CFD-based axial induction factor comparison at 1.25 s.

14 to 17 (the last four elements up to scaled rotor tip of total 17 elements as in Table 9.1), the turbulent wake state is clearly identified by the axial induction factor. CFD-based induction factor derivation by Sivalingam *et al.*[2] is obtained at 1.25 s of the last two cycles and compared against LR-AeroDyn model prediction as shown in Figure 9.48. LR-AeroDyn prediction is within 6% error on these elemental nodes when compared to CFD predictions. Hence, BEM-based engineering models such as LR-AeroDyn and LR-uBEM are able to predict the surge motion responses on the rotor loading reasonably well.

CFD simulation results are exported to Ansys CFD Post to visualize the velocity and vorticity contour profiles. These profiles are plotted in Matlab, which clearly demonstrate how the scaled rotor is interacting with its own wake and the near wake distortion by surge motion frequencies. Figures 9.49 to 9.52 show the velocity contour profiles.

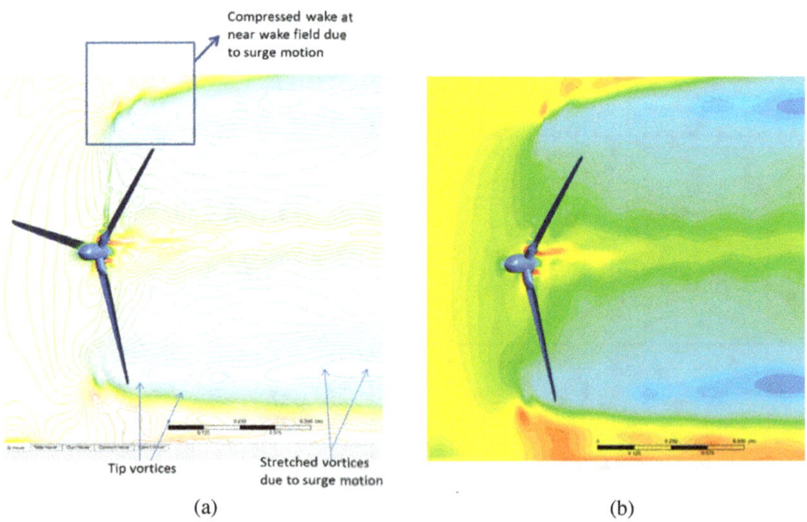

Figure 9.49 SFA10 mean rotor position at the end of 6th cycle of surge motion. (a) Line plot; (b) Contour plot showing tip vertices and compressed and elongated wake in the near wake field by CFD.

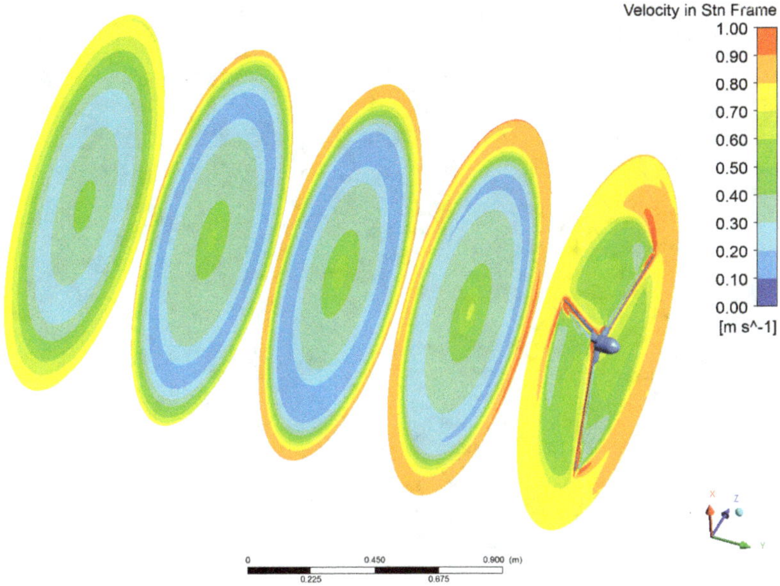

Figure 9.50 SFA10 mean rotor position at the end of 6th cycle of surge motion, near wake field velocity contour plots at 0.5 m intervals from rotor plane in the rotating domain by CFD.

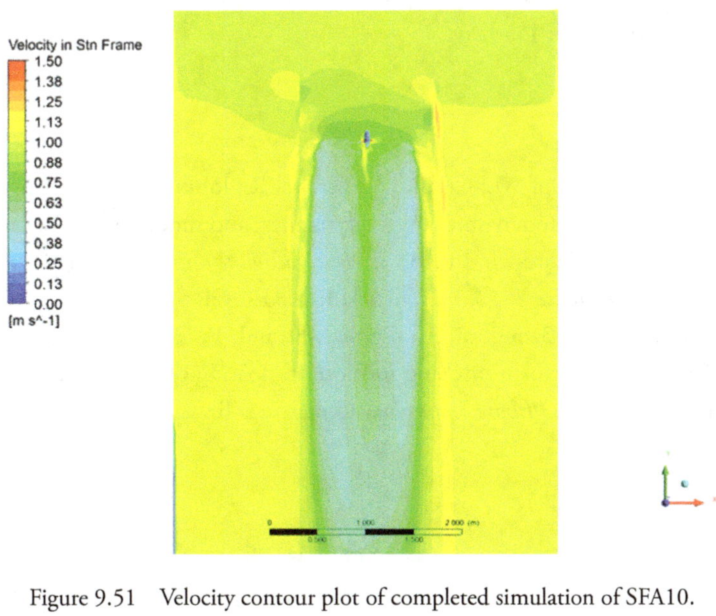

Figure 9.51 Velocity contour plot of completed simulation of SFA10.

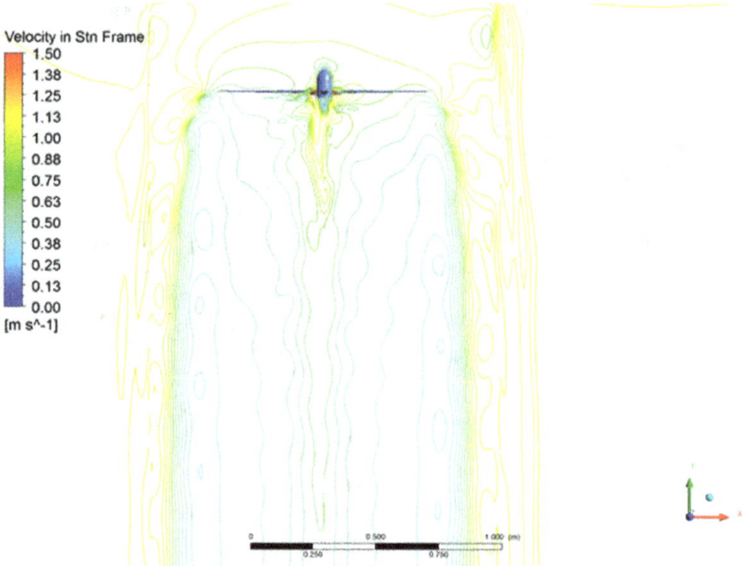

Figure 9.52 Velocity contour line plot of completed simulation of SFA10, showing blade tip vorticity in the downstream.

In addition, these contours display the surge oscillation behavior and corresponding rotor loading.

References

1. Matha, D.; Schlipf, M.; Cordle, A.; Pereira, R.; Jonkman, J. Challenges in simulation of aerodynamics, hydrodynamics, and mooring-line dynamics of floating offshore wind turbines. *International Offshore and Polar Engineering Conference 2011.* Hawaii **2011**, NREL/CP-5000-50544.
2. Sivalingam, K.; Bahuguni, A.; Gulman-Strand, J.; Davies, P.; Nguyen, V. T. Effects of platform pitching motion on floating offshore wind turbine (FOWT) rotor. *Offshore Technology Conference.* Texas **2015**, paper number OTC-25962-MS.

3. Martin, S; Day, S.; Gilmour, C. B. Rotor scaling methodologies for small scale testing of floating wind turbine systems. *34th International Conference on Ocean, Offshore and Arctic Engineering.* Newfoundland **2015**, paper number OMAE2015-41599.
4. Liu, J. C.; Sun, K.; Zhang, J. H.; Zhao, Y. N.; Pan, M. H. Effects of surge on rotor aerodynamics of offshore floating wind turbine. *Advanced Materials Research* **2014**, *1070–1072*, 177–182.
5. de Ridder E.-J.; Otto, W.; Zondervan G.-J., Huijs, F.; Vaz, G. Development of a scaled-down floating wind turbine for offshore basin testing. *33rd International Conference on Ocean, Offshore and Arctic Engineering.* San Francisco **2014**, paper number OMAE2014-23441.
6. Martin, H. R.; Kimball, R. W.; Viselli, A. M.; Goupee, A. J. Methodology for wind/wave basin testing of floating offshore wind turbines. *Journal of Offshore Mechanics and Arctic Engineering* **2014**, *136*, 021902.
7. Burdett, T. A.; van Treuren K. W. Scaling small-scale wind tubrines for wind tunnel testing. *Proceedings of ASME Turbine Expo 2012*, **2012**, 811–820.
8. Whelan, J. I.; Stallard, T. Arguments for modifying the geometry of a scale model rotor. *Proceedings of 9th European Wave and Tidal Energy Conference.* Southampton **2011**.
9. Jonkman, J.; Butterfield, S.; Musial, W.; Scott, G. Definition of a 5-MW reference wind turbine for offshore system development. *Technical Report* **2009**, NREL/TP-500-38060.
10. Kooijman, H. J. T.; Lindenburg, C.; Winkelaar, D.; van der Hooft, E. L. DOWEC 6 MW Pre-Design: Aero-elastic modelling of the DOWEC 6 MW pre-design in PHATAS. *Technical Report* **2003**, ECN, Petten, Netherlands.
11. Martin, S.; Day, A. H. A multi-point performance matched aerofoil design algorithm for a scaled wind turbine rotor model. *Proceedings of the 50th 3AF International Conference on Applied Aerodynamics.* France **2015**.
12. Sivalingam, K.; Martin, S.; Singapore Wala, A. A. Numerical validation of floating offshore wind turbine scaled rotors for surge motion. *Energies* **2018**, *11*, 2578.

13. de Vaal, J. B.; Hansen, M. O. L.; Moan, T. Effect of wind turbine surge motion on rotor thrust and induced velocity. *Wind Energy* **2014**, *17*, 105–121.
14. Micallef, D.; Sant, T. Loading effects on floating offshore horizontal axis wind turbines in surge motion. *Renewable Energy* **2015**, *83*, 737–748.
15. Theodorsen, T. General theory of aerodynamic instability and the mechanism of flutter. *NACA Report* **1935**.
16. von Karman, T.; Sears, W. R. Airfoil theory for non-uniform motion. *Journal of the Aeronautical Sciences* **1938**, *5*, 379–390.
17. Bisplinghoff, R. L.; Ashley, H. *Principles of Aeroelasticity.* Dover Publications **1962**.
18. Leishman, J. G. *Principles of Helicopter Aerodynamics.* Cambridge University Press **2006**.
19. Jonkman, J.; Larsen, T.; Hansen, A.; Nygaard, T.; Maus, K.; Karimirad, M.; Gao, Z.; Moan, T.; Fylling, I.; Nichols, J.; Kohlmeier, M.; Verrgara, J. P.; Merino, D.; Shi, W.; Park, H. Offshore code comparison collaboration within IEA Wind Task 23: Phase IV results regarding floating wind turbine modeling. *Technical Report* **2010**, NREL/CP-500-47534.
20. Moriarty, P. J.; Hansen, A. C. *AeroDyn Theory Manual.* National Renewable Energy Laboratory **2005**.
21. Singapore Wala, A. A. *Aerodynamics Modelling of Floating Offshore Wind Turbines* (PhD Thesis). Nanyang Technological University **2017**.
22. Singapore Wala, A. A.; Ng, E. Y. K.; Bahuguni, A.; Srikanth, N. Quantification and modeling of the dynamic wake effect for floating offshore wind turbines. *Offshore Technology Conference.* Texas **2017**, paper number OTC-27799-MS.
23. Du, Z.; Selig, M. S. A 3-D stall delay model for horizontal axis wind turbine performance prediction. *ASME Wind Energy Symposium.* Nevada **1998**.

24. Singapore Wala, A. A.; Ng, E. Y. K.; Srikanth, N. A Beddoes-Leishman-type model with an optimization-based methodology and airfoil shape parameters. *Wind Energy* **2018**, *21*, 590–603.
25. Sebastian, T. *The Aerodynamics and Near Wake of an Offshore Floating Horizontal Axis Wind Turbine* (PhD Thesis). University of Massachusetts Amherst **2012**.
26. Sebastian, T.; Lackner, M. Offshore floating wind turbines — an aerodynamic perspective. *49th AIAA Aerospace Sciences Meeting including the New Horizons Forum and Aerospace Exposition.* Florida **2011**, AIAA 2011–720.
27. Sebastian, T.; Lackner, M. Analysis of the induction and wake evolution of an offshore floating wind turbine. *Energies* **2012**, *5*, 968–1000.
28. Drela, M. XFOIL: An Analysis and Design System for Low Reynolds Number Airfoils. In: Mueller, T. J., Ed., *Low Reynolds Number Aerodynamics*, Springer **1989**, pp. 1–12.
29. NWTC — AirfoilPrep. *NWTC Information Portal (AirfoilPrep).* **2014**.
30. Jain, A.; Robertson, A. N.; Jonkman, J. M.; Goupee, A. J.; Kimball, R. W.; Swift, A. H. P. FAST code verification of scaling laws for DeepCwind floating wind system tests. *22nd International Offshore and Polar Engineering Conference.* Rhodes **2012**.
31. Doman, D. A.; Murray, R. E.; Pegg, M. J.; Gracie, K.; Johnstone, C. M.; Nevalainen *et al.* Tow-tank testing of a 1/20th scale horizontal axis tidal turbine with uncertainty analysis. *International Journal of Marine Energy* **2015**, *11*, 105–119.

Appendix A1

As an example, the range box plots are given as below for four unsteady case scenarios for cyclic variation of unsteady aerodynamic effects.

SFA1

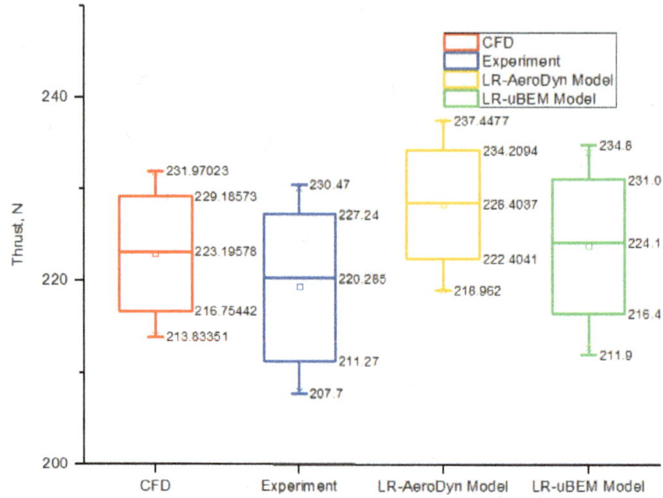
CFD
Experiment
LR-AeroDyn Model
LR-uBEM Model
Thrust, N
240
220
200
231.97023
229.18573
223.19578
216.75442
213.83351
230.47
227.24
220.285
211.27
207.7
237.4477
234.2094
228.4037
222.4041
218.962
234.6
231.0
224.1
216.4
211.9
CFD
Experiment
LR-AeroDyn Model
LR-uBEM Model

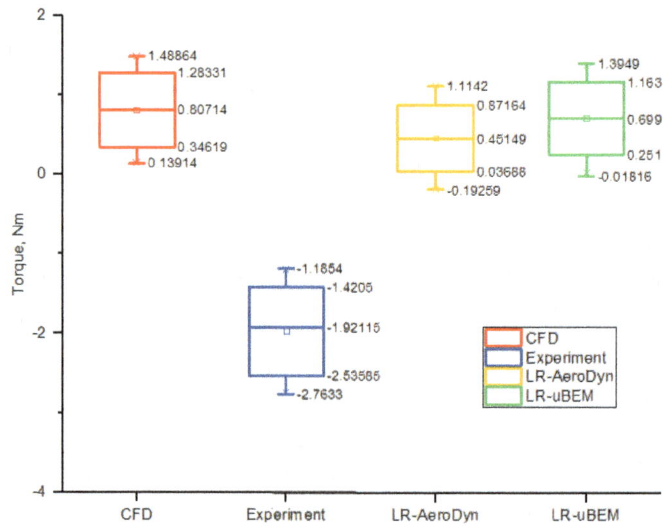
Torque, Nm
2
0
-2
-4
1.48864
1.28331
0.80714
0.34619
0.13914
-1.1854
-1.4205
-1.92115
-2.53585
-2.7633
1.1142
0.87164
0.45149
0.03668
-0.19259
1.3949
1.163
0.699
0.251
-0.01816
CFD
Experiment
LR-AeroDyn
LR-uBEM
CFD
Experiment
LR-AeroDyn
LR-uBEM

SFA2

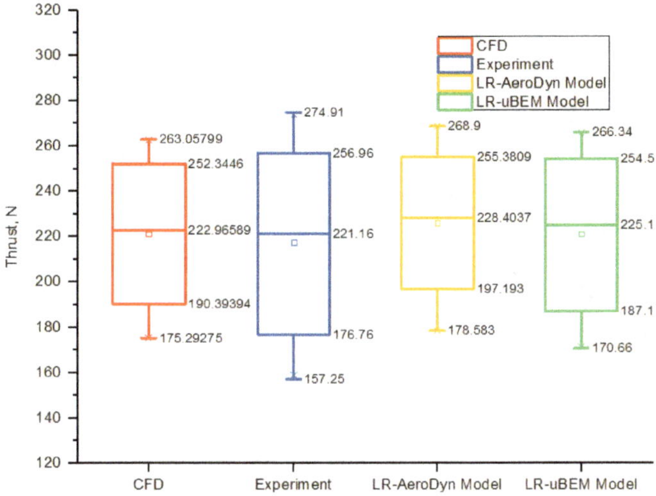

CFD
Experiment
LR-AeroDyn Model
LR-uBEM Model
Thrust, N
263.05799
252.3446
222.96589
190.39394
175.29275
274.91
256.96
221.16
176.76
157.25
268.9
255.3809
228.4037
197.193
178.583
266.34
254.5
225.1
187.1
170.66

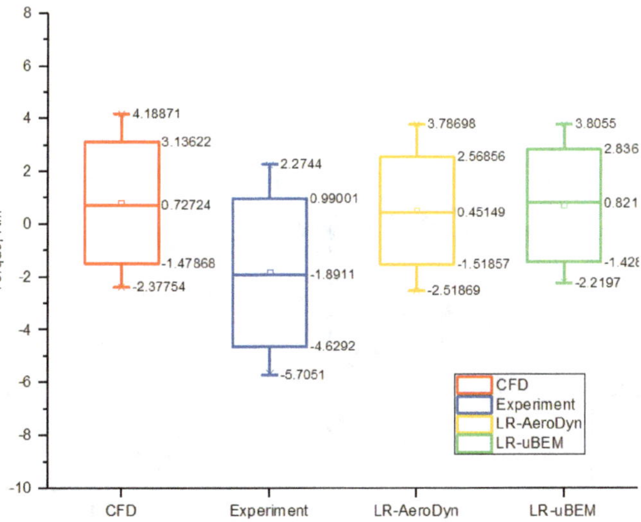

Torque, Nm
4.18871
3.13622
0.72724
-1.47868
-2.37754
2.2744
0.99001
-1.8911
-4.6292
-5.7051
3.78698
2.56856
0.45149
-1.51857
-2.51869
3.8055
2.836
0.821
-2.2197
CFD
Experiment
LR-AeroDyn
LR-uBEM

SFA3

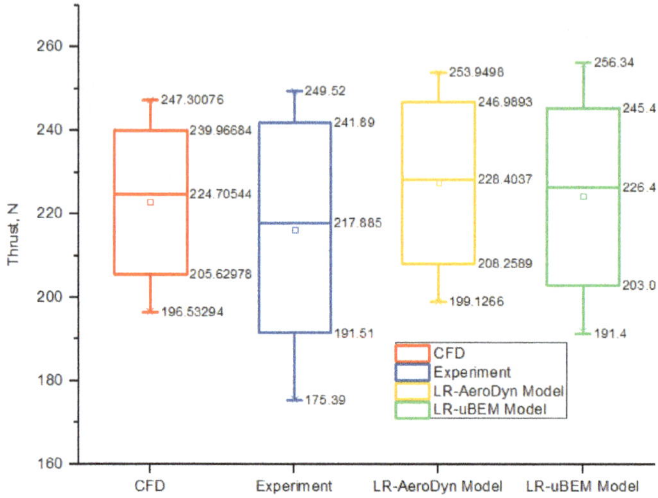

Thrust, N
260
240
220
200
180
160
247.30076
239.96684
224.70544
205.62978
196.53294
249.52
241.89
217.885
191.51
175.39
253.9498
246.9893
228.4037
208.2589
199.1266
256.34
245.4
226.4
203.0
191.4
CFD
Experiment
LR-AeroDyn Model
LR-uBEM Model
CFD
Experiment
LR-AeroDyn Model
LR-uBEM Model

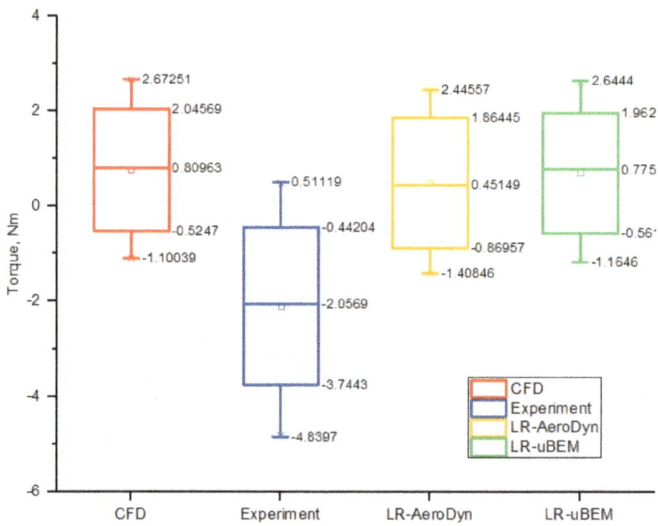

Torque, Nm
4
2
0
-2
-4
-6
2.67251
2.04569
0.80963
-0.5247
-1.10039
0.51119
-0.44204
-2.0569
-3.7443
-4.8397
2.44557
1.86445
0.45149
-0.86957
-1.40846
2.6444
1.962
0.775
-1.1646
CFD
Experiment
LR-AeroDyn
LR-uBEM
CFD
Experiment
LR-AeroDyn
LR-uBEM

SFA4

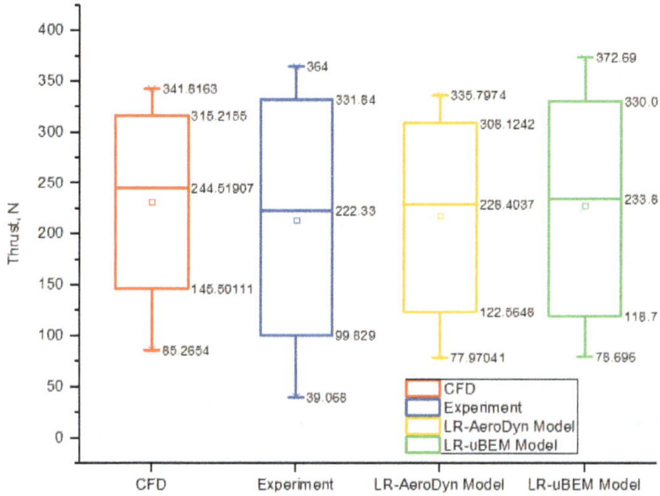

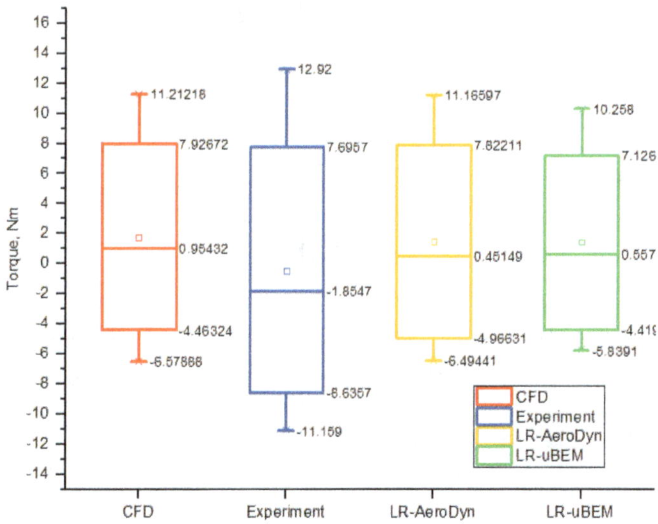

Appendix A2

As an example, axial induction factor for cyclic variation of unsteady aerodynamic effects are shown below for four unsteady case scenarios.

Some blade sections are with more than 0.4 axial induction factors in all the unsteady cases.

SFA1

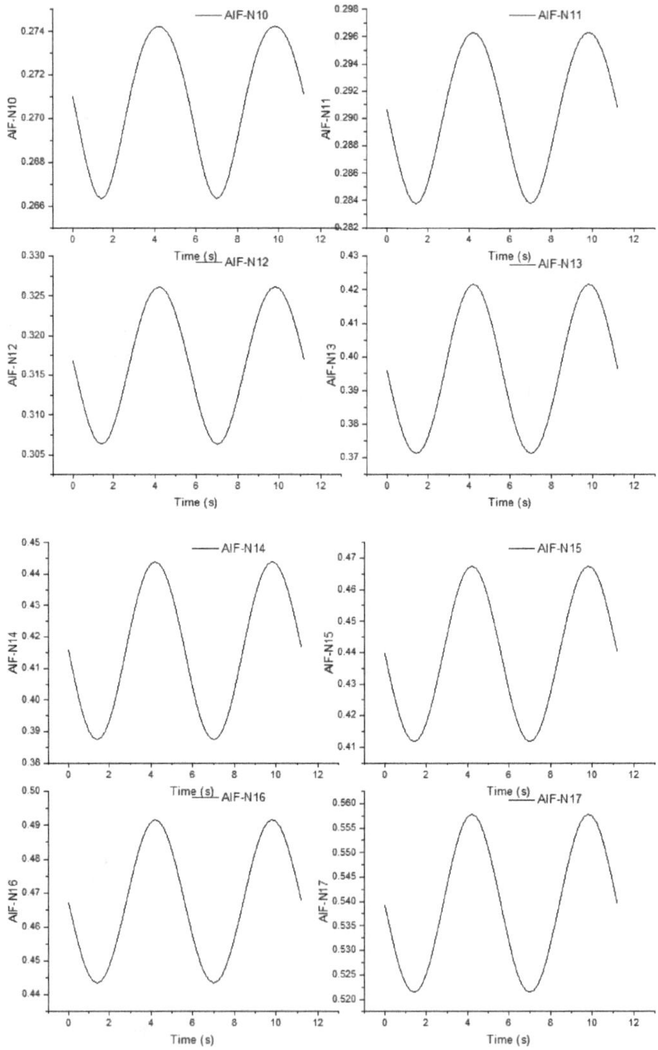

SFA2

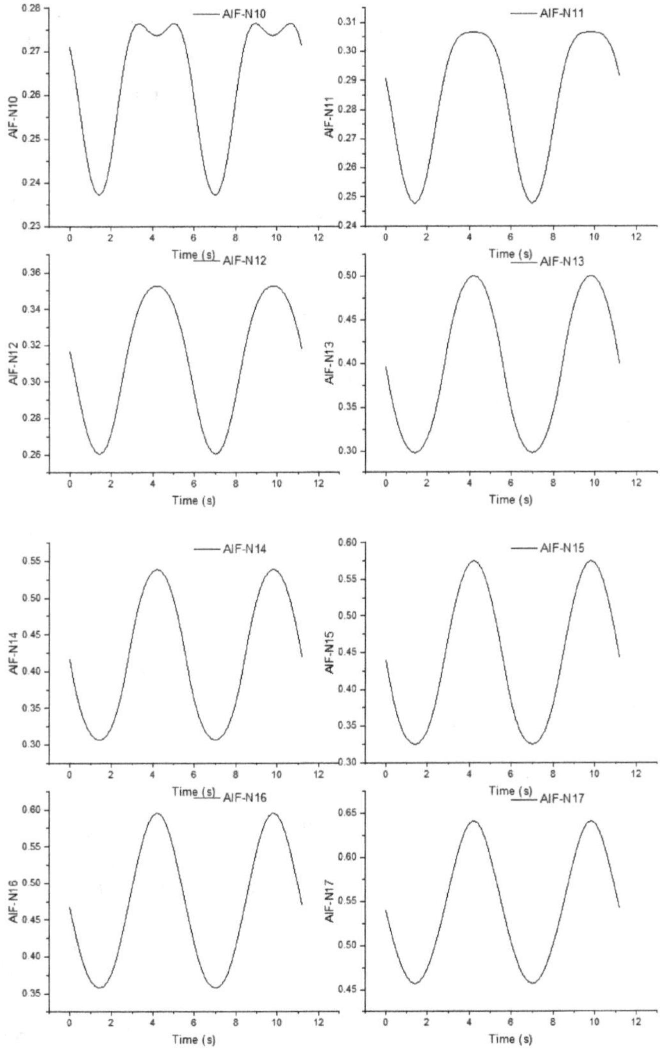
AIF-N10
AIF-N11
AIF-N12
AIF-N13
AIF-N14
AIF-N15
AIF-N16
AIF-N17
Time (s)
0
2
4
6
8
10
12

SFA3

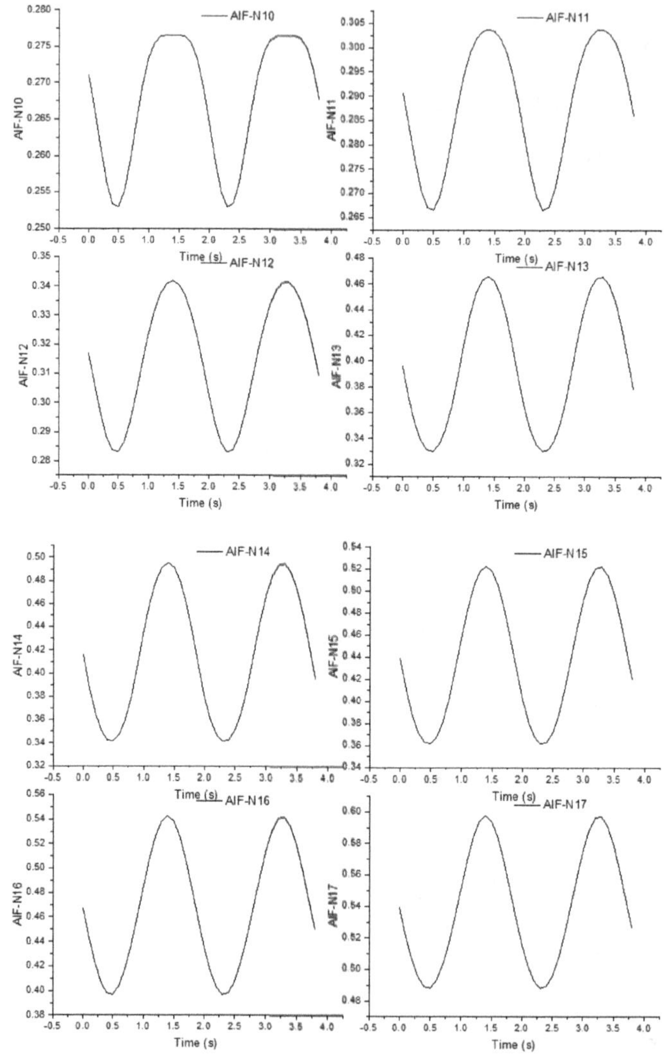
AIF-N10
AIF-N11
AIF-N12
AIF-N13
AIF-N14
AIF-N15
AIF-N16
AIF-N17
Time (s)

SFA4

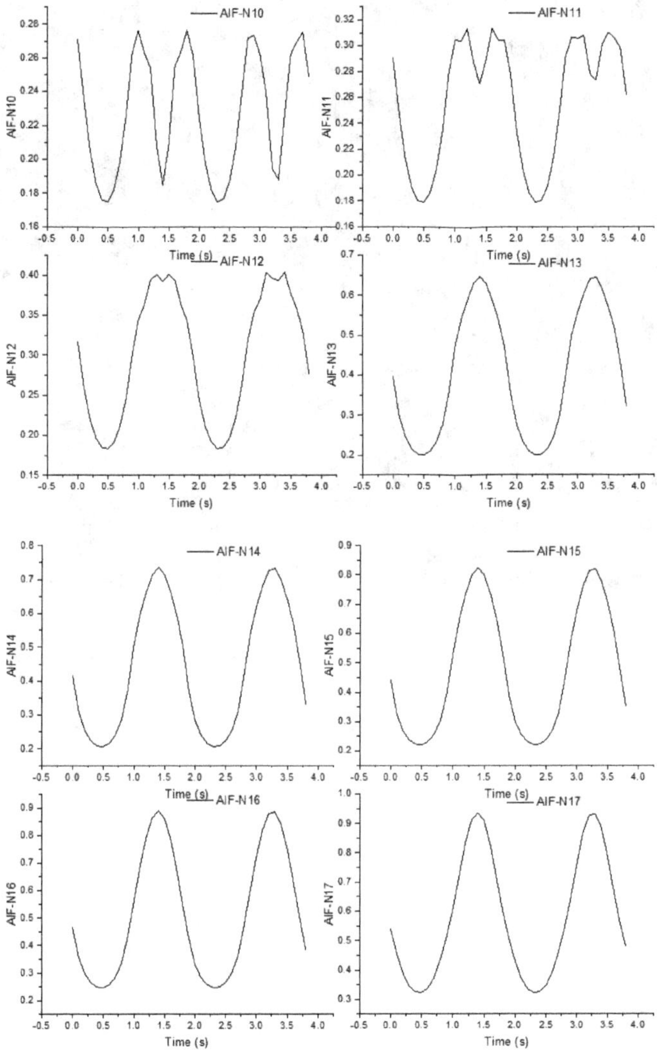
AIF-N10
AIF-N11
AIF-N12
AIF-N13
AIF-N14
AIF-N15
AIF-N16
AIF-N17
Time (s)

Chapter 10

Remaining Useful Life Prediction

10.1 Introduction

Research into offshore wind energy is gaining momentum for its consistent and high wind energy potential compared to onshore wind. According to the EWEA report, by 2016, more than 12.631 GW of offshore wind power is grid connected in Europe.[1] About 25% of the Levelized Cost of Energy (LCoE) is attributed to operation and management (O&M) for the turbines installed offshore or at remote locations, where the turbine accessibility depends on favorable weather conditions (wind, waves, water levels, tidal currents, daylight, and temperatures).[2] Hence, reducing the O&M cost is of paramount importance to lower the power generation cost and make it comparable to fossil fuel-based energy sources. Improving the reliability of critical components or predicting the failures through improved maintenance strategies are the two main approaches to reduce the cost of O&M and in turn LCoE.

The dependency factors that increase the cost of offshore power generation through increased O&M costs are:

- Chances of failed transfer of technicians to the turbine site.
- Chances of technicians to be abandoned at the turbine location because of deterioration in weather conditions.
- Chances that the technicians will be unable to perform the maintenance intervention because of seasickness.

Utility-scale offshore turbines consist of many components that require a huge number of maintenance tasks. In order to minimize the O&M tasks and associated costs,

- The most urgent tasks must be prioritized;
- Vessels, crew and technicians need to be effectively utilized;
- Reduce the spare parts inventory;
- Schedule preventative maintenance to minimize the turbine downtime.

An apt maintenance strategy is based on the nature of the equipment and the conditions in which it operates. Lloyd's Register is capable of modeling all the influencing variables and provides the most advanced predictive maintenance strategy.

The objective of any maintenance strategy is to increase the availability of the wind turbine for power generation. Wind turbine manufacturers and operators prefer condition-based maintenance rather than primitive time-based preventive maintenance. The status and the condition of the components can be obtained through the network-integrated Supervisory Control and Data Acquisition (SCADA) system. The physics-based remaining useful life (RUL) prediction together with condition-based maintenance will be the best strategy that is capable of lowering the labor cost, as the labor cost alone contributes 35% of total repair cost as shown in Figure 10.1.

As shown in Table 10.1, extracted from the Energy Research Centre of the Netherlands (ECN), an optimized O&M strategy will reduce repair

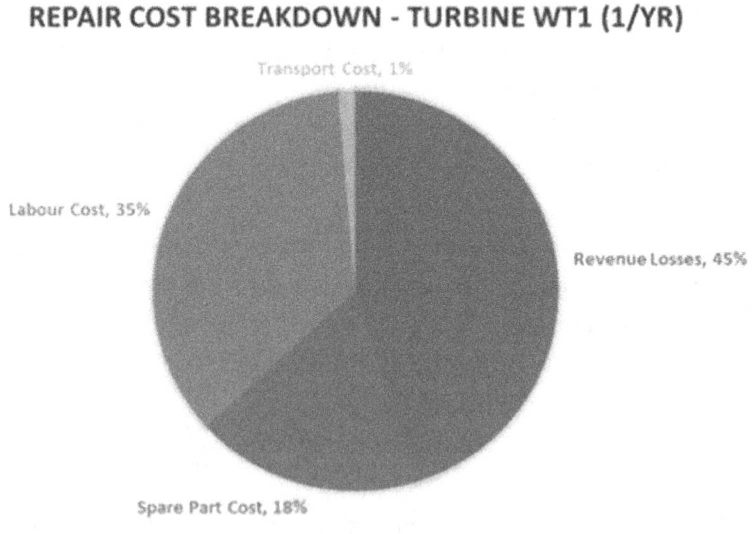

Figure 10.1 Cost breakdown (ECN Report, 2014).

Table 10.1 O&M cost reduction in results of O&M strategy optimization (ECN Report, 2014).

	Updated Scenario	Optimized Scenario	Reduction %
Repair costs (M€/yr)	9.52	8.52	10.43
Rev. losses (M€/yr)	1.64	1.57	4.21
Total effort (M€/yr)	11.16	10.10	9.51
Costs per kWh (c€/kWh)	3.35	3.00	10.58

Assembly	Risk Priority Number (RPN)
Frequency Converter	38.3
Pitch System	33.9
Yaw System	30.8
Gearbox	30.1
Nacelle Auxiliaries	29.0
Control & Comm. System	28.1
Generator	27.6
Main Shaft Set	27.0
Tower	26.0
Power Electrical System	25.0
Foundation	24.6
Cable	24.0
Blade	21.0
Hydraulics System	18.0
Auxiliary Electrical System	17.8
Transition Piece	17.3
Nacelle Structure	16.0
Hub	12.0

Figure 10.2 Critical assemblies ranked by risk priority number.[6]

costs by 10.43% and increases the profit by 4.21% with net savings of EUR 1.06 million per year.

As risk assessment plays a vital role in O&M strategies of offshore wind farms, a comprehensive failure mode and effect analysis (FMEA) is carried out to determine critical assemblies of a generic offshore wind turbine. As shown in Figure 10.2, the most critical component is the frequency or power converter. A novel physics-based methodology has been proposed to predict the RUL of electrical components, especially the power converters of variable speed wind turbines. The proposed methodology has been executed for both fixed and floating wind farm applications by using 5MW NREL virtual wind turbine along with digital twin technology.

Life management and life extension have been proposed by the International Electrotechnical Commission (IEC) wind committee to enrich wind farm operators and owners to decide on end-of-warranty review and its inspections, due diligence, life extension, re-powering, retrofitting of major components and decommissioning. Physics-based RUL prediction will greatly influence those operations with accurate predictions.

10.2 Offshore Wind Turbine Power Converter — Thermal Fatigue Loading Cycle-Based Remaining Useful Life Prediction

As shown in Figure 10.3, power converter Insulated-Gate Bipolar Transistor (IGBT) is constructed with multiple layers of materials with

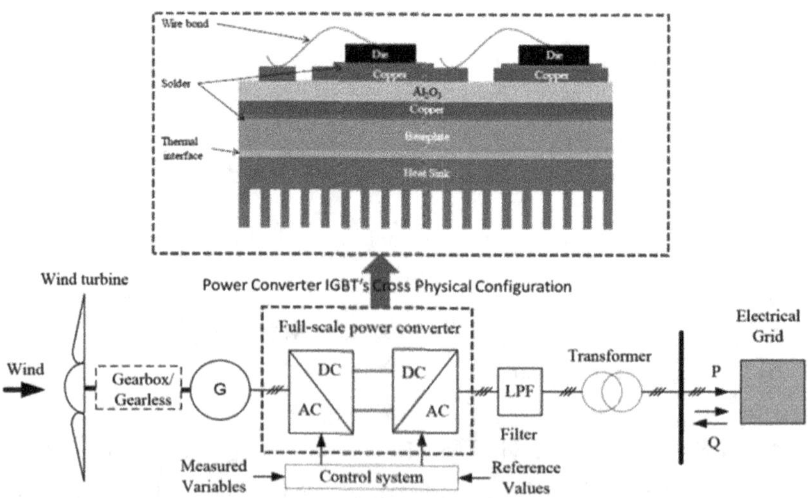

Figure 10.3 Schematic diagram of full-scale power converter and IGBT structural cross-section view.

different thermal conductivity. In Direct Bonded Copper (DBC) IGBT the die is attached to the copper layer via solder. For electrical isolation, a ceramic layer is placed in between the copper layers. For high-power applications such as MW-scale wind turbines, a base plate is typically attached to an air- or water-cooled system. A few stacks of IGBT modules will be used on the generator/rotor side and grid side to form a complete power converter system.

In both onshore and offshore environments, wind speed and temperature variation lead to intricate power loadings (overload condition and system transients) on the power converter which are attributed to high failure rates. Literature on power converters addressing the components and their reliability are limited. The study by Birk and Andresen[3] on power converters identifies the components with high failure rate as IGBT modules, converter control unit and main fans. Different coefficients of thermal expansion for layers of adjacent materials in the IGBT are the primary cause of thermal-driven failure during thermal cycles.[4] The principal modes of failure are bond wire lift-off and solder delamination or crack (at die-to-copper interface and copper-to-baseplate interface).

A typical thermal cycle experienced by a wind turbine power converter is shown in Figure 10.4. As most of the wind turbines are not incorporated

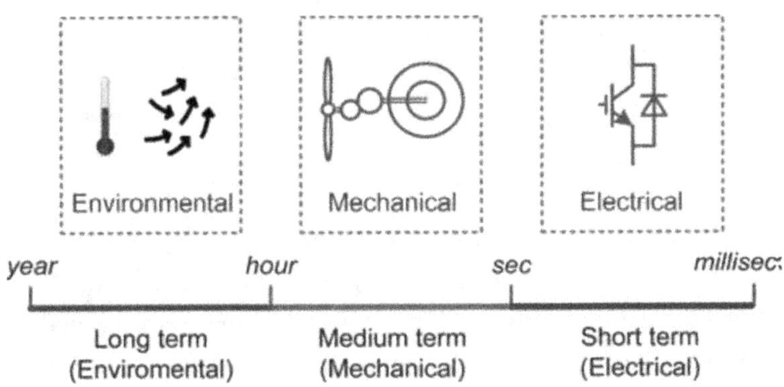

Figure 10.4 Wind power converter thermal cycles.

with condition monitoring for power converter systems, a robust predictive method has to be established to find the junction temperature, temperature fluctuations and their influence on the RUL. With a condition-based monitoring system, it is much easier to predict the damage accumulation and RUL. Mission profile for each category's (long, medium and short term) thermal cycles is different and the approach to predict the damage accumulation also varies accordingly.

10.2.1 *Thermal Loads due to Environmental Conditions*

Thermal loads due to environmental conditions impose long-term thermal cycles dictated by wind speed and ambient temperature. By employing SCADA data, similar to the three-hour interval period shown in Figure 10.5, the damage accumulation due to long-term thermal cycles can be predicted. For floating wind turbines, relative wind speed is used instead of freestream wind speed due to the platform motion.

From the above wind and ambient profiles, loading profile will be generated (in terms of torque and speed) from LR-Aerodyn and LR-uBEM of FAST for lifetime estimation. In a more convenient way, the power loss will be estimated from the SCADA data power values to predict the junction temperature and its fluctuation through a customized thermal model for damage accumulation and RUL estimation. A simple thermal network model is used to compute the junction and case temperature for the IGBT module as shown in Figure 10.6. The values of the thermal network model parameters are provided by the power converter manufacture.

The existing wind farm data-logging interval is five minutes or more and the total damage accumulation for this interval will lead to substantial error if the long-term thermal loading alone is examined. A predicted IGBT temperature is shown in Figure 10.7 for long-term thermal cycle of 5MW NREL wind turbine. Medium- and short-term thermal loading

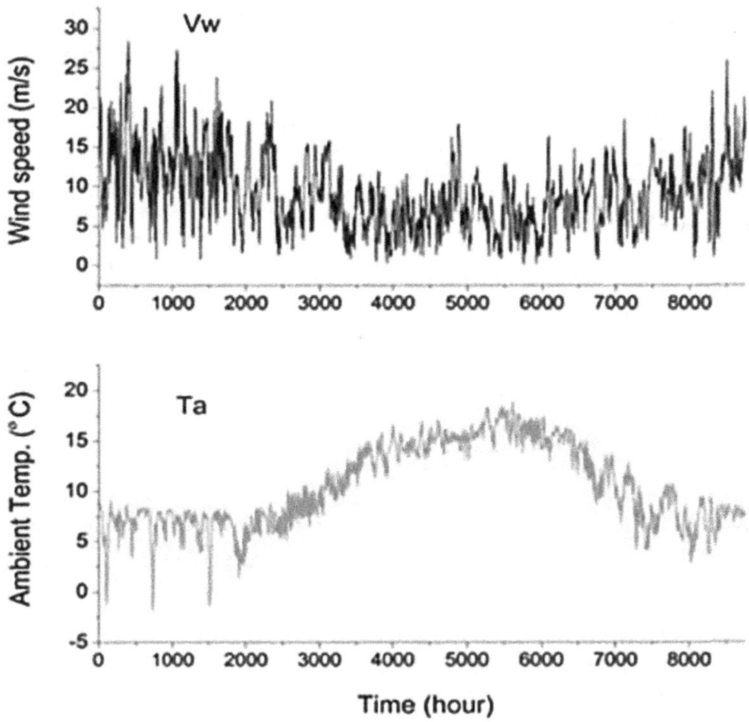

Figure 10.5 One-year mission profile of wind speed and ambient temperature from wind farm/turbine in three-hour averages (Ma *et al.*, 2015).

effects pose significant damage accumulation on the weak points of (chip solder, base plate solder, bond wire) the IGBT structure.

10.2.2 *Thermal Loads due to Mechanical Systems of Wind Turbine*

In order to account for the effect of thermal cycles less than 3 hours in the damage accumulation prediction, the loading profile with time constants of seconds to minutes is established for each wind speed at 2 m/s step for the entire wind spectrum of the whole year. Afterwards, the loading profile generated using FAST and Miner rule is applied for the

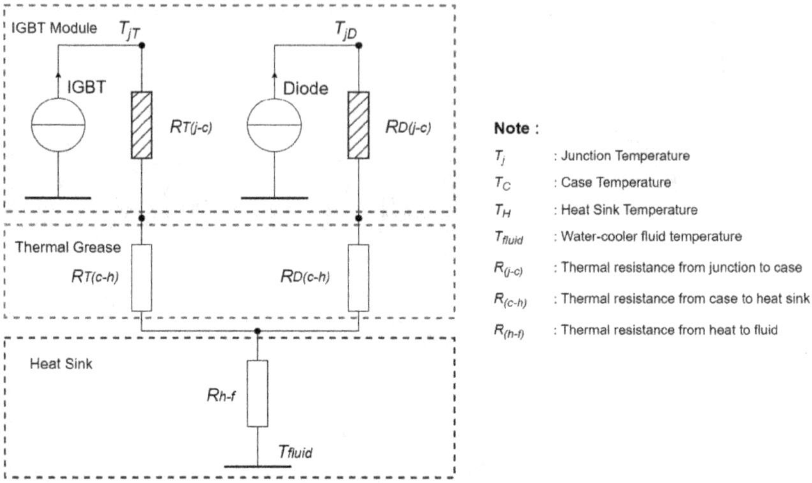

Figure 10.6 Thermal network of power semiconductor devices for the long-term thermal profile generation. Reproduced with permission from Sivalingam, K., Sepulveda, M., Spring, M., Davies, P. A review and methodology development for remaining useful life prediction of offshore fixed and floating wind turbine power converter with digital twin technology perspective. *2018 2nd International Conference on Green Energy and Applications (ICGEA)*, pp. 197–204. Singapore **2018**.

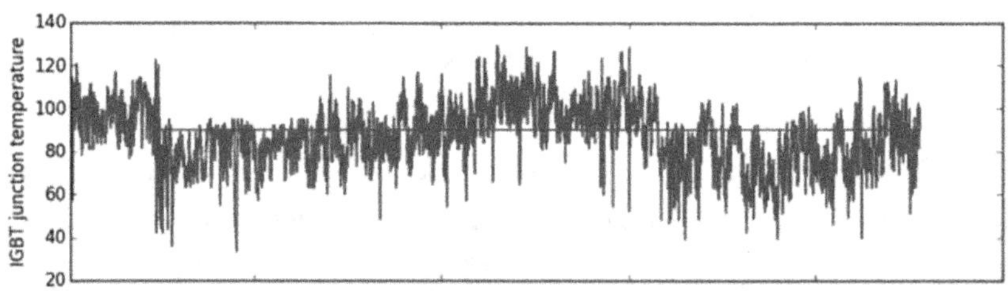

Figure 10.7 Junction temperature and its fluctuation for wind turbine yearly data. Reproduced with permission from Sivalingam, K., Sepulveda, M., Spring, M., Davies, P. A review and methodology development for remaining useful life prediction of offshore fixed and floating wind turbine power converter with digital twin technology perspective. *2018 2nd International Conference on Green Energy and Applications (ICGEA)*, pp. 197–204. Singapore **2018**.

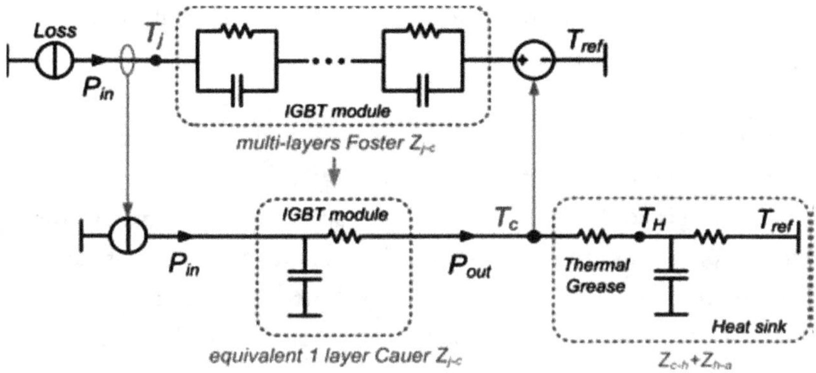

Figure 10.8 Thermal network of power semiconductor devices for medium-term thermal loading. Reproduced with permission from Sivalingam, K., Sepulveda, M., Spring, M., Davies, P. A review and methodology development for remaining useful life prediction of offshore fixed and floating wind turbine power converter with digital twin technology perspective. *2018 2nd International Conference on Green Energy and Applications (ICGEA)*, pp. 197–204. Singapore **2018**.

estimation of total damage accumulation by medium-term thermal cyclic loading. By this approach, the effect of pitch and speed control behaviors can be recognized in the damage accumulation prediction process. Thermal capacitance is included in the thermal network model to capture the thermal transients as in Figure 10.8.

10.2.3 *Thermal Loads due to Electrical Systems of Wind Turbine*

Short-term thermal behavior is attributed to high-frequency current oscillation in the electrical system leading to junction temperature fluctuations. The oscillations are smaller in amplitude and close to the fundamental frequency of the power converter output. Cyclical amplitude is analytically solved as shown in Eq. (10.1):

$$\Delta T_j = P_{loss} \times Z_{th} \times \left(\frac{3}{8f_o}\right) \times 2P_{loss} \times Z_{th} \times \left(\frac{1}{4f_o}\right) \qquad (10.1)$$

where P_{loss} is the loss of power device; Z_{th} is device thermal impedance; and f_o is the fundamental frequency of the converter output.

Previous studies indicate that short-term thermal loading significantly affects the total damage accumulation at rated wind speed conditions and hence it is compelling to include for accurate prediction. All three loading conditions are to be used for the accurate prediction of damage accumulation and RUL. Medium- and short-term loading cycles are found to have greater influence on RUL prediction for FOWT due to its control strategy.

10.3 Physics-based Remaining Useful Life Prediction Methodology

As discussed in the previous sections it is found that the mean junction temperature and temperature cycling are the main cause of temperature-induced stress in the power converter assemblies that affect the reliability or lifetime adversely. A few studies investigated power converter models for damage accumulation prediction which are still to be calibrated for its applicability to offshore wind turbine O&M, i.e., for both fixed-bottom and floating wind turbines.

Until now statistical approaches are largely used for health diagnostics and prognostics of wind turbine components, but they do not completely represent the real condition of the component. Due to recent technology advancements, sensor and data-analytic approaches use operational data such as Condition Monitoring Systems and SCADA to monitor the performance parameters. A physics-based modeling approach accounts for the failure mechanism and determines the system condition using physical laws combined with operational data.[5]

Prior to physics-based modeling, a detailed risk assessment has to be performed by using comprehensive FMEA to identify those failures that have significant impact on the wind turbine operation and cost of

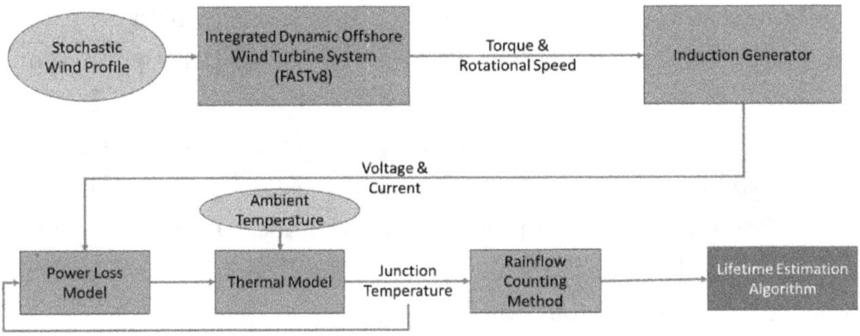

Figure 10.9 RUL prediction flow diagram. Reproduced with permission from Sivalingam, K., Sepulveda, M., Spring, M., Davies, P. A review and methodology development for remaining useful life prediction of offshore fixed and floating wind turbine power converter with digital twin technology perspective. *2018 2nd International Conference on Green Energy and Applications (ICGEA)*, pp. 197–204. Singapore **2018**.

maintainence. A generalized offshore-specific physics-based methodology is implemented to predict the RUL of power converter of wind turbines as shown in Figure 10.9.

Wind profile of the wind turbine is obtained from SCADA data. Then the profile is used in the integrated dynamics offshore wind turbine FAST model which generates torque and speed for the generator model input. A high-level generator model validation is possible by predicting power loss directly from SCADA data and using it as an input to the thermal model to predict the junction temperature. Then, this can be validated with FAST and induction generator model-generated power loss. The flow diagram shown in Figure 10.9 is applicable for both diagnostic (to predict damage accumulation) and prognostic (to predict RUL) approaches.

10.3.1 *Integrated LR-Aerodyn and LR-uBEM Elastic Servo Control Code*

5MW NREL turbine FAST model is used in this study. An accurate and validated aerodynamic model (by Lloyd's Register) for FOWT is used in

FAST. TurbSim code is used to generate turbulent flow field for FAST. Various FAST sub-model files are generated by Matlab-based automation. The details of the electrical drive are not modeled in FAST; instead, the focus is on getting the torque-speed curve as accurate as possible (which effects turbine loads). Though it is possible to extract details of the electrical drive from Matlab Simulink generator models, the most sophisticated built-in Thevenin Equivalent Circuit model for induction machine available in FAST is employed.

10.3.2 *Python-based Induction Generator Model*

A squirrel cage induction generator (SCIG) model is included to extract voltage and current variables to supplement FAST simulations through specific algorithms. SCIG is a three-phase induction machine with three windings in the stator and three or more windings on the rotor (assumed). Generators can be described with a set of equations similar to the motor, simplified to compute voltage and current.

10.3.3 *Python-based Power Loss Prediction Model*

There are three losses in IGBTs and diodes in the power converters. They are conduction losses (P_c), switching losses (P_{sw}) and blocking (or leakage) losses (P_b). Normally blocking losses can be neglected. All the above losses can be predicted accurately by developing a detailed mathematical formulation from power converter data sheets or books. The formulation can be coded in Python.

$$P_{loss} = P_c + P_{sw} + P_b \tag{10.2}$$

10.3.4 *Python-based Thermal Model for Junction and Case Temperature Prediction*

The input for the thermal model is power loss. The python code will predict the junction temperatures of IGBTs and diodes together with case temperature. Junction temperature T_{jm} and junction temperature fluctuation dT_j are given by the following equations.

$$T_{jm\_T/D} = P_{loss} \times \sum_{i=1}^{4} R_{thjc-\frac{T}{D(i)}} + P_{loss} \times \sum_{i=1}^{4} R_{thca-(j)} + T_a \tag{10.3}$$

$$dT_{j\_T/D} = 2P_{loss} \times \sum_{i=1}^{4} R_{thjc-\frac{T}{D(i)}} \times \frac{\left(1 - e^{\frac{t_{on}}{\tau thjc-\frac{T}{D(i)}}}\right)^2}{1 - e^{\frac{t_p}{\tau thjc-\frac{T}{D(i)}}}} \tag{10.4}$$

In Eq. (10.3), R_{thjc} is the thermal resistance from the junction to case of the power module, R_{thca} is the thermal resistance of the air cooling, in which subscripts T and D denote IGBT and the freewheeling diode, whereas subscripts i and j denote four-layer and three-layer Foster structure for power module and air cooling, respectively. P_{loss} is the power loss of each power semiconductor, and T_a is the ambient temperature. In Eq. (10.4), t_{on} denotes the on-state time within each fundamental period of the current at the steady state operation, t_p denotes the fundamental period of the current, and τ denotes the thermal time constant of each Foster layer.

10.3.5 *Python-based Rain Flow Counting Method*

Rainflow counting method is employed to estimate the thermal cycle frequency and amplitude.[6] The method was initially developed for materials science and later applied to power electronics. The local high and low in the data are considered as peaks and valleys and the intermediate range is assumed to be half-cycles. The algorithm pairs the half-cycles to generate full cycles with respect to the mean. The same algorithm is developed in Python.

10.4 Digital Twin Platform

With the profound knowledge of Lloyd's Register's FMEA and RUL prediction experience with fixed-bottom wind turbines, a detailed methodology is developed for offshore fixed and floating platforms. Figure 10.10 shows briefly how the 5MW NREL numerical turbine is used in virtual space for the digital twin technology framework.

A framework is developed for physics-based RUL prediction and for a comparison between fixed and floating offshore 5MW NREL wind turbines to quantify the effect of changes in medium- and short-term thermal cyclic loads on the power converter. A schematic diagram of the digital twin technology platform is shown in Figure 10.11.

For the first three major components in Figure 10.4, virtual sensors are placed (as shown in Figure 10.12) in the 5MW numerical turbine for

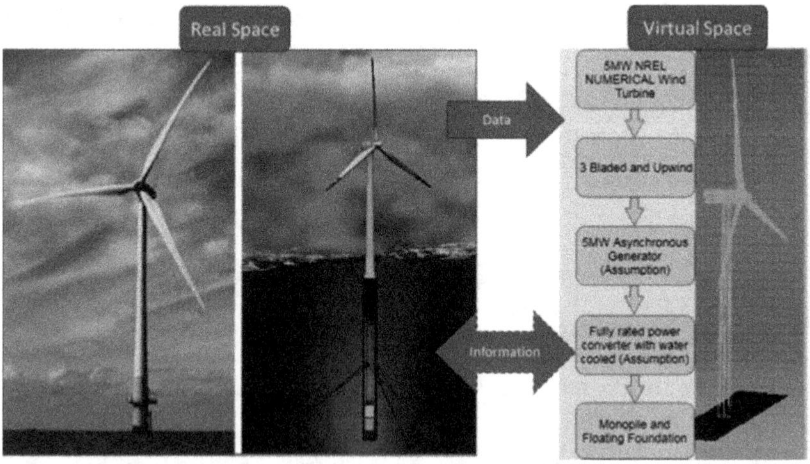

Figure 10.10 5MW wind turbine configuration for fixed and floating applications: schematics of the information-mirroring model.

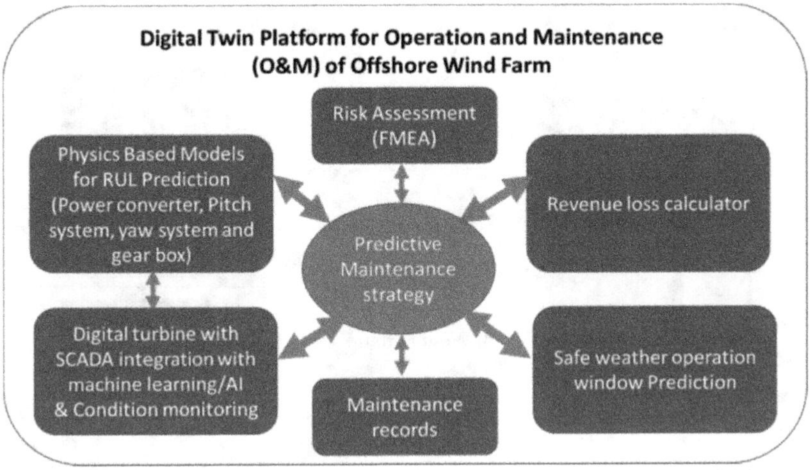

Figure 10.11 Schematic diagram of digital twin architecture for 5MW NREL virtual fixed and floating wind farm O&M.

predicting RUL for optimum O&M strategy comparison, from which is derived the risk of failure for each failure mode. Numerical models used during the wind turbine design phase (5MW NREL FAST model) enable estimates to be made of degradation and remaining life. Examples include replacing statistics of mean time between failures and substituting with mean usage to failure, which may imply counting the number of use cycles for a component or the fatigue damage due to thermal cycles.

Under the digital twin platform, virtual sensors have been derived for locations in the structure without strain gauges using a combination of aeroelastic models and finite element methods. The minimum requirement is to use wind speed, ambient temperature, yaw angle and electrical power generated from SCADA records as shown in Figure 10.12. To support smart maintenance planning and predictive maintenance for wind farm operations, the following are the areas under focus for

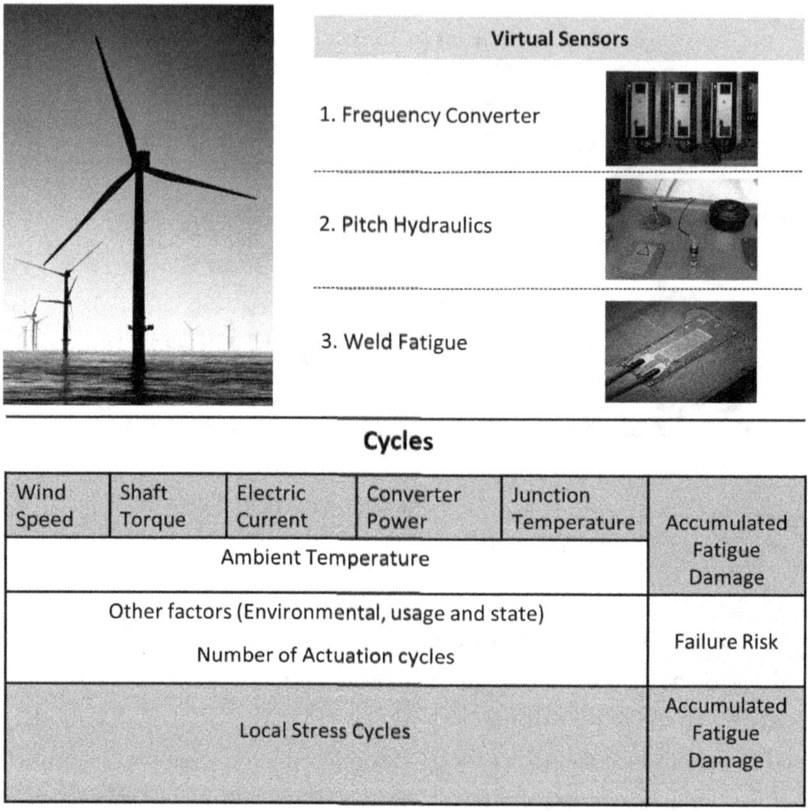

Figure 10.12 Virual sensor details in digital twin technology platform for optimal decision making of maintenance strategy.

fundamental research and development of new technology applied within the digital twin framework:

1. Short-term forecasting at turbine locations — waves, wind, currents, water depth
2. Estimating safe transfer windows — incorporating simplified vessel response models
3. Statistical reliability of offshore fixed structures — incorporating models

4. Turbine in-service loading and accumulation of damage — electronics, structures, mechanical systems, linked with statistical reliability models and environmental forecasts
5. Technology for remote presence at difficult-to-access sites, wearable devices, streamlined exchange of data
6. Synergies between workboat routes and bathymetric surveys, local scour
7. Formal methods for quantitative risk assessment
8. Safety culture, best practices, emergency preparedness, design of systems and testing of procedures

The scope of this section is limited to physics-based methodology for RUL prediction of power converters which is chosen from SEMIKRION by considering all three types of thermal loads. A simple process flow of the approach is given in Figure 10.13 for digital twin platform as a module to it.

For accurate power converter RUL prediction, it is necessary to create a similar environment experienced by the turbine in terms of structural

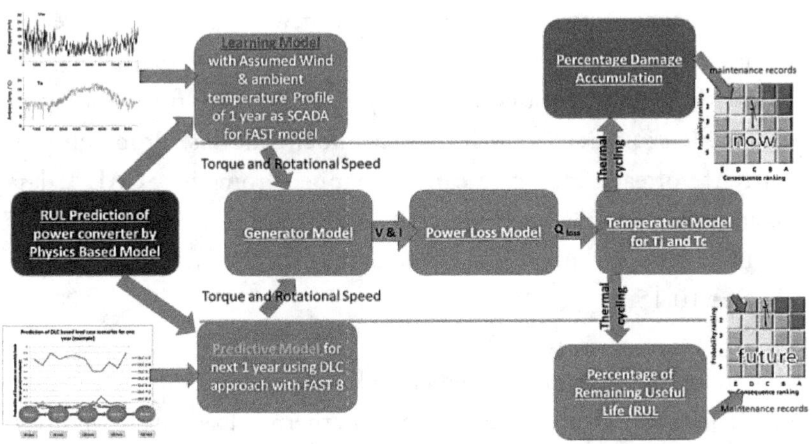

Figure 10.13 Schematic diagram of RUL prediction process flow for power converters.

Table 10.2 Fatigue load case scenarios as per IEC64100-3 for fixed-bottom offshore wind turbines.

Design Situation	DLC	V_{ref} (m/s)
Power production	1.2	42.5
Power production plus occurrence of fault	2.4	42.5
Start up	3.1	42.5
Normal shut down	4.1	42.5
Parked (standing still or idling)	6.4	42.5
Parked and fault detection	7.2	42.5
Transport, assembly, maintenance and repair	8.3	42.5

and electromechanical behavior and wind profile. For the current scope, input load cases for FAST modeling are obtained through IEC 61400-3.[7] As the power converter is subjected to temperature cyclic load, fatigue-based design load cases (DLCs) are chosen. The design scenario shown in Table 10.2 is considered for fixed-bottom offshore wind turbines.

In the case of floating wind turbines, additional load cases are to be included, accounting for the interaction of wind and wave together with mooring line failures. Well-established Monte Carlo Markov Chain (MCMC) technique is employed to generate futuristic event load profile sequences with higher confidence levels. Probability of occurrence of each load case can be mapped from the SCADA data similar to the one shown in Figure 10.14. The future sequence of events for DLCs computed from SCADA data for the next five hours is shown in Figure 10.15.

IGBT junction temperature prediction for the next 5 hours is shown in Figure 10.16 by considering medium-term thermal load in fixed-bottom 5MW NREL turbines.

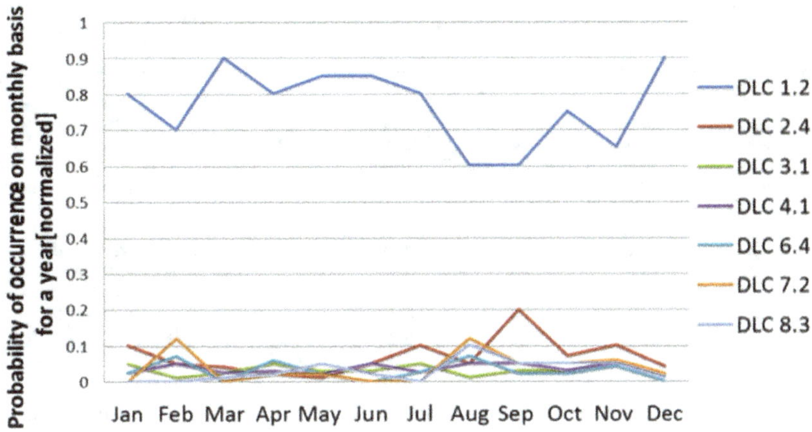

Figure 10.14 Probabilistic occurrence of DLC load cases.

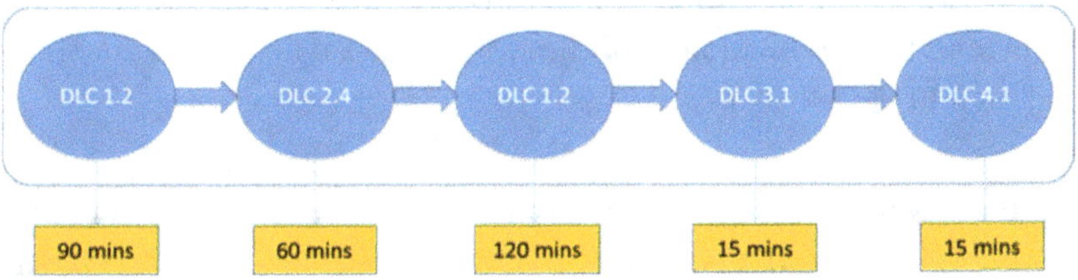

Figure 10.15 Example of future scenario generation by MCMC for 5 hours.

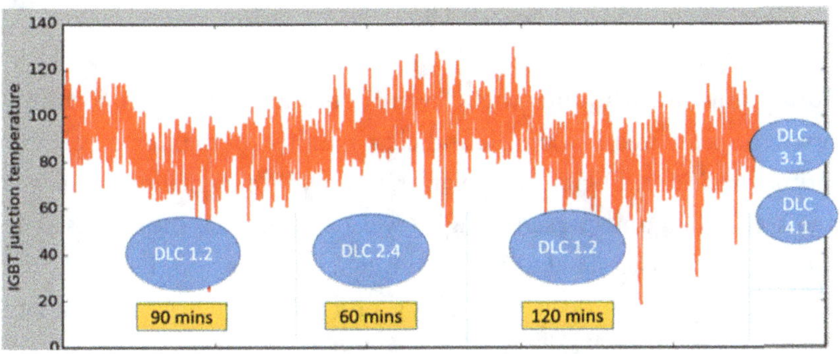

Figure 10.16 IGBT temperature prediction for 5 hours.

Range (°C)		DTj(K)	Tjm(K)	cycle count	Number of cycles to failure (Nf)	Estimated Damage
53.97	59.97	6	253.72	1	6.64259E+13	0.00000000000%
47.97	53.97	6	273	0	9.04297E+12	0.00000000000%
41.98	47.97	5.99	273	0	9.1193E+12	0.00000000000%
35.98	41.98	6	264.195	1	2.16838E+13	0.00000000000%
29.98	35.98	6	265.955	0.5	1.81219E+13	0.00000000000%
23.99	29.98	5.99	263.52	2.5	2.34397E+13	0.00000000001%
17.99	23.99	6	267.892	2	1.49148E+13	0.00000000001%
11.99	17.99	6	254.44	6.5	6.13249E+13	0.00000000001%
8.99	11.99	3	251.11	18	2.92876E+15	0.00000000000%
6	8.99	2.99	256.4	38.5	1.65333E+15	0.00000000000%
3	6	3	266.664	399.5	5.54612E+14	0.00000000007%
1.5	3	1.5	272.6688	4251	1.0091E+16	0.00000000004%
0	1.5	3	273.4109	14402.5	2.85808E+14	0.00000000504%
		total number of cycles		19123		0.00000000520%

Figure 10.17 Damage accumulation prediction for 5 hours based on IGBT temperature.

After IGBT and diode junction temperature prediction for the next 5 hours, rain flow counting algorithm is executed to generate the cycle counts and RUL estimation algorithm will generate damage accumulation as shown in Figure 10.17.

The physics-based RUL prediction of power converter module outcome is shown in Figure 10.18 for fixed-bottom 5MW NREL turbine. Digital twin framework includes the verification and validation process as dynamic data of wind turbine/farm influence the power converter reliability models. The above prediction process can be expanded to the wind farm level as shown in Figure 10.19.

During real-time planning, the relevant information used in algorithms should be presented in a simple and clear manner for making quick and optimal decisions. Digital twin platform converts big data into manageable small data and presents high-level performance indicators that influence the decisions of O&M planning and execution. Figure 10.20 shows the optimum decision-making process in project life cycle for O&M at wind farm level.

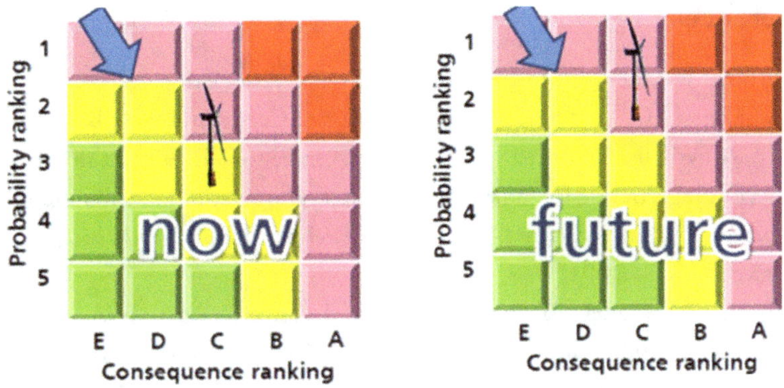

Figure 10.18 Example of physics-based RUL of power converter module output in digital twin platform.

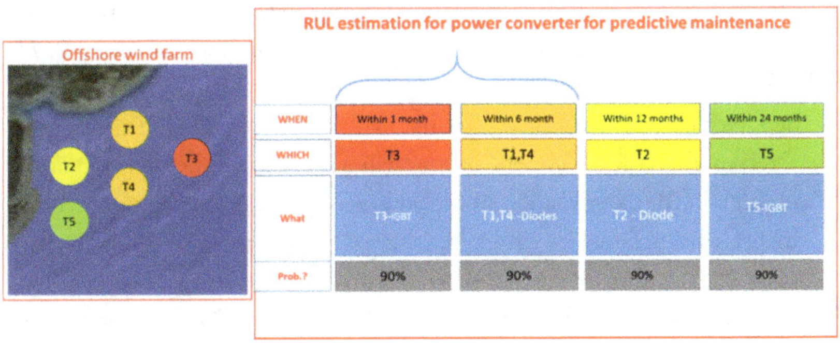

Figure 10.19 Output for wind farm level case of power converter RUL module in digital twin platform for O&M strategy planning.

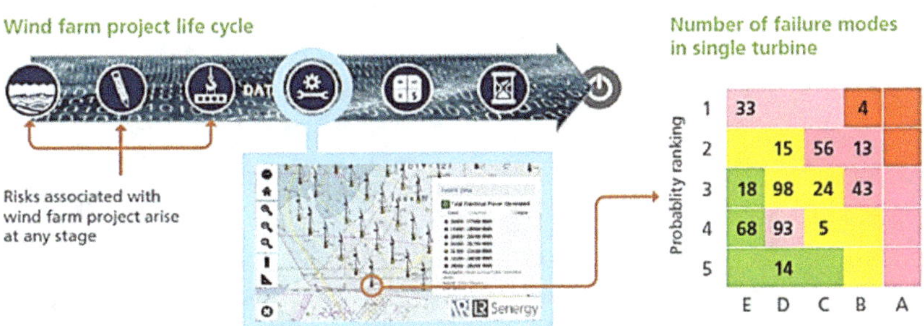

Figure 10.20 Optimal decision making by digital twin for O&M.

References

1. *EWEA Wind Report.* European Wind Energy Association **2016**.
2. van Bussel, G. J.; Henderson, A. R.; Morgan, C. A.; Smith, B.; Barthelmie, R.; Argyriadis, K.; Arena, A.; Niklasson, G.; Peltola, E. State of the art and technology trends for offshore wind energy: operation and maintenance issues. *Proceedings of the Offshore Wind Energy EWEA Special Topic Conference.* Brussels **2001**.
3. Birk, J.; Andresen, B. Parallel-connected converters for optimizing efficiency, reliability and grid harmonics in a wind turbine. *2007 European Conference on Power Electronics and Applications.* Aalborg **2007**.
4. Fischer, K.; Stalin, T.; Ramberg, H.; Thiringer, T.; Wenske, J.; Karlsson, R. Investigation of converter failure in wind turbines. *Elforsk Report* **2012**.
5. Breteler, D.; Kaidis, C.; Tinga, T.; Loendersloot, R. Physics based methodology for wind turbine failure detection, diagnostics & prognostics. *European Wind Energy Association Annual Conference and Exhibition 2015.* Paris **2015**.
6. Sepulveda, M.; Davies, P.; Spring, M. J.; Shek, J. K.; Thies, P. R.; Oterkus, E. Risk assessment of an offshore wind turbine and remaining useful life (RUL) estimation of the power converter. Improving availability by prioritising failures with higher risk to operation. University of Exerter **2016**.
7. *IEC 61400-3:2009. Wind Turbines — Part 3: Design Requirements for Offshore Wind Turbines.* International Electrotechnical Commission **2009**.

Chapter 11

Concluding Remarks

11.1 Summary of Darrieus Rotor Characteristics

One of the goals of this book is to report on a strategy that could be implemented on small wind turbines of power capacity from 10–50 kW to improve the low wind speed characteristics. Most of the previous studies were focused on improving the starting characteristics of the Darrieus rotor, but there exists a clear literature gap on generating power in low wind speed rather than inciting the rotation. The strategy should account for the aspects of commercialization such as low cost and ease of maintenance. The search for the solution begins with a comprehensive review on the existing strategies and their mode of failure to generate considerable torque at low wind speed. Based on the conclusion, four innovative concepts are put forward and subjected to wind tunnel tests. The Hybrid Darrieus-Savonius rotor, while displaying an excellent performance at low wind speed, degrades severely at higher tip-speed ratio (TSR). An innovative solution to transform the Savonius buckets into cylinders was proposed and optimized through a series of computational optimization and wind tunnel validation of the resulting design parameters and their impact on the performance.

The following list summarizes the overall results of the study conducted on Darrieus turbines for low wind speed operation:

- The extensive literature review arrived at an important conclusion that the lift force is not sufficient to enable the Darrieus rotor to generate power from wind speeds of 2 m/s or less.
- The drag force is of paramount importance for low cut-in and sustained rotation to generate power at low wind speed.
- Innovative Darrieus rotor with trailing edge cavity is a potential candidate for sites with average wind speeds of 6–8 m/s as the performance does not drop significantly even at high Reynolds number (Re).
- Thrust load at high winds was a limiting factor for the Savonius turbine to scale up to higher power capacity. The optimized and wind tunnel-validated Telescopic Savonius turbine can be a standalone

option that can withstand up to 50 m/s. No such device exists commercially that can operate under extreme wind speeds.

- The proposed aerodynamic model based on the Double Multiple Streamtube model can predict the blade forces and operating torque of the Adaptive Hybrid Darrieus Turbine (AHDT) close to experimental values for closed configuration.
- The computational studies provide profound knowledge on the flow physics of conventional Darrieus rotor, Hybrid Darrieus-Savonius rotor and Darrieus rotor with large-diameter rotating tower. The optimum AHDT diameter range recommended by the computational fluid dynamics (CFD) results can be used as a rule of thumb for future designs.
- Wind tunnel results arrive at an important conclusion that the larger Savonius rotor in closed configuration (cylinder) has minimal impact on the performance of the Darrieus rotor compared to the smaller-diameter rotor.
- For a two-bladed Darrieus rotor, the optimum ratio of Savonius rotor diameter (D_R) that can be integrated into the Darrieus rotor (D_T) is 1:3, considering the better performance in low wind speed.
- The devised AHDT displays 96.9% higher starting torque than the conventional Darrieus rotor at 3 m/s wind speed. This immense torque at low wind speed will change the perception of Darrieus turbines.
- Apart from torque generation at low wind speed, the Savonius buckets sliding in opposite directions generates enormous resistive torque that can act as an aerodynamic brake. Thus, an implementable solution has been arrived at for the research questions of this study.

The well-focused literature study serves as a consolidated knowledge on various strategies that are investigated on Darrieus rotors and their impact on the power coefficient. Though a large variety of dedicated airfoils are conceived to address specific challenges such as flow curvature, dynamic stall, surface roughness and inexpensive blades, none of the airfoils have been developed with the aim of low wind start-up, forcing the designers to opt for the conventional NACA series. An avenue to modify the conventional airfoils to increase the starting torque has been suggested herein. The two innovative concepts with stepped airfoils and airfoil with trailing edge cavity are the outcomes of the research.

Sophisticated strategies such as trailing edge flaps, morphing blades and blade pitching can generate comparatively higher torque than the conventional airfoils, but the torque is still insufficient to accelerate the rotor. These strategies demand attention from fatigue load reduction in blades and aerodynamic braking at high winds. The complexity at this stage of development eliminates the possibility of incorporating in small wind turbines of power capacity less than 50 kW.

Another group of strategies demands flow-augmenting devices such as diffuser and omnidirectional guide vanes. Though a notable performance enhancement was noted, it is practically impossible to implement for turbines of capacity more than 5 kW, the reasons being economics and the wind load.

11.1.1 *Feasibility Check on Four Innovative Concepts*

A vital conclusion that resulted from the literature study is to exploit the drag force by any possible means to generate a decent power at an average wind speed of 4–6 m/s. The four innovative concepts, i.e., airfoil with steps, Dual Darrieus rotor, Darrieus rotor with Telescopic Savonius rotor, and airfoil with trailing edge cavity are put forward as possible means to create higher drag force.

The KF-N-21 stepped airfoil, though having displayed better starting characteristics than NACA0021, is not favorable due to the shorter operating TSR range with sharp decline in power coefficient. Instability of vortex trapping needs further investigation.

Dual Darrieus rotor displays higher power coefficient than conventional H-rotor till Re 238,556, from thence the difference dips to 21.2% with further increase in Re. The poor starting characteristics and additional blade requirement are not attractive factors.

The trailing edge-modified airfoil can achieve similar power coefficient as conventional airfoils even at higher Re. Further optimization on airfoil and truncation % can be adopted for low-cost turbines, especially for power capacity up to 20 kW.

11.1.2 *Computational and Experimental Studies*

A thorough computational study of both the pressure and velocity fields generated across the wind turbine has been carried out during the optimization process. From this analysis, the distribution of pressure and velocity around the hybrid Darrieus-Savonius wind turbine has been determined with a conclusion on the feasible diameters of the Savonius rotor than can be integrated with the Darrieus rotor. The vorticity plots clearly indicate the wake from the cylinder interacting with blade wake leading to increase in the airfoil drag. A valuable conclusion from the computational study is that the ratio between Darrieus rotor to Savonius rotor lies between 5 and 2 for optimum performance of the hybrid rotor. The experimental study focused on validating the computationally optimized diameters in various configurations in terms of power coefficient and starting characteristics. An optimum diameter ratio of 3 displays better power coefficient for the investigated Re.

11.1.3 *Analysis on Discrepancies between Computational Predictions and Experimental Measurements*

The disagreements between the computationally predicted values and the wind tunnel test results are substantial. Though power coefficient curves of both analyses follow the same pattern, they are offset by a constant value. The deviation is attributed to blade tip loss which have been ignored in the current model. Also, the flow curvature effects have considerable impact on the power performance. Since it is a 2D numerical study, the effects of span-wise flow and the losses associated with that are not accounted for. The discrepancies are observed much at low TSR, due to incapability of CFD to predict at deep stall region causing over-prediction of the turbine performance.

11.2 Potential Progress of Darrieus Turbines

11.2.1 *Design Feasibility of 1 kW Hybrid Darrieus Telescopic Savonius Rotor*

From the experimental results it is evident that the proposed hybrid Darrieus Telescopic Savonius (HDTS) rotor performs better than the conventional Darrieus rotor. The immediate question is whether the proposed rotor is structurally feasible to manufacture and can be implemented commercially. To get an insight on the structural aspects, a 1 kW full-scale turbine has been designed as displayed in Figure 11.1. The Darrieus rotor diameter is 2 m and that of the Telescopic Savonius rotor is 0.8 m. The total height of the rotor is 3.5 m. The symmetric NACA0018 is chosen for maximum performance, as the starting torque is entirely developed by the Savonius rotor. The Savonius buckets are guided by the center shaft through three bucket guides, which transmit the torque developed by the Savonius buckets to the center shaft. For reliability, a polyamide bush is introduced between the sliding members to ensure friction free sliding. The three-stage Telescopic Savonius can be retracted into the housing cylinder or extended by the lead screw actuated by a DC motor. The high rpm DC motor is stepped down through worm gears to impart a reduced rpm for the lead screw. The limit sensors on the top and bottom safely interrupt the power to DC motor and ensure the bucket travel is within the limits. The housing cylinder has a vital function of preventing the freestream wind from inducing resistance to the rotation. Still the retracted buckets inside the housing cylinder will

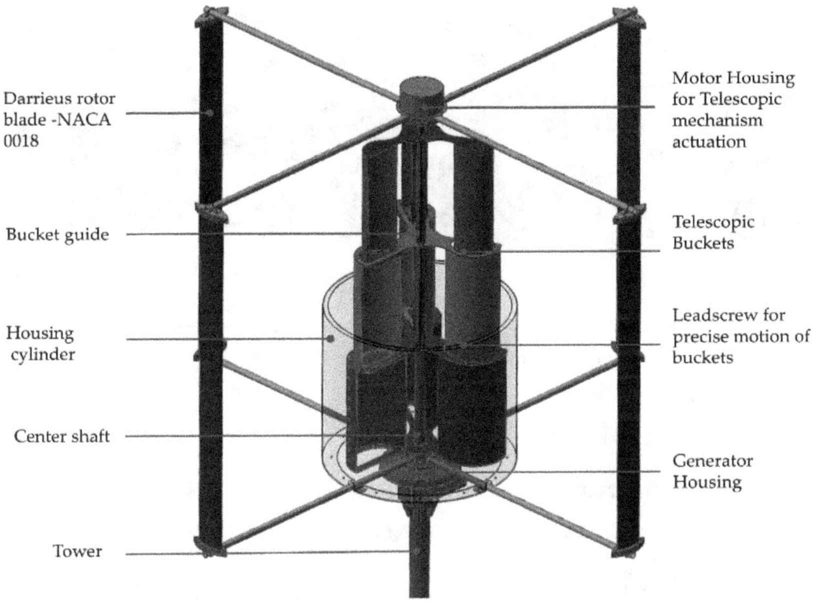

Figure 11.1 1 kW hybrid Telescopic Savonius wind turbine.

create parasitic drag, resulting in a notable torque loss. The major cost elements are identified as DC motor, limit sensors and the lead screw which are standardized and available off-shelf.

11.2.2 *Field Test Comparison of Adaptive Hybrid Darrieus Turbine Configuration*

To ascertain the performance in the field and to quantify the improvement in the starting characteristics, four 0.1 kW turbines of different topology are fabricated, representing the various operating configurations of AHDT as shown in Figure 11.2. These turbines will be installed in low wind speed sites and the performance will be remotely monitored for start-up time and the power coefficient. Since it is cumbersome to incorporate a sliding mechanism in small-scale models, the four configurations representing the operating conditions will suffice to get an insight on comparative performance.

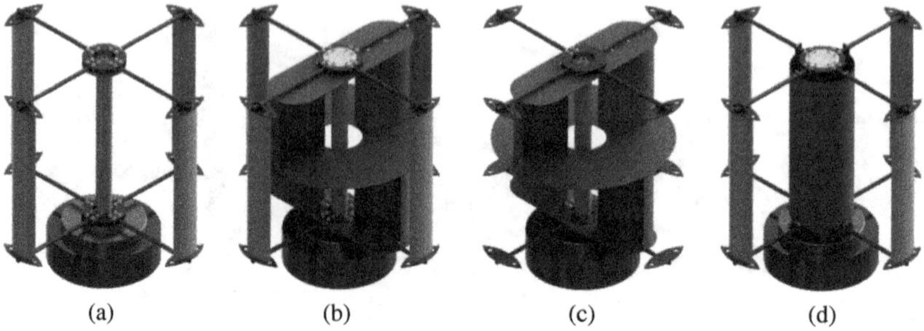

Figure 11.2 (a) Conventional Darrieus. (b) Darrieus with Savonius in open configuration. (c) Conventional Savonius. (d) Darrieus with Savonius in closed configuration.

11.2.3 *Optimization of Darrieus and Savonius Rotors for Adaptive Hybrid Darrieus Turbines*

In the current study, the design parameters of the Darrieus and Savonius rotors are not optimized for AHDT. The Darrieus rotor is evaluated for different solidity, while the rest of the design parameters are constrained. More investigation is recommended to employ various airfoils and helical blades instead of straight blades. The wake structures of straight-bladed Darrieus rotor and helical rotor largely differ. Hence the optimum diameter can vary from the concluded ratio.

The Savonius rotor has plenty of room for further optimization when integrated in AHDT. The overlap ratio, Savonius bucket shape and structural reinforcements as it is employed for aerodynamic braking are a few to mention.

11.3 Recommendations

11.3.1 *Improvements on Aerodynamic Model*

The current proposed model can predict the performance of closed configuration AHDT, while the aerodynamic interaction of AHDT in open Savonius configuration is extremely complex to compute through analytical modeling. The current model does not include the tip loss corrections and dynamic stall models. As a future work these models can be integrated to predict close to the actual values. Aeroelastic models can be included to extend the model to the turbine level.

11.3.2 *Three-dimensional Computational Study*

Though the 2D CFD model employed in the current study is able to capture the overall turbine performance and the optimized Savonius rotor diameter, the power performance must be quantified with blade tip loss, which results in discrepancies with the experimental results. To understand the span-wise flow on the open and closed Savonius rotors, a 3D CFD model will be able to provide better insights on flow interaction between wake and the blades.

11.4 Summary of Floating Offshore Wind Turbines

BEM models are accurate for the prediction of steady load conditions and hence it has been widely employed in the wind industry and certification body for the design and evaluation of wind turbines. Since floating offshore wind turbines (FOWTs) are increasingly important from the power generation perspective, it is necessary to modify the BEM models to be precise enough to predict the unsteady aerodynamics posed by the FOWT. This steered the objectives towards developing and validating the BEM code, which is imperative for the certification of FOWT. Though a handful of previous studies attempted to adapt the BEM codes, there exists an immediate necessity to validate those codes with experimental data. The validation refines the BEM codes further and sheds additional light on the unsteady aerodynamic behaviors of the FOWT. The code validation process includes the complete range of FOWT motions and further comparison with numerical simulation at environmental conditions of interest. The study has been conducted in such a way that the outcome can be immediately inducted into the certification process, further extending the application to maintenance predictions.

The search for the solution begins with an insightful review of the strategies that are currently in practice for the investigation of FOWTs, suitable numerical models that can handle the FOWT motions, numerical code validation process and the extensive use of BEM methodology. The preliminary observations that can be drawn are as listed below.

— Although the FOWTs are subjected to a wide range of motions in all six degrees of freedom, pitch and surge motions were found to have a major influence on the unsteady aerodynamics on the rotor. Hence, the pitch and surge motions are considered to validate the BEM codes.
— Experimental verification with a scaled rotor under similar environmental conditions and its unsteady aerodynamic behavior is indispensable.
— The applicability of developed BEM codes for preventive maintenance through a physics-based approach has been demonstrated.

The existing BEM codes, though well established for the prediction of steady load cases for the fixed-bottom turbines, are not applicable for the FOWT as such. Improvement in the code along with detailed validation at different scenarios are a must. The following list summarizes the overall results on the study conducted on FOWT:

- A high-fidelity CFD method has been developed and implemented for 5MW NREL rotors in FOWT platform pitching motion study for OC3-specified and theoretical wave conditions.
- A novel methodology has been attempted to derive the CFD-based induction factor and was successfully employed in rotor wake state analysis.

- Detailed evaluation of tip loss models was performed at elemental level of the rotor against CFD method predictions.
- The scenarios for scaled rotor experimental tests were confined to emulate the real world FOWT conditions for both static and unsteady aerodynamics in order to validate the numerical model with surge motion.
- A high-fidelity CFD method was developed for the surge motion of the FOWT scaled rotor and validated with experimental data for static and unsteady aerodynamic scenarios.
- A customized BEM-based LR-AeroDyn model, LR-uBEM for scaled rotor unsteady aerodynamics validation was developed and implemented.
- Python-based codes for a physics-based model for remaining useful life prediction of the power converter for FOWT and FOWT wind farm, and solutions for the integration with industrial digital twin platform for O&M were developed. Successfully validated FOWT's BEM model was coupled with FAST model and integrated with digital twin platform.

11.5 Aerodynamic Analysis of Full-Scale 5MW NREL-Based Floating Offshore Wind Turbine

Identifying the critical floating platform motions to validate the BEM codes that can alter the FOWT rotor plane behaviors is critical. After probing through various scenarios a FOWT is subjected to, it is found that pitching and surge motions are causing large motions at the rotor plane level. Hence suitable conditions of OC3 phase IV study with pitching motions have been chosen. A high-fidelity CFD tool has been developed to uncover the aerodynamics at the rotor plane due to the platform pitching motions. The developed CFD model accuracy has been compared with the well-established experimental data available from 5MW NREL wind turbines. CFD-based induction factors have been derived and compared with the BEM-based induction factors for wind speeds up to the rated condition.

The thrust and torque values from the CFD computations show a very good correlation with the BEM-based codes for the below rated wind speeds for most of the blade elements. As anticipated, the CFD prediction of thrust values is low at the blade tips compared to the BEM-based predictions, due to steady state tip loss factors accounted for in BEM. No notable changes are observed in tangential induction factors. The change in thrust values significantly alters the axial induction factors resulting in the deviation between CFD and BEM values. Hence it can be concluded that the CFD computations are far more accurate and vary accordingly on each blade element compared to BEM predictions. The dynamic effects of the platform pitching motion on the rotor for OC3 phase IV case 5.1 are studied with rigid body assumption in CFD and compared with FAST predictions. The value of the induction factor indicates the operating state of the rotor. For the OC3 phase IV case 5.1, the rotor operates in the windmill state as the induction factor is less than 0.4. Through CFD-based dynamic mesh motion methodology, the induction factor is also found to be less than 0.4, indicating that the CFD-based approach is able to predict the aerodynamic effects caused by the pitching motion precisely. Validity of the BEM for FOWT application has been checked by comparing the induction factors of CFD and BEM, indicating that BEM can be refined further for predicting fine variation of cyclic loads. The detailed CFD approach provides insight on the flow pattern and the flow separation regions on the blades. An imperative finding from the CFD study is that the pitching motions did not induce any flow separation on the blades.

A theoretical wave height of 12 m is specified to create a turbulent wake state at the rotor, in order to compare the predictions of FAST BEM methods and high-fidelity CFD. The elemental values of the blade such as thrust are almost similar, except at the blade tip due to tip loss models in BEM. It is also evident that the chosen CFD model is capable of handling single moving reference frame (MRF) scheme at a time along with a moving mesh for translating motions. Also, to determine the effect of larger pitching motions, it is necessary to include the multiple MRF, which limits the CFD application. The boundary layer effects can be well documented through CFD with a fine resolution mesh. Thus, by poring over the CFD results, it is apparent that the CFD methodologies employed present a clear picture on the flow pattern and related aerodynamic effects at various operating states of the rotor.

11.6 Numerical Validation of Scaled Floating Offshore Wind Turbine Rotor

The unsteady aerodynamic loading due to surge motion has been investigated through numerical codes such as LR-AeroDyn and LR-uBEM and compared with the experimental results of the scaled rotor. The CFD results are studied in detail for experimental scenarios of interest. In almost all the selected unsteady scenarios, numerical codes predict mean thrust values within 10% error level against the experimental results. Thrust varies between 4% and 110% respectively for the maximum and minimum surge oscillations and the CFD investigation reveals that the windmill changes its operating state due to the strong interaction of rotor with its own wake instantaneously at higher amplitudes and surge frequencies. For higher frequency surge motions, LR-uBEM is able to predict more precisely than LR-AeroDyn, which has a difference of 10–15%.

11.7 Methodology Development and Implementation of Remaining Useful Life for Floating Offshore Wind Turbine Power Converter

Predictive maintenance is critical especially for FOWT incurring a substantial cost for maintenance and to reduce the downtime. The power converter, being crucial for FOWT operation, has been chosen for physics-based RUL prediction for better maintenance under O&M operation. The prediction of damage accumulation and RUL through validated L-Aerodyn-/LR-uBEM-based models are successfully implemented in the digital twin framework along with FAST. The above predictive methodology applies well to fixed and floating offshore wind turbine power converters by accounting for medium- and short-term thermal transient loadings together as well as long-term thermal loading. The models are integrated into a complete digital twin platform framework for optimum predictive maintenance strategy.

Index

www.ingramcontent.com/pod-product-compliance
Lightning Source LLC
LaVergne TN
LVHW010325220926
848484LV00004BB/841

9789811272486